Mathematical Sciences Research Institute
Publications 4

# Mathematical Sciences Research Institute Publications

| | |
|---|---|
| *Volume 1* | D. Freed and K. Uhlenbeck: Instantons and Four-Manifolds |
| *Volume 2* | S.S. Chern (ed.): Seminar on Nonlinear Partial Differential Equations |
| *Volume 3* | J. Lepowsky, S. Mandelstam, and I.M. Singer (eds.): Vertex Operators in Mathematics and Physics |
| *Volume 4* | V. Kac: Infinite Dimensional Groups with Applications |
| *Forthcoming* | S.S. Chern and P. Griffiths: Essays on Exterior Differential Systems |
| | C.C. Moore (ed.): Group Representations, Ergodic Theory, Operator Algebras, and Mathematical Physics: Proceedings of a Conference in Honor of G.W. Mackey |

# Infinite Dimensional Groups with Applications

Edited by V. Kac

Springer-Verlag
New York Berlin Heidelberg Tokyo

V. Kac
Department of Mathematics
Massachusetts Institute of Technology
Cambridge, MA 02139
U.S.A.

Mathematical Sciences Research Institute
1000 Centennial Drive
Berkeley, CA 94720
U.S.A.

AMS Subject Classification: 22E65, 22E70, 17B65

Library of Congress Cataloging-in-Publication Data
Main entry under title:
Infinite dimensional groups with applications.
(Mathematical Sciences Research Institute publications; 4)
Bibliography: p.
1. Lie groups—Addresses, essays, lectures.
2. Lie algebras—Addresses, essays, lectures.
I. Kac, Victor G. II. Series
QA387.I565 1985 512'.55 85-17382

The Mathematical Sciences Research Institute wishes to acknowledge support from the National Science Foundation.



Printed and bound by R. R. Donnelley & Sons, Harrisonburg, Virginia.
Printed in the United States of America.

9 8 7 6 5 4 3 2 1

ISBN 0-387-96216-6 Springer-Verlag New York Berlin Heidelberg Tokyo
ISBN 3-540-96216-6 Springer-Verlag Berlin Heidelberg New York Tokyo

DEDICATED TO THE MEMORY OF

**CLAUDE CHEVALLEY**

(11 February 1909 - 28 June 1984)

# PREFACE

This volume records most of the talks given at the Conference on Infinite-dimensional Groups held at the Mathematical Sciences Research Institute at Berkeley, California, May 10–May 15, 1984, as a part of the special program on Kac–Moody Lie algebras. The purpose of the conference was to review recent developments of the theory of infinite-dimensional groups and its applications. The present collection concentrates on three very active, interrelated directions of the field: general Kac–Moody groups, gauge groups (especially loop groups) and diffeomorphism groups.

I would like to express my thanks to the MSRI for sponsoring the meeting, to Ms. Faye Yeager for excellent typing, to the authors for their manuscripts, and to Springer-Verlag for publishing this volume.

V. Kac

# INFINITE DIMENSIONAL GROUPS WITH APPLICATIONS

## CONTENTS

## INFINITE DIMENSIONAL GROUPS WITH APPLICATIONS

### CONTENTS (Cont'd)

# THE LIE GROUP STRUCTURE OF DIFFEOMORPHISM GROUPS AND INVERTIBLE FOURIER INTEGRAL OPERATORS, WITH APPLICATIONS*

By

Malcolm Adams,[1] Tudor Ratiu,[2] and Rudolf Schmid[3]

## Abstract

This is a survey of basic facts about the differentiable structure of infinite dimensional Lie groups. The groups of diffeomorphisms and of invertible Fourier integral operators on a compact manifold have a structure which is weaker than that of a Lie group in the classical sense. This differentiable structure is called ILH (inverse limit of Hilbert) Lie group. We indicate applications to the well-posedness problem, to hydrodynamics, plasma physics, general relativity, quantum field theory, and completely integrable PDE's.

---

*Part of this material has been presented as a lecture by Rudolf Schmid at the Conference on Infinite Dimensional Lie Groups, MSRI, Berkeley, May 10-15, 1984.

[1]Institute for Advanced Study, Princeton, NJ 08540. Research partially supported by NSF postdoctoral fellowship MCS 82-11332 while at the University of California, Berkeley.

[2]University of Arizona, Department of Mathematics, Tucson, Arizona 85721, and Mathematical Sciences Research Institute, Berkeley. Research supported by NSF postdoctoral fellowship MCS 83-11674.

[3]Yale University, Department of Mathematics, New Haven, Connecticut 06520, and Mathematical Sciences Research Institute, Berkeley. Research partially supported by NSF grants MCS 83-01124 and MCS 81-20790.

## Table of Contents

## Introduction

Infinite dimensional Lie groups play an increasingly important role in pure as well as applied mathematics. This paper is a survey of a class of examples which prove to be quite valuable in the study of various partial differential equations arising from physics, namely the diffeomorphism groups and various extensions thereof, including our recent work on Fourier integral operators. We describe here one way in which these groups can be construed as Lie groups and discuss several of the applications to mathematical physics.

To define infinite dimensional Lie groups one needs a notion of smooth infinite dimensional manifold. Of course the definition follows that of finite dimensional manifolds except that the model space $\mathbb{R}^n$ is replaced by an infinite dimensional vector space. Since the differential calculus of functions on $\mathbb{R}^n$ does not generalize to any vector space we need to assume some additional structure on the model space. Difficulties arise in choosing the model space general enough to include interesting examples while still restricted enough to allow a meaningful differential calculus. These difficulties are already present in such natural examples as diffeomorphism groups.

The basic idea for the construction of a manifold structure on spaces of mappings dates back to a paper of Eells [1958]. He introduced a $C^\infty$ Banach manifold structure for the set of continuous maps from a compact topological space into a finite dimensional $C^\infty$ manifold. His method has been exemplary for all later attempts because it represents the most natural construction of charts on spaces of mappings. Abraham [1961], Smale, and Palais [1968] considered the corresponding problem for the space $C^r(M,N)$, $1 \leqslant r < \infty$ of $C^r$ maps from a compact manifold M into a Banach manifold N. It turned out that if M is a compact $C^r$ manifold and N a finite dimensional $C^{r+s+2}$ manifold, then $C^r(M,N)$ is a Banach manifold of class $C^s$. In particular the space $\mathscr{D}^r(M)$ of $C^r$-diffeomorphisms of a smooth compact manifold M is open in $C^r(M,M)$ and thus is a smooth Banach manifold. Unfortunately this is not a Lie group in the classical sense because the composition map

$$c: \mathscr{D}^r(M) \times \mathscr{D}^r(M) \longrightarrow \mathscr{D}^r(M)$$

is merely continuous, not $C^\infty$. On the other hand it is easy to see that the composition

$$c: \mathscr{D}^{s+r}(M) \times \mathscr{D}^r(M) \longrightarrow \mathscr{D}^r(M)$$

is of class $C^s$. This suggests that the limit space, $\mathscr{D}^\infty(M)$, should be a Lie group. A first attempt to prove this was done by Leslie [1967] who tried to endow $\mathscr{D}^\infty(M)$ with a Fréchet manifold structure. As will be explained in §4, there are many inequivalent ways to define smoothness for maps between Fréchet spaces. It turns out that the concept of smoothness used by Leslie is too weak for making $\mathscr{D}^\infty(M)$ into a Lie group and the statements in his paper were only much later proved by Gutknecht [1977] by using very technical convergence structures on spaces of maps between Fréchet spaces.

Parallel to all these developments in the study of manifolds of maps, the seminal paper of Arnold [1966] shifts the emphasis from the study of manifolds to applications of Lie groups in mechanics. There, he outlines an original program for the study of mechanical systems whose configuration spaces are Lie groups. Ignoring all technical details, Arnold treats $\mathscr{D}^\infty(M)$ as a Lie group and sketches a geometrical program for the study of hydrodynamics that has been expanding ever since. In this way, Arnold ties the Lie group structure of $\mathscr{D}^\infty(M)$ to important questions in hydrodynamics and motivates much of the later research in infinite dimensional Lie groups.

The second, successful, attempt to make $\mathscr{D}^\infty(M)$ into a Lie group was done by Omori [1970] in 1968. He regards $\mathscr{D}^\infty(M)$ as the inverse limit of the Banach manifolds $\mathscr{D}^r(M)$. Instead of worrying about a differential calculus for Fréchet spaces, Omori defines a generalized manifold structure on $\mathscr{D}^\infty(M)$ and calls it an ILB (inverse limit of Banach) manifold. In this sense, $\mathscr{D}^\infty(M)$ becomes a smooth Lie group. In the course of his work, Omori defines an abstract structure of ILB manifolds and Lie groups, faithfully tailored on the only existing example of that time, the diffeomorphism groups. His original paper was soon followed by Ebin and Marsden [1970] who besides clarifying, enlarging, and completing Omori's work raise the question of ILB Lie subgroups of $\mathcal{D}^{\infty}(M)$, if M has additional structure. Among the many subgroups of $\mathcal{D}^{\infty}(M)$ they consider are the groups of volume preserving and symplectic diffeomorphisms which turn out to be of crucial importance in fluid dynamics and plasma physics. Ebin and Marsden [1970] also give a startling application of the Lie group structure of the group of volume preserving diffeomorphisms by proving for the first time the well-posedness of the initial boundary value problem for Euler's equations for an ideal, homogeneous, incompressible fluid on a compact region with boundary (or a compact Riemannian manifold with boundary). In order to get this remarkable result, Ebin and Marsden move further away from $\mathcal{D}^{\infty}(M)$ by completely shifting the emphasis to the study of $\mathcal{D}^{r}(M)$ and never using in a significant manner the $C^{\infty}$ diffeomorphisms. We adopt here the same point of view and refer the reader to Ebin and Marsden [1970], Marsden, Ebin, Fischer [1972], and Omori [1974] for many of the technical details glossed over in the present review article.

Omori and his collaborators [1980-83] are trying to construct a Fréchet Lie group structure on the space of invertible Fourier integral operators of order zero. So far they give to this space the structure of a topological group. We enlarge the concept of ILB Lie groups in the spirit of Omori, Ebin, and Marsden to what, it seems to us, a weaker and more natural definition. In this new enlarged sense we can handle both the diffeomorphisms and the full class of invertible formal Fourier integral operators. We hope that some of the techniques described below will find their way in the theory of Kac-Moody Lie groups.

We want to close this brief historical sketch with a few comments on another line of research in infinite dimensional Lie groups. The original question raised by Leslie [1967] on the Fréchet Lie group structure of $\mathcal{D}^{\infty}(M)$ has sparked a certain amount of interest, in spite of the success of the Omori-Ebin-Marsden approach. We refer the reader to the excellent review article on this subject of Milnor [1983] and to Hamilton [1982] for the many intricacies of the Fréchet differential calculus. We shall not adopt this point of view in the present paper.

The plan of this review article is the following. In sections 1 to 4 we move from classical Banach Lie groups to Fréchet Lie groups and finally to ILB Lie groups. The remaining bulk of the paper is mainly concerned with the description of various examples of ILB Lie groups and the sketch of some of their applications. We are unequal in our emphasis. While we discuss to some extent many of the methods used to show that a given group has an ILB structure, we only mention some of the applications, ignoring all technical details and being content with a formal "believable" argument and a quote to the relevant references. The applications we mention all center around the Hamiltonian character of certain PDE's. Roughly speaking the philosophy of these methods relies on the fact that certain groups are configuration spaces and/or symmetry groups for a dynamical system, that corresponding momentum maps are important conserved quantities, and that the material-spatial-convective picture in continuum mechanics has group theoretical underpinnings. Moreover, what is easy to show in one picture, might be difficult in another and hence it is important to be able to freely pass from one continuum mechanical description to another; these passages are all canonical, since they are equivariant momentum maps.

We shall use in this paper non-canonical Hamiltonian formulations of PDE's. Classically, if M is the configuration space for a problem, then the cotangent bundle $T^*M$ is the phase space and Hamilton's equations for H are given by $\dot{F} = \{F,H\}$ for any function $F: T^*M \longrightarrow \mathbb{R}$. The Poisson bracket here is given by the canonical symplectic form on $T^*M$. If one replaces $T^*M$ by the dual $\mathfrak{g}^*$ of a Lie algebra $\mathfrak{g}$, one still has a Poisson bracket, called the Lie-Poisson bracket (a term coined by Marsden and Weinstein [1983])

$$\{F,G\}(\alpha) = \left\langle \alpha, \left[\frac{\delta F}{\delta \alpha}, \frac{\delta G}{\delta \alpha}\right]\right\rangle$$

where $F,G: \mathfrak{g}^* \longrightarrow \mathbb{R}$ and the functional derivatives $\delta F/\delta\alpha$, $\delta G/\delta\alpha$ are elements of $\mathfrak{g}$ representing the Fréchet derivatives of F and G:

$$DF(\alpha)\cdot \beta = \langle \beta, \frac{\delta F}{\delta \alpha}\rangle.$$

Here $\langle , \rangle$ is the pairing of $\mathfrak{g}$ with $\mathfrak{g}^*$. Thus for $H\colon \mathfrak{g}^* \longrightarrow \mathbb{R}$ the non-canonical Hamilton equations are given by $\dot{F} = \{F,H\}$, where $\{,\}$ is the Lie-Poisson bracket. The coadjoint orbits of the Lie group $G$ of $\mathfrak{g}$ in $\mathfrak{g}^*$ are symplectic manifolds by the Kirillov-Kostant-Souriau theorem and their Poisson brackets coincide with the restrictions of the Lie-Poisson bracket to the orbits. For an introduction see Marsden et al. [1983] and for the general theory of Poisson manifolds, see Weinstein [1983].

**Acknowledgement**

We want to express our gratitude to Jerry Marsden for his numerous remarks which considerably improved our exposition. Many thanks to Ernst Binz and Dan Freed for their remarks on §8 and 9, respectively.

## 1. Banach Lie groups

Classically, by a Lie group G one means a group which is also a smooth ($C^\infty$) finite dimensional manifold, the group structure and manifold structure being compatible in the sense that the group operation (product) $\mu: (g_1, g_2) \longmapsto g_1 \cdot g_2$ is a smooth ($C^\infty$) map of manifolds $\mu: G \times G \longrightarrow G$. The implicit function theorem implies that the inversion $\nu: g \longmapsto g^{-1}$ is also a smooth map of manifolds $\nu: G \longrightarrow G$. This definition makes perfectly good sense in the category of infinite dimensional smooth manifolds which are modelled on Banach spaces, and there is a well-developed theory which parallels to some extent the theory of finite dimensional Lie groups; see e.g. Bourbaki [1975], and for Banach manifolds in general see e.g. Lang [1972], Choquet-Bruhat et al. [1982], Abraham, Marsden and Ratiu [1983]. For instance, some fundamental properties of classical Lie group theory are valid also in this case.

(1) The Lie algebra $\mathfrak{g} = T_eG$ of a Lie group G is a complete invariant of the local structure of the group.

(2) The exponential map exp: $\mathfrak{g} \longrightarrow G$ is a $C^\infty$ diffeomorphism from a neighborhood of zero in $\mathfrak{g}$ onto a neighborhood of the identity in G, hence it defines a local chart of G around the identity. Moreover, all one-parameter subgroups of G are of the form exp $t\xi$, for some $\xi \in \mathfrak{g}$.

There are, however, two classical theorems which do not have analogs in the category of Banach Lie groups.

(1) If G is a finite dimensional Lie group with Lie algebra $\mathfrak{g}$ and H is a closed subgroup, then H is a Lie subgroup and its Lie algebra is given by $\mathcal{H} = \{\xi \in \mathfrak{g} \mid \exp t\xi \in H \text{ for all } t \in \mathbb{R}\}$. Only half of this statement is true if G is a Banach Lie group. Namely, if H is a closed subgroup of the Banach Lie group G with Banach Lie algebra $\mathfrak{g}$, the set $\mathcal{H} = \{\xi \in \mathfrak{g} \mid \exp t\xi \in H \text{ for all } t \in \mathbb{R}\}$ is a closed Banach Lie subalgebra of $\mathfrak{g}$. If in addition H is locally compact, then H is a finite dimensional closed Lie subgroup of G with

Lie algebra $\mathcal{H}$; the proof of this statement is done as in the finite dimensional case. But in general, closedness of H does not imply that H is a Lie subgroup as the following counterexample of Bourbaki [1975] shows. Regard the Hilbert space $\ell^2(\mathbb{R})$ as an additive Hilbert Lie group and denote its elements by $(a_1, a_2, \ldots)$. Let $G_n = \{a_m \in (1/m)\, \mathbb{Z} \mid 1 \leq m \leq n\}$ and observe that $G_n$ is a closed Lie subgroup of $\ell^2(\mathbb{R})$ for all $n \in \mathbb{N}$. Consequently $H = \bigcap_{n \in \mathbb{N}} G_n$ is a closed subgroup. One shows that H is totally disconnected and not discrete and therefore it cannot be a submanifold, hence a Lie subgroup, of $\ell^2(\mathbb{R})$.

(2) If $\mathfrak{g}$ is a finite dimensional Lie algebra, there exists a connected Lie group G who has $\mathfrak{g}$ as its Lie algebra. This statement is false in the category of Banach Lie groups. We shall discuss a counterexample below, after introducing loop groups.

**Examples:**

(A) The infinite dimensional vector space $C^\infty(M)$ of all smooth functions on a compact manifold M is a group under pointwise addition (i.e. $\mu: C^\infty(M) \times C^\infty(M) \longrightarrow C^\infty(M)$; $\mu(f,g) = f + g$ and $\nu: C^\infty(M) \longrightarrow C^\infty(M)$; $\nu(f) = -f$.) As for any vector space, we formally have $T_e C^\infty(M) = C^\infty(M)$, so the Lie algebra of $C^\infty(M)$ coincides with $C^\infty(M)$ and the bracket is trivial $[f,g] = 0$. The space $C^\infty(M)$ is *not* a Banach Lie group since spaces of $C^\infty$-functions form a Fréchet space rather than a Banach space. To get a Banach Lie group we can complete $C^\infty(M)$ in either the uniform $C^k$-topology to $C^k(M)$, $0 \leq k < \infty$, or the Sobolev $H^s$-topology to $H^s(M)$, $s \geq 0$. Then the vector spaces $C^k(M)$ and $H^s(M)$ are Banach Lie groups.

**Application:**

The vector group $C^\infty(\mathbb{R}^3)$ and the Banach Lie groups $C^k(\mathbb{R}^3)$ and $H^s(\mathbb{R}^3)$ are related to the gauge group of electromagnetism in the sense that Maxwell's equations are invariant under the gauge transformation of the vector potential $A \longmapsto A + \nabla\varphi$, $\varphi \in C^\infty(\mathbb{R}^3)$ (for details see e.g. Marsden et al., [1983] and §7).

(B) The infinite dimensional vector space $C^\infty(M,\mathbb{R}\setminus\{0\})$ of smooth, nowhere vanishing functions on M is a group under pointwise multiplication; $\mu(f,g) = f\cdot g$ and $\nu(f) = f^{-1}$. This group is abelian and formally its Lie algebra is $C^\infty(M)$ with trivial bracket. If M is compact, then $C^k(M,\mathbb{R}\setminus\{0\})$, $k \geq 0$, is a Banach Lie group. For $H^s(M,\mathbb{R}\setminus\{0\})$ we need M compact as well but even then $H^s$ need not be closed under pointwise multiplication. This requires in addition $s > (\dim M)/2$ (Adams [1975], Palais [1965]). Then $H^s(M,\mathbb{R}\setminus\{0\})$ is a Hilbert Lie group. This example generalizes to the case where $\mathbb{R}\setminus\{0\}$ is replaced by any finite dimensional Lie group G. Then $C^k(M,G)$ is a Banach Lie group and $H^s(M,G)$ is a Hilbert Lie group if $s > (\dim M)/2$, under pointwise multiplication $\mu(f,g)(x) = f(x)\cdot g(x)$, $x \in M$, the latter product $\cdot$ taken in G, and $\nu(f)(x) = f(x)^{-1}$. The Lie algebra is $C^k(M,\mathfrak{g})$ and $H^s(M,\mathfrak{g})$ respectively with bracket $[\xi,\eta](x) = [\xi(x),\eta(x)]$, $x \in M$, the latter bracket taken in $\mathfrak{g}$, the Lie algebra of G. Moreover, the exponential map $\exp\colon \mathfrak{g} \longrightarrow G$ defines the map $\mathrm{Exp}\colon H^s(M,\mathfrak{g}) \longrightarrow H^s(M,G)$, $\mathrm{Exp}(\xi) \equiv \exp\circ\xi$, $\xi \in H^s(M,\mathfrak{g})$, which is a local diffeomorphism; i.e. Exp is the exponential map of the Hilbert Lie group $H^s(M,G)$. The same holds for $C^k(M,G)$.

<u>**Applications**</u>:

(1) If $M = S^1$, the circle, then $C^k(S^1,G) = \Omega^k(G)$ is the loop group of G, (Garland [1980], Pressley and Segal [1984]). Its Lie algebra is the loop algebra $C^k(S^1,\mathfrak{g})$, where $\mathfrak{g}$ is the Lie algebra of G.

(2) If G is the structure group of a principal fiber bundle $\pi\colon P \longrightarrow M$, then the group of gauge transformations $\mathcal{G} = \{\varphi \in C^\infty(P,G) \mid \varphi(p\cdot g) = g^{-1}\varphi(p)g\}$ plays an important role in the geometry of quantum field theories, (Singer [1978], [1980], Mitter [1980], and §9).

As already mentioned, not every Banach Lie algebra has an underlying Banach Lie group. Central extensions of loop algebras provide such counterexamples. In order to have a corresponding loop group extension a certain integrality condition must be satisfied. Let

G be a compact, connected, simply connected Lie group with Lie algebra $\mathfrak{g}$ and denote by $\Omega(G)$ the Hilbert Lie group of $H^s$-loops in G and by $\Omega(\mathfrak{g})$ the Hilbert Lie algebra of $H^s$-loops in $\mathfrak{g}$ for $s > 1/2$. Then $\Omega(\mathfrak{g})$ is the Lie algebra of $\Omega(G)$. Let

$$(1) \qquad 0 \longrightarrow \mathbb{R} \longrightarrow \tilde{\Omega}(\mathfrak{g}) \longrightarrow \Omega(\mathfrak{g}) \longrightarrow 0$$

be a Lie algebra extension defined by a Lie algebra 2-cocycle w. Then w defines a left invariant closed 2-form on $\Omega(G)$, which we will denote also by w. The following theorem can be found in Pressley and Segal [1984].

**Theorem.**

(i) The Lie algebra extension (1) corresponds to a group extension

$$1 \longrightarrow S^1 \longrightarrow \tilde{\Omega}(G) \longrightarrow \Omega(G) \longrightarrow e$$

if and only if the differential form $w/2\pi$ represents an integral cohomology class on $\Omega(G)$. The group extension is then unique.

(ii) If $\lambda w$ is not integral for any $\lambda \in \mathbb{R}$, $\lambda \neq 0$, then $\tilde{\Omega}(\mathfrak{g})$ does not correspond to any Lie group.

(iii) The cocycle w satisfies the integrality condition if and only if $\langle h_\alpha, h_\alpha \rangle$ is an even integer for any coroot $h_\alpha$ of the group G.

Note that (i) implies (ii) for if there is a Lie group with Lie algebra $\tilde{\Omega}(\mathfrak{g})$ it must be an extension of $\Omega(G)$. For example, choose $G = SU(2,\mathbb{C}) \times SU(2,\mathbb{C})$ and choose on the first factor of $\mathfrak{g} = su(2,\mathbb{C}) \times su(2,\mathbb{C})$ the standard cocycle (trace) and on the second factor any irrational multiple of the standard cocycle. Then the resulting cocycle satisfies hypothesis (ii) and therefore $\tilde{\Omega}(\mathfrak{g})$ has no underlying Banach Lie group.

Many of the groups that arise in physics as configuration spaces, symmetry groups, or gauge groups cannot be given a local

Banach structure and a more general concept of Lie groups is necessary.

## 2. Diffeomorphism groups

Amongst the most important "classical" examples of infinite dimensional groups are the diffeomorphism groups of manifolds. Their differentiable structure is not that of a Banach Lie group.

Let M be a compact manifold and denote by $\mathcal{D}^s(M)$ the set of all $H^s$-diffeomorphisms on M, $s > (\dim M)/2$. $\mathcal{D}^s(M)$ is a group under composition, i.e. $\mu: \mathcal{D}^s(M) \times \mathcal{D}^s(M) \longrightarrow \mathcal{D}^s(M)$: $\mu(f,g) = f \circ g$ and $\nu$: $\mathcal{D}^s(M) \longrightarrow \mathcal{D}^s(M)$, $\nu(f) = f^{-1}$; the unit element is $e = id$, the identity map. $\mathcal{D}^s(M)$ is a *smooth* ($C^\infty$) Hilbert manifold whose tangent space at $f \in \mathcal{D}^s(M)$ equals

$$T_f\mathcal{D}^s(M) = \{X_f: M \longrightarrow TM \mid X_f \text{ is } H^s \text{ and } \pi \circ X_f = f\},$$

the vector space of all $H^s$-vector fields covering f. A chart of f in $\mathcal{D}^s(M)$ is obtained by the exponential map of a Riemannian metric on M; see for example Ebin-Marsden [1970], Palais [1965], [1968], Marsden-Ebin-Fischer [1972]. The manifold $\mathcal{D}^s(M)$ is not, however, a Banach Lie group, since the group operation is differentiable only in the following restricted sense.

The derivative at $f \in \mathcal{D}^s(M)$ of right multiplication $R_g$: $\mathcal{D}^s(M) \longrightarrow \mathcal{D}^s(M)$, $R_g(f) = f \circ g$, given by

$$X_f \in T_f\mathcal{D}^s(M) \longmapsto X_f \circ g \in T_{f \circ g}\mathcal{D}^s(M),$$

is again right multiplication, so an easy inductive argument proves that $R_g$ is $C^\infty$ for any $g \in \mathcal{D}^s(M)$. On the other hand, the derivative at $f \in \mathcal{D}^s(M)$ of left multiplication $L_g$: $\mathcal{D}^{s+k}(M) \longrightarrow \mathcal{D}^s(M)$, $k \geq 0$, $L_g(f) = g \circ f$ is given by

$$X_f \in T_f\mathcal{D}^{s+k}(M) \longmapsto Tg \circ X_f \in T_{g \circ f}\mathcal{D}^s(M),$$

where $Tg$: $TM \longrightarrow TM$ denotes the derivative (tangent map) of the diffeomorphism g. But if g is of class $H^{s+k}$, Tg is only of class

$H^{s+k-1}$ and thus the above process (taking higher derivatives) can be repeated only k times. In this way it is shown that if $g \in \mathcal{D}^{s+k}(M)$, $L_g$: $\mathcal{D}^s(M) \longrightarrow \mathcal{D}^s(M)$ is $C^k$, $k \geq 0$. More generally, composition $(f,g) \in \mathcal{D}^{s+k}(M) \times \mathcal{D}^s(M) \longmapsto f \circ g \in \mathcal{D}^s(M)$ is a $C^k$ map. Therefore, the group operation $\mu$ is *not* smooth, but only continuous. Similarly, the inverse map $\nu$: $f \longmapsto f^{-1}$ is continuous when regarded as a map from $\mathcal{D}^s(M)$ to $\mathcal{D}^s(M)$, but is $C^k$ if regarded as a map $\nu$: $\mathcal{D}^{s+k}(M) \longrightarrow \mathcal{D}^s(M)$.

The tangent manifold $T(\mathcal{D}^s(M))$ can be identified with the set of all $H^s$ mappings from M to TM that cover diffeomorphisms and it is again an infinite dimensional Banach manifold. A vector field X on $\mathcal{D}^s(M)$ is a map X: $\mathcal{D}^s(M) \longrightarrow T(\mathcal{D}^s(M))$ such that $X(f) \in T_f(\mathcal{D}^s(M))$, i.e. $\pi \circ X(f) = f$. $T_e(\mathcal{D}^s(M))$ is the space $\mathfrak{X}^s(M)$ of $H^s$ vector fields on M.

Recall now that for a Lie group G one gives a Lie algebra structure to $T_eG$ by extending $\xi, \eta \in T_eG$ to left invariant vector fields $X_\xi, X_\eta$ on G and defining $[\xi,\eta] = [X_\xi, X_\eta](e)$ where the bracket on the right hand side is the Lie bracket of vector fields. This procedure is not possible in the case of $\mathcal{D}^s(M)$. Recall that multiplication $\mu$: $\mathcal{D}^{s+k}(M) \times \mathcal{D}^s(M) \longrightarrow \mathcal{D}^s(M)$ is a $C^k$ map. Taking the derivative in the second variable we have TL: $\mathcal{D}^{s+k}(M) \times T_e(\mathcal{D}^s(M)) \longrightarrow T(\mathcal{D}^s(M))$ is $C^{k-1}$ so we can't use TL to translate $\xi \in T_e(\mathcal{D}^s(M))$ around to get a continuous vector field. On the other hand

$$\mathrm{TR}: T_e(\mathcal{D}^{s+k}(M)) \times \mathcal{D}^s(M) \longrightarrow T(\mathcal{D}^s(M))$$

is a priori $C^{k-1}$ and in fact turns out to be $C^k$. Thus if $\xi \in T_e(\mathcal{D}^{s+1}(M))$, the right invariant vector field $Y_\xi$, given by $Y_\xi(f) = T_eR_f(\xi) = \xi \circ f$ is a $C^1$ vector field on $\mathcal{D}^s(M)$. Thus, if $\xi, \eta \in T_e(\mathcal{D}^{s+1}(M))$ we can define the bracket $[\xi,\eta] \in T_e(\mathcal{D}^s(M))$ by $[\xi,\eta] = -[Y_\xi, Y_\eta](e)$, $Y_\xi, Y_\eta$ being the right invariant $C^1$-vector fields on $\mathcal{D}^s(M)$ and the bracket on the right hand side being the Lie bracket of vector fields on $\mathcal{D}^s(M)$. (Note $[Y_\xi, Y_\eta]$ is a $C^0$-vector field on $\mathcal{D}^s(M)$.) The minus sign is included because on Lie groups if $\xi, \eta \in T_eG$ and $Y_\xi, Y_\eta$ are the right invariant vector fields for $\xi$

and $\eta$ then $[\xi,\eta] = -[Y_\xi,Y_\eta](e)$. Using this definition of bracket on $T_e(\mathcal{D}^{s+1}(M)) \approx \mathfrak{X}^{s+1}(M)$ it turns out that $[\xi,\eta]$ is just the negative of the Lie bracket of $\xi$ and $\eta$ considered as vector fields on M. Note that for $\xi,\eta \in \mathfrak{X}^{s+1}(M)$, $[\xi,\eta] \in \mathfrak{X}^s(M)$, i.e. one derivative is lost so the "Lie algebra" is not closed under bracket.

To obtain a Lie algebra of vector fields on M which is closed under the bracket, one has to consider $\mathfrak{X}^\infty(M)$ the space of $C^\infty$-vector fields. This is formally the Lie algebra of the "Lie" group $\mathcal{D}^\infty(M)$ of $C^\infty$-diffeomorphism on M. The problem in this case is that $\mathcal{D}^\infty(M)$ is *not* a Banach manifold because $\mathfrak{X}^\infty(M)$ is not a Banach space but only a Fréchet space (a complete metrizable topological vector space, whose topology cannot be defined by a single norm but by a family of seminorms) (Palais [1965], [1968]). There is no canonical extension of the classical differential calculus on Banach spaces to the case of Fréchet spaces, e.g. one has various, nonequivalent, choices even for the definition of a $C^1$-map on a Fréchet space and for all these notions of differentiability there is no classical inverse function theorem (Keller [1974], Schmid [1983], Hamilton [1982]).

The fact that any Fréchet space is the inverse limit of Banach spaces (so in particular $\mathfrak{X}^\infty(M) = \varprojlim \mathfrak{X}^s(M)$) and the properties of the diffeomorphism groups lead to the abstract concept of ILH-Lie groups discussed in §4 (Omori [1970]).

## 3. The exponential mapping

One of the most important constructions connecting the Lie group G to its Lie algebra $\mathfrak{g}$ is the exponential mapping exp: $\mathfrak{g} \longrightarrow G$ defined in the following manner. For $\xi \in \mathfrak{g}$, let $\varphi_\xi(t)$ be the integral curve of the left invariant vector field $X_\xi$ on G, passing through e at $t = 0$. Then one defines $\exp \xi = \varphi_\xi(1)$. A remarkable property of exp: $\mathfrak{g} \longrightarrow G$ is that it is a diffeomorphism of a neighborhood of zero in $\mathfrak{g}$ onto a neighborhood of the identity in G, if G is a Banach Lie group. Thus inverting exp onto its range where it is a diffeomorphism, one obtains a chart at the identity in G. Now left translating this chart at every point one gets an atlas in which many formulas involving group operations become particularly simple.

It is also proved that using the right invariant vector field $Y_\xi$ determined by $\xi$ gives the same exponential map. Besides being of practical importance by providing an atlas, the exponential mapping provides the key to many of the classical results in Lie group theory. For instance, using the exponential map, it can be shown that any closed subgroup H of a finite dimensional Lie group G is a Lie subgroup, and that any Lie subalgebra $\mathfrak{h}$ of $\mathfrak{g}$ determines a connected Lie subgroup H of G, namely H is generated by exp $\mathfrak{g}$. For any finite dimensional Lie algebra $\mathfrak{g}$ there is a connected, simply connected Lie group G, whose Lie algebra is isomorphic to $\mathfrak{g}$.

Unfortunately, for diffeomorphism groups the power of the exponential map is greatly reduced by a number of pathologies. In complete analogy with the classical case,

$$\exp: T_e(\mathcal{D}^s(M)) = \mathcal{X}^s(M) \longrightarrow \mathcal{D}^s(M)$$

is defined as follows. For $\xi \in \mathcal{X}^s(M)$, let $\varphi_t \in \mathcal{D}^s(M)$ be its flow. Then the curve $t \longmapsto \varphi_t$ is an integral curve of the right invariant vector field $Y_\xi$ on $\mathcal{D}^s(M)$ determined by $\xi$. Thus the exponential map is given by

$$\exp(\xi) = \varphi_1.$$

The map exp is continuous, but unlike the case of Banach Lie groups, it is *not* $C^1$; moreover, there is no neighborhood of the identity in $\mathcal{D}^s(M)$ onto which exp maps surjectively. (For counterexamples see e.g. Hamilton [1982].) This means in practice that, in spite of the fact that G and $\mathfrak{g}$ are known, the construction of charts on G requires additional information.

These facts are important to keep in mind because in physical applications various subgroups of $\mathcal{D}^s(M)$ and subalgebras of $\mathcal{X}^\infty(M)$ play an important role. In view of the pathologies mentioned above, we cannot conclude that they are Lie subgroups, or are given by a corresponding Lie subgroup respectively. Other special arguments are needed and often one must explicitly construct local charts.

The lack of a good exponential map also causes enormous

difficulties in the representation theory of diffeomorphism groups. For $M = S^1$, the circle, see Goodman and Wallach [1984], and also their contribution in these proceedings.

The same constructions as for $H^s$-diffeomorphisms can be done for $C^k$-diffeomorphisms and the same results hold for the "Lie" group $\mathcal{D}^k(M)$ with "Lie" algebra $\mathcal{X}^k(M)$ (Omori [1974], Palais [1965]).

## 4. ILH-Lie groups

A collection of groups $\{G^\infty, G^s \mid s \geq s_0\}$ is called an *ILH-Lie group* (inverse limit of Hilbert) if:

(i) each $G^s$ is a Hilbert manifold of class $C^{k(s)}$, modeled on the Hilbert space $E^s$, where the order of diffentiability $k(s)$ tends to $\infty$ as $s \longrightarrow \infty$;

(ii) for each $s \geq s_0$, there are linear continuous, dense inclusions $E^{s+1} \hookrightarrow E^s$ and dense inclusions of class $C^{k(s)}$, $G^{s+1} \hookrightarrow G^s$;

(iii) each $G^s$ is a topological group and $G^\infty = \varprojlim G^s$ is a topological group with the inverse limit topology;

(iv) if $(U^s, \varphi^s)$ is a chart on $G^s$, then $(U^s \cap G^t, \varphi^s | U^s \cap G^t)$ is a chart on $G^t$, for all $t \geq s$;

(v) group multiplication $\mu: G^\infty \times G^\infty \longrightarrow G^\infty$ can be extended to a $C^k$-map $\mu: G^{s+k} \times G^s \longrightarrow G^s$ for any $s$ such that $k \leq k(s)$;

(vi) inversion $\nu: G^\infty \longrightarrow G^\infty$ can be extended to a $C^k$-map $\nu: G^{s+k} \longrightarrow G^s$, for any $s$ satisfying $k \leq k(s)$;

(vii) right multiplication $R_g$ by $g \in G^s$ is a $C^{k(s)}$-map $R_g: G^s \longrightarrow G^s$.

If the manifolds are Banach manifolds rather than Hilbert manifolds then $\{G^\infty, G^s \mid s \geq s_0\}$ is an ILB-Lie group.

A collection of vector spaces $\{g^\infty, g^s \mid s \geq s_0\}$ is called an *ILH(ILB)-Lie algebra* if

(i) each $g^s$ is a Hilbert (Banach)-space and for each $s \geq s_0$ there are linear, continuous, dense inclusions $g^{s+1} \hookrightarrow g^s$ and $g^\infty = \varprojlim g^s$ is a Fréchet space with the inverse limit topology;

(ii) there exist bilinear, continuous, antisymmetric maps $[\ ,\ ]$: $g^{s+2} \times g^{t+2} \longrightarrow g^{\min(s,t)}$, for all $s,t \geq s_0$, which satisfy the Jacobi identity on $g^{\min(s,t,r)}$ for elements in $g^{s+4} \times g^{t+4} \times g^{r+4}$.

If $\{G^\infty, G^s \mid s \geq s_0\}$ is an ILH(ILB)-Lie group, put $g^s \equiv T_eG^s$ and $g^\infty = \varprojlim g^s$. Then $\{g^\infty, g^s \mid s \geq s_0\}$ is the ILH(ILB)-Lie algebra of the ILH(ILB)-Lie group $\{G^\infty, G^s \mid s \geq s_0\}$.

The classical examples of ILH-Lie groups are the diffeomorphism groups

$$\{\mathcal{D}^\infty(M), \mathcal{D}^s(M) \mid s > (\dim M)/2\}$$

with ILH-Lie algebras

$$\{\mathfrak{X}^\infty(M), \mathfrak{X}^s(M) \mid s > (\dim M)/2\}.$$

The properties in the definitions are proved in Ebin [1968], Ebin-Marsden [1970], Omori [1974], Marsden-Ebin-Fischer [1972].

<u>**Remark**</u>: We noted in §2 that for $G^s = \mathcal{D}^s(M)$ the bracket is actually defined on $g^{s+1} \times g^{t+1} \longrightarrow g^{\min(s,t)}$. This is because TR: $T_e(\mathcal{D}^{s+k}(M)) \times \mathcal{D}^s(M) \longrightarrow T(\mathcal{D}^s(M))$ is actually $C^k$, not just $C^{k-1}$. Since we don't require this to hold for our general ILH manifolds property (ii) above is appropriately adjusted.

The last section sketches a proof of the fact that the group of invertible Fourier integral operators on a compact manifold is an ILH-Lie group.

The terminology of ILH(ILB)-Lie groups was introduced by Omori [1970] in order to study the $C^\infty$-diffeomorphism groups. It turns out, as we shall see in the last section, that Omori's definition is too restrictive for the Fourier integral operators. This is the reason why our definition of ILH(ILB)-Lie groups given at the beginning of this section is weaker than Omori's. Moreover, in contrast with Omori, we shift the emphasis from the limit space $G^\infty$ to the tower of spaces $G^s$, the properties of $G^\infty$ being naturally induced by the properties of all the $G^s$.

There are two main points of view about the treatment of $G^\infty$. One point of view is to consider $G^\infty$ as a Fréchet manifold modeled on the Fréchet space $E^\infty$; the charts are $(U^s \cap G^\infty, \varphi^s | U^s \cap G^\infty)$ for $s \geq s_0$. For this to make sense, we need a calculus in Fréchet spaces and in particular a definition for a map to be $C^k$. Unfortunately there are many inequivalent ways to define this concept. For example, if E and F are Fréchet spaces, U is open in E, differentiability of a map $f: U \longrightarrow F$ is defined in the usual manner. However, in contrast with the definitions for Banach spaces, f is said to be $C^1$, if $Df(x)\cdot h$ is jointly continuous in $(x,h) \in U \times E$. The map f is said to be $C^k$ if $D^kf(x)\cdot(h_1,\ldots,h_k)$ exists and is jointly continuous in $(x,h_1,\ldots,h_k) \in U \times E \times \ldots \times E$; see Hamilton [1982] for a development of this differential calculus. The reason for changing the usual definition for a $C^k$ map is due to the fact that the vector space L(E,F) of continuous linear maps between Fréchet spaces is not a Fréchet space. Moreover, if E and F are Fréchet spaces there is no vector space topology on the space of k-linear maps $L^k(E,F)$ such that the evaluation map ev: $L^k(E,F) \times E^k \longrightarrow F$, $ev(u,h_1,\ldots,h_k) = u(h_1,\ldots,h_k)$ is continuous; see Keller [1974]. Thus, one can define a map $g: X \longrightarrow L^k(E,F)$, X a locally convex space, to be continuous if $\tilde{g}$: $X \times E^k \longrightarrow F$, $\tilde{g}(x,h_1,\ldots,h_k) = g(x)(h_1,\ldots,h_k)$ is continuous. This defines a convergence structure (pseudo topology) on $L^k(E,F)$. It is the coarsest convergence structure on $L^k(E,F)$ which makes the evaluation map ev: $L^k(E,F) \times E^k \longrightarrow F$ continuous. A map $f: U \subseteq E \longrightarrow F$ is called $C^k$, with respect to this convergence structure, if $D^kf: U \subseteq E \longrightarrow L^k(E,F)$ exists and is continuous. Using different convergence structures on $L^k(E,F)$ leads to different, inequivalent notions of $C^k$

differentiability. This point of view of differentiability is developed in Keller [1974]; see also Schmid [1978] and Michor [1980]. In this way, the definitions above and the theorems derived from them imitate closely the usual differential calculus, with the exception that the classical inverse function theorem and implicit function theorem don't hold; see Hamilton [1982] and Schmid [1983]. But whatever differential calculus one uses, one is faced with pathologies in the theory of Fréchet manifolds. This raises the question whether the whole technical apparatus is really worth using for the specific problem at hand.

There exists, however, an entirely different point of view in which the central role is not played by the Fréchet Lie group $G^{\infty}$, but by the tower of Banach manifolds $G^s$ which have $G^{\infty}$ as limit. Suppose, for example, that one wants to prove some kind of short time existence theorem of a vector field on $G^{\infty}$. Instead of worrying about the Fréchet differentiable structure, extend the vector field in question to all $G^s$ and use there the standard theorem of local existence of integral curves. Then prove a regularity theorem which bounds below the time of existence for each $G^s$, the bound being independent of s. This enables the passage $s \longrightarrow \infty$ and one gets a local existence result in $G^{\infty}$. The advantage of this approach lies in the fact that one can appeal to the powerful elliptic theory both for existence and regularity.

We shall devote the rest of the paper to specific examples of ILH-Lie groups and comment on their applications to physical problems.

## 5. Volume preserving diffeomorphisms and incompressible hydrodynamics

Let $(M,\mu)$ be a compact (finite dimensional) manifold, $\partial M \neq \emptyset$, $\mu$ a volume element on M. The set

$$\mathcal{D}^s_{\mu}(M) = \{\varphi \in \mathcal{D}^s(M) \mid \varphi^*\mu = \mu\}$$

of volume preserving diffeomorphisms on M of class $H^s$, $s > (\dim M)/2 + 1$, is a subgroup of $\mathcal{D}^s(M)$ for each s. It is shown in Ebin and

Marsden [1970] and Omori [1974] that $\{\mathcal{D}^{\infty}_{\mu}(M), \mathcal{D}^{s}_{\mu}(M) \mid s > (\dim M)/2 + 1\}$ is an ILH-Lie subgroup of the ILH-Lie group $\{\mathcal{D}^{\infty}(M), \mathcal{D}^{s}(M) \mid s > (\dim M)/2 + 1\}$ with ILH-Lie algebra $\{\mathcal{X}^{\infty}_{\mu}(M), \mathcal{X}^{s}_{\mu}(M) \mid s > (\dim M)/2 + 1\}$, where

$$\mathcal{X}^{s}_{\mu}(M) = \{\xi \in \mathcal{X}^{s}(M) \mid \operatorname{div} \xi = 0\},$$

the divergence free vector fields of class $H^s$ on M with respect to the volume form $\mu$.

We shall give the proof that $\mathcal{D}^{s}_{\mu}(M)$ is a closed smooth submanifold of $\mathcal{D}^{s}(M)$ if $\partial M = \emptyset$ for illustrative purposes; this proof is a typical example in the theory of diffeomorphism groups. The idea is very simple and is based on two key facts. The first uses the Hodge decomposition theorem to insure that the affine subspace

$$[\mu]_s = \mu + d(H^{s+1}(\Omega^{n-1}(M)))$$

is closed in $H^s(\Omega^n(M))$; we denote by $H^s(\Omega^k(M))$ the exterior k-forms on M of class $H^s$. Define the map

$$F: \mathcal{D}^{s}(M) \longrightarrow [\mu]_s, \text{ by}$$

$$F(\varphi) = \varphi^*\mu.$$

The definition is correct, i.e. $\varphi^*\mu \in [\mu]_s$, since by the change of variables theorem $\int_M (\varphi^*\mu - \mu) = 0$, so that by the de Rham theorem

$\varphi^*\mu - \mu = d\alpha$ with $\alpha \in H^{s+1}(\Omega^{n-1}(M))$. The second key fact in the proof is that F is a $C^\infty$ map by the $\omega$-lemma. Moreover,

$$T_\varphi F(V) = \varphi^*(L_{V\circ\varphi^{-1}}\mu),$$

for $V \in T_\varphi \mathcal{D}^{s}(M)$; $L_X$ denotes the Lie derivative with respect to the vector field X. But since $\mu$ is a volume form, $d\mu = 0$, so that

$$T_\varphi F(V) = d i_{V\circ\varphi^{-1}}\mu,$$

where $i_X$ denotes the operation of interior product of a form with the vector X. But since $\mu$ is non-degenerate, the map

$$X \in \mathfrak{X}^s_\mu(M) \longmapsto i_X\mu \in H^s(\Omega^{n-1}(M))$$

is an isomorphism, so that $T_\varphi F$ is onto. Thus F is a submersion and $F^{-1}(\mu)$ is a smooth closed submanifold of $\mathcal{D}^s(M)$.

If $\partial M \neq \emptyset$ one needs to work with de Rham type theorems due to Duff [1952]; see Ebin and Marsden [1970]. Note that the above proof gives the charts only implicitly.

The main application of $\mathcal{D}^s_\mu(M)$ is based on the fact that it is the configuration space of an incompressible, homogeneous, ideal fluid. Let M be a compact orientable, finite dimensional Riemannian manifold and let $\mu$ be the Riemannian volume. Since $\mathcal{D}^s_\mu(M)$ admits the smooth weak Riemannian metric

$$(U,V)_\eta = \int_M \langle U(m),V(m)\rangle \ (\eta(m))\mu(m)$$

for $U,V \in T_\eta\mathcal{D}^s_\mu(M)$, and $\langle\ ,\ \rangle$ the metric on M, one can talk about geodesics on $\mathcal{D}^s_\mu(M)$. It turns out that this metric is right invariant and that the spray is smooth. This remarkable fact found by Ebin and Marsden [1970] has as consequence the local existence and uniqueness of geodesics on $\mathcal{D}^s_\mu(M)$ and their smooth dependence on initial conditions. Moreover, if $V_t$ is an integral curve of the spray on $\mathcal{D}^s_\mu(M)$ and $\eta_t = \pi\circ V_t$, then

$$v_t = V_t\circ\eta_t^{-1}$$

satisfies the Euler equations:

$$\text{(5.1)} \qquad \begin{cases} \dfrac{\partial v}{\partial t} + \nabla_v v = -\nabla p \\ \operatorname{div} v = 0, \end{cases}$$

where $\nabla_v$ is the covariant derivative of the metric on M. The time dependent vector field $v_t$ represents the Eulerian (or spatial) velocity of the fluid, whereas the solution $V_t$ of the geodesic spray equations represents the material velocity of the fluid. In this way, the well-posedness for the initial value problem of the Euler equations is equivalent to the local existence and uniqueness of solutions for the geodesic spray on $\mathcal{D}^s_\mu(M)$. The dependence of the solutions of the Euler equations on initial conditions is continuous, because the pull-back $v_t = V_t \circ \eta_t^{-1}$ involves right composition with an inverse as well as left translation and both operations are continuous but not smooth. This theorem, whose proof was sketched above, is due to Ebin and Marsden [1970]. The relationship $v_t = V_t \circ \eta_t^{-1}$ represents the momentum map of the right action of the group $\mathcal{D}^s_\mu(M)$ on the weak symplectic manifold $T\mathcal{D}^s_\mu(M)$, the symplectic form being naturally induced by the weak metric from the canonical one on the cotangent bundle.

Equivariant momentum maps are canonical, i.e. preserve Poisson brackets. This suggests that the Euler equations are Hamiltonian themselves, without recourse to the material picture and the associated geodesic spray. This is indeed the case as shown by Marsden and Weinstein [1983]. The Lie algebra $\mathfrak{X}^\infty_\mu(M)$ is weakly paired with itself by the $L^2$-inner product; the pairing is weakly non-degenerate due to the Hodge decomposition of any vector field in a $L^2$ orthogonal sum of a gradient and a divergence free vector field tangent to $\partial M$. Thus, as a "dual" of a Lie algebra, $\mathfrak{X}^\infty_\mu(M)$ carries the Lie-Poisson bracket

$$\text{(5.2)} \qquad \{F,G\}(v) = -\int_M \left\langle v, \left[\frac{\delta F}{\delta v}, \frac{\delta G}{\delta v}\right]\right\rangle \mu,$$

where the functional derivative $\delta F/\delta v$ is the divergence free vector field tangent to $\partial M$ representing the Fréchet derivative of F, if it exists, i.e.

$$DF(v)\cdot v' = \int_M \left\langle \frac{\delta F}{\delta v}, v' \right\rangle \mu.$$

With respect to the kinetic energy Hamiltonian

(5.3)
$$H(v) = (1/2) \int_M |v|^2 \mu$$

the Euler equations are Hamiltonian, i.e. are equivalent to

(5.4)
$$\dot{F} = \{F,H\}$$

for any $F: \mathfrak{X}^\infty_\mu(M) \longrightarrow \mathbb{R}$. To check this, note first that $\frac{\delta H}{\delta v} = v$ by (5.3), so that a straightforward computation yields

(5.5)
$$\dot{F} = \{F,H\}(v) = -\int_M \langle \nabla_v v, \frac{\delta F}{\delta v} \rangle \mu.$$

Now write $\nabla_v v$ as a divergence free part tangent to $\partial M$ plus a gradient (the Helmholtz-Hodge decomposition), i.e.

$$\nabla_v v = X - \nabla p$$

for some $X \in \mathfrak{X}^\infty_\mu(M)$ and $p \in C^\infty(M)$ uniquely determined by the Neumann problem

$$\Delta p = -\mathrm{div}\, (\nabla_v v)$$

$$\frac{\partial p}{\partial n} = -(\nabla_v v) \cdot n \text{ on } \partial M,$$

where n is the outward unit normal on $\partial M$. Returning to (5.5) we see that $\frac{\partial v}{\partial t} + X = 0$ or

$$\frac{\partial v}{\partial t} + \nabla_v v = -\nabla p$$

which is Euler's equation (5.1).

This derivation of the Euler equations might seem a lucky guess; it is not, and for the relevant theory see Marsden and Weinstein [1983] and Marsden, Ratiu, and Weinstein [1984a,b], where the central role of the momentum map $v_t = V_t \circ \eta_t^{-1}$ is also explained. The Hamiltonian setting above has also a lot of consequences. Kelvin's circulation theorem appears as a symmetry of the problem. Using the degeneracy of the Lie-Poisson bracket, important Liapunov stability results for stationary solutions can be obtained; see Arnold [1965], [1969] and Holm et al [1985], Abarbanel et al [1985].

## 6. Semidirect product diffeomorphism groups and compressible hydrodynamics

In this section we shall outline briefly an application of the diffeomorphism group to compressible hydrodynamics. Let M be a compact Riemannian manifold with Riemannian volume dX. Imagine M filled with compressible fluid in motion. Given the mass density $\rho_0(X)$ of the fluid in material coordinates, assuming that shocks do not occur and, moreover, that the particle paths are $C^\infty$, one concludes that the configuration space of the fluid motion is the group $\mathcal{D}^\infty(M)$. Thus, as in the previous section, passing to particle paths of class $H^s$, $s > (\dim M)/2 + 1$, one can describe this mechanical system, which is a vector field on the tangent bundle $T\mathcal{D}^s(M)$, as a second order equation on $\mathcal{D}^s(M)$. It turns out that this second order equation is not the geodesic spray of a metric, unlike the case of incompressible fluid. Using the weighted Hodge decomposition as a main tool and the geometric ideas presented in the previous section, Marsden [1976] proved the local existence and uniqueness of solutions $(v,\rho)$ for $s > (\dim M)/2 + 1$; here v and $\rho$ denote the Eulerian velocity field and density.

Let $t \longmapsto \eta_t(X) = x(t)$ denote the particle path of the material point X and let $V_t(X) = d\eta_t(X)/dt$ denote the material velocity of the fluid (which is a vector field covering $\eta_t$). The passage from the Lagrangian $(V_t,\rho_0)$ to the Eulerian $(v_t,\rho)$ description is given in classical fluid mechanics by the map $v_t = V_t \circ \eta_t^{-1}$, $(\rho_t \circ \eta_t)J(\eta_t) = \rho_0$ where $J(\eta_t) = \frac{dx(t)}{dX}$ denotes the Jacobian of $\eta_t$. In differential forms language, if dx denotes the Riemannian volume in the Eulerian description, i.e. $\eta_t^*\, dx = J(\eta_t)dX$, we have that

$\eta_t^*(\rho_t dx) = (\rho_t \circ \eta_t)J(\eta_t)dX = \rho_0 dX$. Taking the time derivative of this relation one gets the well-known differential form of the conservation of mass law:

$$\frac{\partial \rho}{\partial t} + \operatorname{div}(\rho v) = 0, \tag{6.1}$$

where div is taken with respect to the volume dx. The remaining Euler equations

$$\frac{\partial v}{\partial t} + \nabla_v v = -(1/\rho)\, \nabla p \tag{6.2}$$

represent the spatial balance of momentum. For isentropic fluids the pressure p is determined in terms of the internal energy density $w(\rho)$ by $p'(p) = \rho^2 w'(\rho)$ and in addition it is assumed that $p'(\rho) > 0$ (positive sound speed); the latter condition makes the system (6.1), (6.2) well posed with boundary condition that v is parallel to $\partial M$.

Note that whereas in Lagrangian description the variable was the material velocity V, in Eulerian coordinates, the variables are the spatial velocity v and spatial density $\rho$ and the material density $\rho_0$ becomes an initial condition for $\rho$. Thinking now of $\rho_0$ as a parameter, i.e. letting $\rho$ have any initial condition, gives the dynamic variables $(M,\rho)$, where $M = \rho v$ is the spatial momentum density of the fluid. Thus, the phase space of the Eulerian isentropic fluid motion is $\mathfrak{X}^\infty(M) \times C^\infty(M,\mathbb{R})$.

The total energy is given by

$$H(M,\rho) = (1/2) \int_M (|M|^2/\rho)dx + \int_M \rho w(\rho)dx. \tag{6.3}$$

It can be easily checked directly, that with respect to the bracket

$$\begin{aligned}\{F,G\}(M,\rho) = &\int_M \left\langle M, \left[\frac{\delta F}{\delta M},\frac{\delta G}{\delta M}\right]\right\rangle dx \\ &- \int_M \rho\left(L_{\delta F/\delta M} \frac{\delta G}{\delta \rho} - L_{\delta G/\delta M} \frac{\delta F}{\delta \rho}\right)dx,\end{aligned} \tag{6.4}$$

where $\langle\ ,\ \rangle$ is the metric on M, $|\ |$ its norm, and $L_u$ the Lie derivative with respect to the vector field u, the equations (6.1), (6.2) are Hamiltonian with Hamiltonian function given by (6.3).

The bracket (6.4) is in fact the Lie-Poisson bracket on the dual of the semidirect product ILH Lie algebra $\mathfrak{X}^\infty(M) \ltimes C^\infty(M,\mathbb{R})$, the action of $\mathfrak{X}^\infty(M)$ on $C^\infty(M,\mathbb{R})$ being given by minus the Lie derivative. As in the previous section, we identify the dual $(\mathfrak{X}^\infty(M) \ltimes C^\infty(M,\mathbb{R}))^*$ only via a weakly nondegenerate pairing with $\mathfrak{X}^\infty(M) \times C^\infty(M,\mathbb{R})$, the phase space of Eulerian isentropic fluid motion: $\mathfrak{X}^\infty(M)$ and $C^\infty(M,\mathbb{R})$ are identified with themselves via the $L^2$-pairing with respect to the metric $\langle\ ,\ \rangle$ and the volume dx.

The situation described in this example is quite general. The key observation is that the total energy (6.3) written in Lagrangian coordinates is invariant under the action of the subgroup

$$\mathcal{D}^\infty(M)_{\rho_0} = \{\varphi \in \mathcal{D}^\infty(M) \mid \rho_0 = (\rho_0 \circ \varphi^{-1}) J(\varphi^{-1})\}.$$

This then leads to a momentum map and a reduction. The semidirect product reduction theorem (see Guillemin and Sternberg [1984], Ratiu [1981], and Marsden, Ratiu and Weinstein [1984a,b]) deals exactly with this situation. The outline of this result is the following. Let G be a Lie group and V a representation space of G. Denote by $G_a$ the isotropy of the induced representation of G on the dual V* at $a \in V^*$. Then $G_a \backslash T^*G$, the right coset space of T*G by the lift of the right translation of $G_a$ on G, is isomorphic as a Poisson manifold to the union of the orbits $\mathcal{O}_{(\mu,a)}$ in the dual of the semidirect product Lie algebra $\mathfrak{g} \ltimes V$, for all $\mu \in \mathfrak{g}^*$. Moreover, if $H_a$ is a family of Hamitonians on T*G compatible with the relevant actions (and thus, in particular, invariant under $G_a$), then this family induces a Hamiltonian H on $(\mathfrak{g} \ltimes V)^*$. In the example above we have $G = \mathcal{D}^\infty(M)$, $V = C^\infty(M,\mathbb{R})$, the representation is push-forward, $a = \rho_0 dX$, $H_a$ is the Hamiltonian (6.3) written in Lagrangian coordinates, and H is given by (6.3). For other applications of this theorem to magnetohydrodynamics and elasticity, see Marsden, Ratiu and Weinstein [1984a].

## 7. Symplectomorphisms and plasma physics

Let $(M,\omega)$ be a finite dimensional symplectic manifold without boundary. The set

$$\mathcal{D}^s_\omega(M) = \{\varphi \in \mathcal{D}^s(M) \mid \varphi^*\omega = \omega\}$$

of canonical diffeomorphisms of class $H^s$ (symplectomorphisms) on M is a subgroup of $\mathcal{D}^s(M)$. Proceeding as in section 5, it is shown that if M is compact, $\{\mathcal{D}^\infty_\omega(M), \mathcal{D}^s_\omega(M) \mid s > (\dim M)/2\}$ is an ILH-Lie subgroup of the ILH-Lie group $\{\mathcal{D}^\infty(M), \mathcal{D}^s(M) \mid s > (\dim M)/2\}$ with ILH-Lie algebra $\{\mathcal{X}^\infty_\omega(M), \mathcal{X}^s_\omega(M) \mid s > (\dim M)/2\}$; we denote by $\mathcal{X}^s_\omega(M) = \{\xi \in \mathcal{X}^s(M) \mid L_\xi\omega = 0\}$, the locally Hamiltonian vector fields of class $H^s$. See Ebin and Marsden [1970] and Omori [1974]. Again the charts on $\mathcal{D}^s_\omega(M)$ are obtained only implicitly. However, in this case, there is a direct method due to Weinstein [1971] to construct explicit charts using Poincaré's generating function. In Schmid [1978] it is shown directly using the Γ-differentiability, that $\mathcal{D}^\infty_\omega(M)$ is a Fréchet Lie group. If $\partial M \neq \emptyset$, due to the fact that the boundary conditions for $\xi$ and $i_\xi\omega$ are unrelated, there is no theorem stating that the symplectomorphism group is an ILH-Lie group.

The group of canonical transformations appears everywhere in mechanics. We refer to Arnold [1978], Abraham and Marsden [1978], and Choquet-Bruhat et al [1982] for Hamiltonian mechanics and to Chernoff and Marsden [1974] for the infinite dimensional situation. We shall limit ourselves here to a relatively recent application of the group of canonical transformation and its Lie algebra to the fundamental equations of plasma physics due to Marsden and Weinstein [1982].

A plasma is a collection of charged particles of various species (electrons, protons etc.) moving in $\mathbb{R}^3$. For simplicity one assumes that there is only one species of particles of charge q and mass m, and it is useful to approximate their positions $x \in \mathbb{R}^3$ and velocities v by a density on phase space which may be a smooth function $f(x,v,t)$. Denote by $E(x,t)$ and $B(x,t)$ the electric and magnetic fields respectively generated by the motion of the charged particles. Then the motion of the plasma can be described by the

*Maxwell-Vlasov* equations:

$$(7.1)\quad \begin{cases} \partial f/\partial t + v\cdot\partial f/\partial x + (q/m)\,(E + v \times B)\partial f/\partial v = 0 \\ \\ \partial B/\partial t = -\ \mathrm{curl}\ E \\ \\ \partial E/\partial t = \mathrm{curl}\ B - J_f, \text{ where the current density is} \\ \quad J_f = q\cdot\int vf(x,v,t)dv \\ \\ \mathrm{div}\ E = \rho_f, \text{ where the charge density is} \\ \quad \rho_f = q\int f(x,v,t)dv \\ \\ \mathrm{div}\ B = 0 \end{cases}$$

We think of this system of coupled, non-linear evolution equations as an initial value problem for the variables f, E and B.

In the limit case where B = 0 the Maxwell-Vlasov system reduces to a single equation of the field variable f(x,v,t)

$$(7.2)\qquad \frac{\partial f}{\partial t} + v \cdot \frac{\partial f}{\partial v} - \frac{q}{m}\frac{\partial \varphi_f}{\partial x} \cdot \frac{\partial f}{\partial x} = 0,$$

the Poisson-Vlasov equation, where the scalar potential $\varphi_f$ is defined by $\Delta\varphi_f = -\rho_f$. One can show that f(x,v,t) evolves in time by a canonical transformation $\eta_t$ of $\mathbb{R}^6$, i.e. $f(x,v,t) = \eta_t^* f(x,v,t_0)$ where $\eta_t \in \mathcal{D}_\omega(\mathbb{R}^6)$ and $\omega$ is the canonical symplectic structure on $\mathbb{R}^6$. If we identify any Hamiltonian vector field $X_h(x,v)$ on $\mathbb{R}^6$ with its Hamiltonian $h(x,v)$: $\mathbb{R}^6 \longrightarrow \mathbb{R}$ we get a Lie algebra isomorphism of $\mathcal{X}_\omega^\infty(\mathbb{R}^6)$ with $C^\infty(\mathbb{R}^6)$ via $[X_h,X_g] = -X_{\{h,g\}}$ where $\{h,g\}$ is the canonical Poisson bracket of functions on $\mathbb{R}^6$. Moreover, if we identify the dual of this Lie algebra via the $L^2$-pairing $\langle h,f\rangle = \int h(x,v)f(x,v)dxdv$ with itself, we can regard the plasma density f(x,v) as an element of $\mathfrak{g}^* = C^\infty(\mathbb{R}^6)$. Now the Poisson-Vlasov equation can be written in Lie-Poisson form on $C^\infty(\mathbb{R}^6)$ i.e. in the form

$$\dot{F} = \{F,H\}$$

where

$$(7.3) \qquad H(f) = \frac{1}{2} \int mv^2 f(x,v,t)dxdv + \frac{1}{2} \int \varphi_f \rho_f dx$$

is the total energy and $\{\ ,\ \}$ is the Lie-Poisson bracket on $C^\infty(\mathbb{R}^6)$, namely

$$\{F,G\}(f) = \int f \ \{\frac{\delta F}{\delta f},\frac{\delta G}{\delta f}\}dxdv.$$

Maxwell's equations $\dot{E}$ = curl B and $\dot{B}$ = -curl E are Hamilton's equations on the cotangent space $T^*\mathfrak{A}$ with respect to the canonical symplectic structure and the Hamiltonian $H(E,B) = \frac{1}{2} \int (|E|^2 + |B|^2)dx$ is the total field energy. Here the configuration space $\mathfrak{A} = \{A: \mathbb{R}^3 \longrightarrow \mathbb{R}^3\}$ is the space of vector potentials on $\mathbb{R}^3$, and B = curl A for some $A \in \mathfrak{A}$; then $(A,E) \in T^*\mathfrak{A}$. The invariance of Maxwell's equations under the gauge transformations $A \longmapsto A + \nabla\varphi$ for any $\varphi \in C^\infty(\mathbb{R}^3)$ leads via the reduction procedure to the two remaining Maxwell equations div E = $\rho$ and div B = 0.

Next we turn our attention to the full Maxwell-Vlasov equations (7.1) and remark that the same symmetry group $C^\infty(\mathbb{R}^3)$ acts on the space of plasma densities which leads to a momentum map on the coupled phase space $C^\infty(\mathbb{R}^6) \times T^*\mathfrak{A}$. Reducing by this symmetry one obtains a reduced phase space with a Poisson structure with respect to which the Maxwell-Vlasov equations are Hamiltonian, i.e. of the form $\dot{F} = \{F,H\}$ where

$$(7.4) \quad H(f,E,B) = \frac{1}{2}\int mv^2 f(x,v)dxdv + \frac{1}{2}\int |B(x)|^2 dx + \frac{1}{2}\int |E(x)|^2 dx$$

is the total energy of the plasma. The noncanonical Poisson bracket turns out to be the following. For any two functions F and G of the field variables (f,E,B) we have:

$$\{F,G\}(f,E,B) = \int f\left\{\frac{\delta F}{\delta f},\frac{\delta G}{\delta f}\right\}dx\,dv + \int\left(\frac{\delta F}{\delta E}\cdot \operatorname{curl}\frac{\delta G}{\delta B} - \frac{\delta G}{\delta E}\cdot \operatorname{curl}\frac{\delta F}{\delta B}\right)dx$$

$$+ \int\left(\frac{\delta F}{\delta E}\cdot\frac{\partial f}{\partial v}\frac{\delta G}{\delta f} - \frac{\delta G}{\delta E}\cdot\frac{\partial f}{\partial v}\frac{\delta F}{\delta f}\right)dx\,dv$$

$$+ \int fB\cdot\left(\frac{\partial}{\partial v}\frac{\delta F}{\delta f}\times\frac{\partial}{\partial v}\frac{\delta G}{\delta f}\right)dx\,dv.$$

Notice that the coupling of the plasma and electromagnetic fields appears in the Poisson structure rather than in the Hamiltonian, and is produced by the action of the infinite dimensional (Banach) gauge group on the uncoupled phase spaces of matter and fields.

In this example where a dynamical system of infinitely many degrees of freedom (the Maxwell-Vlasov equations) is described as a Hamiltonian system, infinite dimensional Lie groups appear as configuration space as well as symmetry groups. Similar structures were found for multifluid plasmas; see Marsden et al. [1983] and references therein.

**Remarks on the use of Hamiltonian formulation**

Having seen in the first three sections how diffeomorphism groups and their Lie algebras appear in the Hamiltonian formulation of fluid and plasma systems, we briefly comment on the possible applications of this Hamiltonian formalism for interacting systems. Some possible applications are:

- stability calculations
- normal mode calculations
- variational formulations
- perturbation theory
- classical-quantum relationship
- qualitative studies of periodic orbits and invariant tori
- search for conserved quantities and, for special equations on special coadjoint orbits, complete integrability
- applications of the Poincaré-Melnikov-Arnold method.

Few of these programs have been caried out.

Recently stability results for fluid and plasma systems were proved in Holm et al. [1985], and Abarbanel et al. [1985] by developing a method due to Arnold [1965], [1969]. This method relies heavily on the noncanonical Hamiltonian formalism. One finds that there are Casimir functionals that Poisson-commute with all functionals of the given dynamical variables and these enable one, when added as conserved quantities to the energy, to obtain convexity estimates, bounding the growth of perturbations. These estimates yield nonlinear stability, in the presence of a long-time existence theorem. These methods combined with the recent advances in the Hamiltonian structure of fluids and plasmas were used in the aforementioned papers to obtain nonlinear stability criteria of equilibria for a variety of fluid and plasma systems such as stratified Euler and Boussinesq flow, adiabatic fluids, MHD (magnetohydrodynamics), multifluid plasmas, and the Maxwell-Vlasov equations.

For normal mode calculations the Hamiltonian structure of plasma systems has been used in Spencer and Schmid [1984]. There, the Hamiltonian structure for fluid electrodynamics in the Coulomb case is used to derive the Bohm-Gross and ion-acoustic dispersion relations and the mode coupling in an unmagnetized homogeneous Coulomb plasma.

The Poincaré-Melnikov-Arnold method is used in proving the existence of Smale horseshoes in the phase portrait of vector fields. We refer the reader to Guckenheimer and Holmes [1983] and Holmes and Marsden [1983] for applications and further references.

## 8. <u>Diffeomorphism groups in general relativity</u>

In general relativity, vacuum spacetimes are Lorentz 4-manifolds whose metric g has vanishing Ricci curvature Ric(g). The equations Ric(g) = 0 are the vacuum Einstein field equations. The full symmetry group of the problem is the diffeomorphism group of spacetime.

In the 3+1 formulation of general relativity one wishes to split up the Lorentz manifold V, in an arbitrary manner, into "space" and "time". In this manner one is lead to the consideration of the manifold of spacelike embeddings of a three-dimensional oriented manifold M into V and the problem of solving Ric(g) = 0 converts to an initial value problem for a metric on M with g prescribed at $t = 0$. Thus, one regards the 3-metric as evolving in the infinite dimensional space $\mathcal{M}^s$ of $H^s$-metrics on M. This evolution turns out to be Hamiltonian and is commonly called the ADM formalism; for a concise treatment, see Fischer and Marsden [1972], [1979]. The diffeomorphism group $\mathcal{D}^s(M)$ plays the role of a symmetry group, thought of as the totality of coordinate transformations of space. Associated to this symmetry is the conservation law $\delta\pi = 0$, where $\pi$ is the conjugate momentum to the 3-metric in $T^*\mathcal{M}^s$.

The quotient space $\mathcal{M}^s/\mathcal{D}^s(M)$ which is relevant to this system is not a manifold but a stratified space as shown by Fischer [1970]. To avoid (or resolve) these singularities, the following procedure is used by Binz [1979]. Let $(\mathbb{R}^n,\langle,\rangle)$ be a fixed Euclidean space and $E^s(M,\mathbb{R}^n)$ the space of all $H^s$-embeddings of M into $\mathbb{R}^n$. Pulling back the scalar product $\langle,\rangle$ by any $j \in E^s(M,\mathbb{R}^n)$ yields the smooth map $m\colon E^s(M,\mathbb{R}^n) \longrightarrow \mathcal{M}^s$; $m(j) = j^*\langle,\rangle$. If n is large enough the map m is surjective by Nash's theorem. Right composition of $\mathcal{D}^s(M)$ on $E^s(M,\mathbb{R}^n)$ given by $j\circ\varphi$, for any $j \in E^s(M,\mathbb{R}^n)$ and $\varphi \in \mathcal{D}^s(M)$, turns $E^s(M,\mathbb{R}^n)$ into a smooth principal bundle with structure group $\mathcal{D}^s(M)$ and base manifold $U^s(M,\mathbb{R}^n) = E^s(M,\mathbb{R}^n)/\mathcal{D}^s(M)$, the $H^s$ manifold consisting of all submanifolds of $\mathbb{R}^n$ of diffeomorphism type M; see Binz and Fischer [1978]. The map m defines the commuting diagram

$$\begin{array}{ccc} E^s(M,\mathbb{R}^n) & \xrightarrow{\ m\ } & \mathcal{M}^s \\ \downarrow & & \downarrow \\ U^s(M,\mathbb{R}^n) & \xrightarrow{\ \bar{m}\ } & \mathcal{M}^s/\mathcal{D}^s(M) \end{array}$$

The fibers of m are $H^s$-manifolds (Binz [1984]). The Hamiltonian system for Einstein's evolution equation on $\mathcal{M}^s$ lifts to $E^s(M,\mathbb{R}^n)$. (This can be done in different ways, see also Kuchar [1976].) The solution curves on $E^s(M,\mathbb{R}^n)$ and hence those in the quotient $U^s(M,\mathbb{R}^n)$ (which is now a manifold) are mapped by m and $\bar{m}$ onto those of $\mathcal{M}^s$ and $\mathcal{M}^s/\mathcal{D}^s(M)$, respectively.

The study of the solution space $\mathcal{E}$ of Einstein's equations is a central problem in general relativity. Fischer and Marsden [1972] found general sufficient conditions for $\mathcal{E}$ to be a Sobolev (or Fréchet) manifold in terms of the Cauchy data for vacuum spacetimes and Moncrief [1980] showed that these are equivalent to the requirement of the absence of Killing fields.

If the metric g has non-trivial Killing fields, which is the most interesting case, $\mathcal{E}$ is no longer a smooth manifold, it has conical singularities in the neighborhood of spacetime metrics with non-trivial Killing vector fields and the reduced phase space $\mathcal{E}$/(gauge transformations) is considerably more complicated. Combined with an appropriate slice theorem and the reduction methods using momentum maps it was shown in Arms, Marsden and Moncrief [1982], that $\mathcal{E}$/(gauge transformations) is a stratified symplectic manifold. This structure theorem for the space of solutions is closely related to difficulties with perturbation series found originally by Brill and Deser [1968] and which led to the ideas of linearization stability (Fischer and Marsden [1973]). In order to make the slice theorem work one needs exact conditions on the topologies on the spaces $\mathfrak{M}^s$ and $\mathfrak{D}^s(M)$; they are in fact formulated in general for Sobolev $W^{s,p}$-spaces, see Arms et al. [1982] and Isenberg and Marsden [1982].

Another important theorem using infinite dimensional analysis was the important Positive Mass Theorem which states that any non-trivial perturbation of Minkowski space leads to a spacetime with strictly positive mass (or internal gravitational energy). The idea of the proof is to show that on the space $\mathcal{E}$ of solutions of Einstein's equations, the mass functional has a non-degenerate critical point at Minkowski (flat) space. Choquet-Bruhat and Marsden [1976] used a Morse lemma of Tromba [1976] to make the non-degenerate critical point idea (of Brill and Deser [1968]) work. The mass theorem was subsequently proved by Schoen and Yau [1979] using methods of minimal surfaces. Another proof has been given by Witten [1981] (see also Parker and Taubes [1982]) using spinors.

## 9. Gauge groups and quantum field theories

The diffeomorphism subgroups that arise in gauge theories as gauge groups behave nicely because they are isomorphic to subgroups of Hilbert Lie groups.

One considers a principal G-bundle $\pi: P \longrightarrow M$. The group $\mathfrak{G}$ of gauge transformations of P is defined by

$$\mathfrak{G} = \{\varphi \in \mathfrak{D}^{\infty}(P) \mid \varphi(p \cdot g) = \varphi(p) \cdot g,\ \pi\varphi(p) = \pi(p),\ p \in P,\ g \in G\}.$$

$\mathfrak{G}$ is a group under composition, hence a subgroup of the diffeomorphism group $\mathfrak{D}^{\infty}(P)$. Since a gauge transformation $\varphi$ preserves fibers we can realize each such $\varphi \in \mathfrak{G}$ via $\varphi(p) = p \cdot \tau(p)$ where $\tau: P \longrightarrow G$ satisfies $\tau(p \cdot g) = g^{-1}\ \tau(p)g$, for $p \in P$, $g \in G$. Let

$$\mathcal{C} \equiv \{\tau: P \longrightarrow G \mid \tau(p \cdot g) = g^{-1}\ \tau(p)\ g,\ p \in P,\ g \in G\}.$$

$\mathcal{C}$ is a group under pointwise multiplication, hence a subgroup of $C^{\infty}(P,G)$, which extends to a Hilbert Lie group if equipped with the $H^s$-Sobolev topology, (see example B, §1). We give $\mathcal{C}$ the induced topology and extend it to a Hilbert Lie group denoted by $\mathcal{C}^s$. There is a natural group isomorphism $\mathcal{C} \longrightarrow \mathfrak{G}$, $\tau \longmapsto \varphi$, defined by $\varphi(p) = p \cdot \tau(p)$, $p \in P$. This allows one to extend $\mathfrak{G}$ to a Hilbert Lie group $\mathfrak{G}^s$.

Let $\mathfrak{A}$ denote the affine space of connections on P. To each $A \in \mathfrak{A}$ one associates its covariant differential $D_A$ and its curvature two-form $F_A$. When A is regarded as a potential, i.e. as a Lie algebra valued one-form on P, then $F_A = D_A A = dA + (1/2)[A,A]$ is called the field strength of A. For $A \in \mathfrak{A}$ and $\varphi \in \mathfrak{G}$, the pulled back connection $\varphi^* A$ is the gauge transform of A. Thus $\mathfrak{G}$ acts as a transformation group on $\mathfrak{A}$. Topologizing $\mathfrak{A}$ with the $H^s$-topology, the $\mathfrak{G}^{s+1}$ action on $\mathfrak{A}^s$ is $C^{\infty}$. Lifting this action to the cotangent bundle $T^*\mathfrak{A}$, one gets the momentum map $J(A,\pi) = D_A^*\pi$, $(A, \pi) \in T^*\mathfrak{A}$. For a principal bundle whose base space M is a 4-dimensional spacetime with a compact spacelike Cauchy surface, e.g. $M = T^3 \times \mathbb{R}$, the space $T^*\mathfrak{A}$ is the appropriate phase space for a

Hamiltonian formulation of the Yang-Mills equations $D_A^* F_A = 0$. The space $T^*\mathfrak{A}/\mathfrak{G}$ was studied by Mitter [1980]. The solution set of the Yang-Mills equations is in 1-1 correspondence with the constraint set $J^{-1}(0) \subset T^*\mathfrak{A}$ of the momentum map J. The manifold structure of this set has an elegant complexity and has been studied by Arms [1981] and Moncrief [1980]. In the latter paper it is shown that $J^{-1}(0)$ is a submanifold except near points which represent symmetrical (or reducible) solutions of the Yang-Mills equations. At such points the structure is described by a quadratic form and together with the linearized constraint this is sufficient to describe the singularities in the solution set at symmetric points. Moreover, it is shown that the Yang-Mills dynamical system is Hamiltonian on the reduced phase space $J^{-1}(0)/\mathfrak{G}$ with respect to the energy $H = (1/2)(\|\pi\|^2 + \|F_A\|^2)$ (see also Nill [1983]). Arms [1981] showed that the Cauchy problem for the Yang-Mills equations is well posed and that linearization stability can be guaranteed when the fields are not too symmetrical. Similar results were found for coupled fields such as Einstein-Yang-Mills and Yang-Mills-Higgs fields; see Arms, Marsden and Moncrief [1982].

The gauge group $\mathfrak{G}$ plays the role of a symmetry group in general quantum field theories. One considers a principal G-bundle over the base space M of an oriented Riemannian 4-manifold and G a compact nonabelian Lie group, e.g. $M = S^4$, $G = SU(N)$. A connection one form $A \in \mathfrak{A}$ is considered as a vector potential and its curvature $F_A$ as the associated field. The $\mathfrak{G}$ action on $\mathfrak{A}$ is given by $D_{\varphi A} = \varphi D_A \varphi^{-1}$ and $F_{\varphi A} = \varphi F_A \varphi^{-1}$. From this it follows that the action function $S(A) = \|F_A\|^2$ is gauge invariant, $S(\varphi A) = S(A)$, i.e. it is the same for gauge equivalent potentials. In other words, the action S(A) depends on the orbit space $\mathfrak{A}/\mathfrak{G}$.

In the Feynman approach to quantum field theory one wants to compute

$$\int_{\mathfrak{A}} e^{-S(A)} f(A)\ \mathcal{D}(A) \Big/ \int_{\mathfrak{A}} e^{-S(A)}\ \mathcal{D}(A)$$

for certain functionals f(A). One of the difficulties is, that the integrand of the numerator may be constant on $\mathfrak{G}$-orbits and these have infinite measures. One should really integrate over the orbit space $\mathfrak{A}/\mathfrak{G}$. The physicists get around this difficulty by choosing a particular gauge, i.e. choosing a continuous section s: $\mathfrak{A}/\mathfrak{G} \longrightarrow \mathfrak{A}$ and then integrate over $s(\mathfrak{A}/\mathfrak{G})$ with a weight factor given by the Jacobian change of variables from one orbit to the other. This introduces the Faddeev-Popov ghosts as the integral of a measure along the fibers. Singer [1978] showed that such a gauge fixing is globally not possible, i.e. there is no global slice.

$\mathfrak{A}/\mathfrak{G}$ is in general not a manifold since the $\mathfrak{G}$ action on $\mathfrak{A}$ is not free. However, if Z denotes the set of constant gauge transformations with values in the center of $\mathfrak{G}$ and $\tilde{\mathfrak{G}} = \mathfrak{G}/Z$, then $\tilde{\mathfrak{G}}$ does act freely on $\mathfrak{B} = \{B \in \mathfrak{A} \mid B \text{ an irreducible connection}\}$. Then the space $\mathcal{N} = \mathfrak{B}/\tilde{\mathfrak{G}}$ is a nice infinite dimensional manifold; see Singer [1978], [1980] and Mitter [1980]. The topology of $\mathcal{N}$ is determined by local slices. $\mathcal{N}$ is an infinite dimensional Riemannian manifold whose tangent space at each point is isomorphic to the local slice. Moreover, $\mathfrak{B}$ is a principal bundle over $\mathcal{N}$ with group $\tilde{\mathfrak{G}}$ and there is a natural connection on $\mathfrak{B}$, locally given by $F_{\mu\nu}(B) = 2(D_B^* D_B)^{-1} b_{\mu\nu}^*$, for $B \in \mathfrak{B}$, where $b_{\mu\nu}^* = \sum_i [\mu_i, \nu_i]$, at each point. From this curvature field one can compute the characteristic classes; see Singer [1980].

There is another property of connections which is invariant under the $\mathfrak{G}$-action on $\mathfrak{A}$, and this is the notion of self-dual connections, self-dual Yang-Mills fields, or instantons. On a 4-manifold M a connection A is self-dual if $F_A = * F_A$ (where $*$ is the Hodge star operator). In fact, since the star operator is conformally invariant on 2-forms, the property of self-duality of a connection is invariant under the larger group of transformations of a principal bundle consisting of those which induce conformal transformations on the base space M. It is shown in Atiyah, Hitchin, and Singer [1978] that for a compact self-dual Riemannian 4-manifold M with positive scalar curvature and principal G-bundle P over M with G a compact semisimple Lie group, the moduli space $\mathfrak{M} = \{[A] \in \mathcal{N} \mid A \text{ self-dual}\}$ of irreducible self-dual connections modulo gauge transformations (if nonempty) is a smooth finite dimensional submanifold of $\mathcal{N}$ and $\dim \mathcal{M} = p_1(\mathrm{ad}) - (1/2)(\dim G)(\chi - \tau)$, where $p_1(\mathrm{ad})$ is the first Pontrjagin class of the adjoint bundle, $\chi$ is the Euler characteristic of M, and $\tau$ is the signature of M. The proof of this important result involves a careful study of the $H^s$-Sobolev topologies on the infinite dimensional spaces involved, the Atiyah-Singer index theorem for the Dirac operator and the inverse and implicit function theorems for Banach spaces.

One problem with the moduli space $\mathcal{M}$ is, that it is not compact, but the analytic work of Uhlenbeck [1982] contains the detailed analysis at infinity in the moduli space and here the bubble theorem gives the exact structure at infinity. Over a compact region $\Omega \subset M$, of an oriented 4-manifold M, self-dual connections absolutely minimize the Yang-Mills action integral

$$\mathcal{YM}(A) = \frac{1}{2}\int_\Omega \|F_A\|^2$$

and it is shown by Uhlenbeck [1982] that for a sufficiently small constant $\epsilon$ the set

$$\{[A] \in \mathcal{M} \mid A \text{ self-dual}, \frac{1}{2}\int_\Omega \|F_A\|^2 < \epsilon\}$$

is compact. For noncompact regions this is true for the modified functional

$$\mathcal{YM}_{(p)}(A) = \frac{1}{2}\int_\Omega \|F_A\|^p, \; p > 2.$$

The constant $\epsilon$ has to be relatively small, for example if we take $\Omega = M = S^4$ then on a line bundle with instanton number (index) k the energy of a self-dual connection is $4\pi^2 k$, so we must have $\epsilon < 4\pi^2 k$. (See Lawson [1983] and Uhlenbeck and Freed [1984], §8 for details.) Using implicit function theorems on Banach spaces, Taubes [1983] constructed self-dual solutions for $k = 1$ to the Yang-Mills equations on a compact 4-manifold with positive definite intersection form, induced by the standard t'Hooft instantons on the

4-sphere $S^4$. In a subsequent paper Taubes [1984] constructed self-dual solutions for k sufficiently large on manifolds with indefinite intersection matrix. Moreover, regularity results and stability estimates were found for the Yang-Mills equations; see e.g. Uhlenbeck [1982], Lawson [1983], Taubes [1983].

Another application of diffeomorphism groups to quantum theory is given in Goldin and Sharp [1983]. There, unitary representations of semidirect products of diffeomorphism groups with vector spaces of tensor fields are used to describe many different quantum systems, such as n-particle Bose and Fermi systems, infinite free Bose gas, a quantum particle with internal degrees of freedom both with integral and 1/2 integral spin. From the local currents of these representations the generators of the symmetries are constructed.

## 10. <u>The globally Hamiltonian vector fields and quantomorphism groups</u>

In this section we want to discuss some examples to the problem of finding a Lie group for a given Lie algebra in infinite dimensions. We have already seen in §7 that the ILH-Lie algebra $\mathcal{X}^\infty_\omega(M)$ of all locally Hamiltonian vector fields has as underlying ILH-Lie group $\mathcal{D}^\infty_\omega(M)$ the group of canonical transformations (symplectomorphisms). In Ratiu and Schmid [1981], the Lie groups for the globally Hamiltonian vector fields and for the infinitesimal quantomorphisms have been found. We describe these results below.

**10.1.** Let $(M,\omega)$ be a compact symplectic manifold. Then the symplectomorphisms $\mathcal{D}^\infty_\omega(M)$ are an ILH-Lie group with ILH-Lie algebra $\mathcal{X}^\infty_\omega(M)$, the locally Hamiltonian vector fields. The commutator algebra is $[\mathcal{X}^\infty_\omega(M),\mathcal{X}^\infty_\omega(M)] = \mathcal{H}^\infty_\omega(M)$, the *globally* Hamiltonian vector fields on M. (Recall that a globally Hamiltonian vector field $X_f$ corresponds to the one form df of a function f under the isomorphism between one forms and vector fields defined by $\omega$.) Then it is shown that the commutator subgroup $[\mathcal{D}^\infty_\omega(M)_0,\mathcal{D}^\infty_\omega(M)_0]$ of the identity component $\mathcal{D}^\infty_\omega(M)_0$ in $\mathcal{D}^\infty_\omega(M)$ is an ILH-Lie subgroup of $\mathcal{D}^\infty_\omega(M)$ with ILH-Lie algebra $\mathcal{H}^\infty_\omega(M)$ (under a homotopy condition on $\mathcal{D}^\infty_\omega(M)_0$). Note that this is only true as ILH-Lie groups; it is not true that $\mathcal{H}^s_\omega(M)$ equals $[\mathfrak{X}^s_\omega(M),\mathfrak{X}^s_\omega(M)]$, due to the loss of derivatives. Nevertheless an explicit construction is given for the tower of groups $\mathfrak{H}^s(M)$ of the Lie algebra tower of $H^s$-globally Hamiltonian vector fields $\mathcal{H}^s_\omega(M)$.

10.2. As we have already seen in the example of plasma physics (see section 7) another Lie algebra associated with a symplectic manifold $(M,\omega)$ is of special interest, namely the space of smooth functions $\mathcal{C}^\infty_\omega(M)$ with bracket $[f,g] = -\{f,g\}$, the Poisson bracket being defined by $\omega$; $\{f,g\} = \omega(X_f,X_g)$. For $X_f$, $X_g \in \mathcal{H}^\infty_\omega(M)$ the generating Hamiltonian for their bracket $[X_f,X_g]$ is $-\{f,g\}$, so the map $f \longmapsto X_f$ is a Lie algebra homomorphism of $\mathcal{C}^\infty_\omega(M)$ into $\mathcal{H}^\infty_\omega(M)$. Since $\omega$ is nondegenerate, the center of $\mathcal{C}^\infty_\omega(M)$ consists of constant functions (assuming M is connected). In other words the sequence

$$0 \longrightarrow \mathbb{R} \longrightarrow \mathcal{C}^\infty_\omega(M) \longrightarrow \mathcal{H}^\infty_\omega(M) \longrightarrow 0$$

is exact, i.e. $\mathcal{C}^\infty_\omega(M)$ is a central extension of $\mathcal{H}^\infty_\omega(M)$. An obvious guess for the Lie group of $\mathcal{C}^\infty_\omega(M)$ is to look for a one-dimensional central extension of $[\mathcal{D}^\infty_\omega(M)_0,\mathcal{D}^\infty_\omega(M)_0]$ and to realize this extension explicitly. This is done in the $H^s$-topologies, i.e. in the ILH-category, as follows.

Let $(N,\theta)$ be a compact, exact, contact manifold. Assume that the foliation $\mathcal{F}$ of N defined by the fundamental vector field defines the symplectic manifold $(M = N/\mathcal{F},\omega)$ such that N is a principal circle bundle over M with connection form $\theta$ and curvature $\omega$, i.e. N is the quantizing manifold of M whose automorphism group $\mathcal{D}^s_\theta(N) = \{\eta \in \mathcal{D}^s(N),\ \eta^*\theta = \theta\}$ is the quantomorphism group of $(N,\theta)$. Its identity component $\mathcal{D}^s_\theta(N)_0$ is the central extension by $S^1$ of the $H^s$ completion of $[\mathcal{D}^\infty_\omega(M)_0,\mathcal{D}^\infty_\omega(M)_0]$. The ILH-Lie algebra $\mathfrak{X}^\infty_\theta(N) = \{X \in \mathfrak{X}^\infty(N) \mid L_X\theta = 0\}$ of $\mathcal{D}^\infty_\theta(N)$ is isomorphic to the Lie algebra $\mathcal{C}^\infty_\omega(M)$. But as in the finite dimensional situation there is another ILH-Lie group with the same ILH-Lie algebra, $S^1 \times [\mathcal{D}^\infty_\omega(M)_0,\mathcal{D}^\infty_\omega(M)_0]$, which has as Lie algebra $\mathcal{C}^\infty_\omega(M)$ too.

## 11. The group of homogeneous symplectomorphisms of $T^*M \setminus \{0\}$

The symplectic manifold most widely used is the cotangent bundle $T^*M$ of a manifold M. But the symplectic form on $T^*M$ is exact, i.e. it is the differential of a canonical one-form $\theta$ on $T^*M$. Thus one can ask about the structure of diffeomorphisms of $T^*M$ that preserve $\theta$. It is well-known that a diffeomorphism $\varphi: T^*M \longrightarrow T^*M$ satisfying $\varphi^*\theta = \theta$ is necessarily a lift, i.e. $\varphi = T^*\eta$, for $\eta \in \mathcal{D}^\infty(M)$. In order to avoid this trivial situation, consider $T^*M \setminus \{0\}$, and its diffeomorphisms preserving $\theta$. Then it is known that $\varphi^*\theta = \theta$ if and only if $\varphi$ is symplectic and homogeneous of degree one, i.e. $\varphi(\tau\alpha_m) = \tau\varphi(\alpha_m)$ for all $\tau > 0$ and $\alpha_m \in T^*_mM$. Now we can consider the group $\mathcal{D}^s_\theta(T^*M \setminus \{0\})$ of homogeneous symplectomorphisms of $T^*M \setminus \{0\}$. But right away we are faced with the problem of non-compactness of $T^*M \setminus \{0\}$. We shall sketch below, following Ratiu and Schmid [1981] how $\{\mathcal{D}^\infty_\theta(T^*M \setminus \{0\}), \mathcal{D}^s_\theta(T^*M \setminus \{0\}) \mid s \geq \dim M + 1/2\}$ is an ILH-Lie group with ILH-Lie algebra $\{\mathcal{S}^\infty(T^*M \setminus \{0\}), \mathcal{S}^{s+1}(T^*M \setminus \{0\}) \mid s > \dim M + 1/2\}$, where $\mathcal{S}^s(T^*M \setminus \{0\}) = \{H: T^*M \setminus \{0\} \longrightarrow \mathbb{R} \mid H$ is of class $H^s$ and homogeneous of degree one$\}$, with the Poisson bracket as Lie algebra bracket. Note the gain in one derivative at the Lie algebra level. The basic idea of the ensuing discussion is that $\mathcal{D}^s_\theta(T^*M \setminus \{0\})$ is algebraically isomorphic to the group of all $H^s$ contact transformations of the cosphere bundle of M, which is a compact manifold if M is. We start by recalling the relevant facts.

The multiplicative group of strictly positive reals $\mathbb{R}_+$ acts smoothly on $T^*M \setminus \{0\}$ by $\alpha_x \longmapsto \tau\alpha_x$, $\tau > 0$, $\alpha_x \in T^*_xM$, $\alpha_x \neq 0$. This action is free and proper and therefore $\pi: T^*M \setminus \{0\} \longrightarrow Q \equiv (T^*M \setminus \{0\})/\mathbb{R}_+$ is a smooth principal fiber bundle over Q, the *cosphere bundle* of M. Note that Q is compact (supposing M is) and odd-dimensional. Q carries no canonical contact one-form but for each global section $\sigma: Q \longrightarrow T^*M \setminus \{0\}$ we can define an exact contact one-form $\theta_\sigma$ on Q by $\theta_\sigma = \sigma^*\theta$. Such global sections exist in abundance; for example, any Riemannian metric on M identifies $T^*M$ with TM and Q with the unit sphere bundle. Then the usual inclusion of the sphere bundle into TM gives a section $\sigma$. The section $\sigma$ is uniquely determined by a smooth function $f_\sigma$:

$T^*M\setminus\{0\} \longrightarrow \mathbb{R}_+$ defined by $\sigma(\pi(\alpha_x)) = f_\sigma(\alpha_x)\alpha_x$. In other words, $f_\sigma$ measures how far from the section $\sigma$ an element $\alpha_x \in T^*M\setminus\{0\}$ lies. The function $f_\sigma$ is homogeneous of degree -1 and $\pi^*\theta_\sigma = f_\sigma\theta$.

An $H^{s+1}$ *contact transformation* on Q is a diffeomorphism $\varphi \in \mathcal{D}^{s+1}(Q)$ such that for any two sections $\sigma,\zeta$: $Q \longrightarrow T^*M\setminus\{0\}$, there exists an $H^{s+1}$ function $h_{\sigma\zeta}: Q \longrightarrow \mathbb{R}_+$ satisfying $\varphi^*\theta_\sigma = h_{\sigma\zeta}\theta_\zeta$. Equivalently, $\varphi \in \mathcal{D}^{s+1}(Q)$ is an $H^{s+1}$ contact transformation if and only if for each global section $\sigma$ there exists an $H^{s+1}$ function $h_\sigma: Q \longrightarrow \mathbb{R}_+$ such that $\varphi^*\theta_\sigma = h_\sigma\theta_\sigma$. The function $h_\sigma$ is uniquely determined by $\sigma$, namely $h_\sigma = \langle\varphi^*\theta_\sigma, E_\sigma\rangle$, where $E_\sigma$ is the Reeb vector field on Q determined by the contact structure $\theta_\sigma$ and $\langle,\rangle$ denotes the pairing between vector fields and one forms. ($E_\sigma$ is the unique vector field satisfying $\langle\theta_\sigma, E_\sigma\rangle = 1$ and $i_{E_\sigma}(d\theta_\sigma) = 0$, where $i_{E_\sigma}(d\theta_\sigma)$ denotes the interior product of $E_\sigma$ with $d\theta_\sigma$; in local coordinates $(x^1,\ldots,x^{n-1},y^1,\ldots,y^{n-1},t)$ on Q, where $\theta_\sigma = \sum_{i=1}^{n-1} y^i dx^i + dt$ we have $E_\sigma = \frac{\partial}{\partial t}$.) Therefore the group of $H^{s+1}$-contact transformations on Q is isomorphic to the group

$$\mathrm{Con}_\sigma^{s+1}(Q) = \{(\varphi,h) \in \mathcal{D}^{s+1}(Q) \ltimes H^{s+1}(Q,\mathbb{R}\setminus\{0\}) \mid \varphi^*\theta_\sigma = h\theta_\sigma\}$$

for any fixed but arbitrary global section $\sigma$, where $\mathcal{D}^{s+1}(Q) \ltimes H^{s+1}(Q,\mathbb{R}\setminus\{0\})$ is the semidirect product of the Lie groups $\mathcal{D}^{s+1}(Q)$ and $H^{s+1}(Q,\mathbb{R}\setminus\{0\})$ ($H^{s+1}(Q,\mathbb{R}\setminus\{0\})$ as a multiplicative group) with composition law $(\varphi_1,h_1)\cdot(\varphi_2,h_2) = ((\varphi_1\circ\varphi_2), h_2(h_1\circ\varphi_2))$. Omori [1974] has shown that $\mathrm{Con}_\sigma^{s+1}(Q)$ is a closed Lie subgroup of the Lie group $\mathcal{D}^{s+1}(Q) \ltimes H^{s+1}(Q,\mathbb{R}\setminus\{0\})$. The Lie algebra of $\mathcal{D}^{s+1}(Q) \ltimes H^{s+1}(Q,\mathbb{R}\setminus\{0\})$ is the semidirect product $\mathfrak{X}^{s+1}(Q) \ltimes H^{s+1}(Q,\mathbb{R})$ of $H^{s+1}$-vector fields and $H^{s+1}$-functions, with bracket $[(X,f),(Y,g)] = ([X,Y], X(g) - Y(f))$. The Lie algebra of $\mathrm{Con}_\sigma^{s+1}(Q)$ is $\mathrm{con}_\sigma^{s+1}(Q) = \{(Y,g) \in \mathfrak{X}^{s+1}(Q) \ltimes H^{s+1}(Q,\mathbb{R}) \mid L_Y\theta_\sigma = g\theta_\sigma\}$ ($L_Y$ denoting the Lie derivative along the vector field Y).

In Ratiu and Schmid [1981] (theorem 4.1) it is shown that the group $\mathcal{D}_\theta^{s+1}(T^*M\setminus\{0\})$ is isomorphic (as a group) to the Lie group $\mathrm{Con}_\sigma^{s+1}(Q)$. The isomorphism is given by $\Phi$: $\mathcal{D}_\theta^{s+1}(T^*M\setminus\{0\}) \longrightarrow \mathrm{Con}_\sigma^{s+1}(Q)$, $\Phi(\eta) = (\varphi,h)$ where $\varphi$ is defined by $\varphi\circ\pi = \pi\circ\eta$ and h by $h\circ\pi = (f_\sigma\circ\eta)/f_\sigma$, $\sigma(\pi(\alpha_x)) = f_\sigma(\alpha_x)\alpha_x$. The inverse of $\Phi$ is given by $\Phi^{-1}(\varphi,h) = (\sigma\circ\varphi\circ\pi)/(h\circ\pi)\cdot f_\sigma$.

Since $\mathrm{Con}_\sigma^{s+1}(Q)$ and $\mathrm{Con}_\varsigma^{s+1}(Q)$ are isomorphic as ILH-Lie groups for any two global sections $\sigma$ and $\varsigma$, the isomorphism $\Phi$ determines an ILH-Lie group structure on $\mathcal{D}_\theta^{s+1}(T^*M\setminus\{0\})$ which is independent of $\sigma$ (or independent of the Riemannian metric if $\sigma$ is induced from such). Furthermore, the Lie algebra $\mathcal{X}_\theta^{s+1}(T^*M\setminus\{0\}) = \{Y \in \mathcal{X}^{s+1}(T^*M\setminus\{0\}) \mid L_Y\theta = 0\}$ of $\mathcal{D}_\theta^{s+1}(T^*M\setminus\{0\})$ is isomorphic to $\mathcal{S}^{s+2}(T^*M\setminus\{0\}) = \{H \in H^{s+2}(T^*M\setminus\{0\},\mathbb{R}) \mid H \text{ homogeneous of degree one}\}$. This is because $L_Y\theta = 0$ if and only if Y is globally Hamiltonian, homogeneous of degree zero, with Hamiltonian function $H = \theta(Y)$, homogeneous of degree one. Moreover, the Lie algebras $\mathcal{X}_\theta^{s+1}(T^*M\setminus\{0\})$ and $\mathrm{con}_\sigma^{s+1}(Q)$ are isomorphic via $T_e\Phi$: $\mathcal{X}_\theta^{s+1}(T^*M\setminus\{0\}) \longrightarrow \mathrm{con}_\sigma^{s+1}(Q)$. Explicitly, $T_e\Phi(X_H) = (X,k)$, where X is uniquely defined by $T\pi\circ X_H = X\circ\pi$ and k by $k\circ\pi = \{f_\sigma,H\}/f_\sigma$, ($\{,\}$ is the canonical Poisson bracket on $T^*M$). The map $H \longmapsto H\circ\sigma$ is an isomorphism from $\mathcal{S}^{s+2}(T^*M\setminus\{0\})$ onto $H^{s+2}(Q,\mathbb{R})$, with inverse j: $H^{s+2}(Q,\mathbb{R}) \longrightarrow \mathcal{S}^{s+2}(T^*M\setminus\{0\})$, where j(f) is the extension to $T^*M\setminus\{0\}$ by homogeneity of degree one of $f\circ\sigma^{-1}$: $\sigma(Q) \subset (T^*M\setminus\{0\}) \longrightarrow \mathbb{R}$, for $f \in H^{s+2}(Q,\mathbb{R})$. The composition of these two isomorphisms with $T_e\Phi^{-1}$ gives an isomorphism

$$F:\quad \mathrm{con}_\sigma^{s+1}(Q) \xrightarrow{T_e\Phi^{-1}} \mathcal{X}_\theta^{s+1}(T^*M\setminus\{0\}) \longrightarrow \mathcal{S}^{s+2}(T^*M\setminus\{0\}) \xrightarrow{j^{-1}} H^{s+2}(Q,\mathbb{R}),$$

$F(X,k) = \theta_\sigma(X)$. In the condition $L_X\theta_\sigma = k\theta_\sigma$, the function k is uniquely determined by X, namely, $k = E_\sigma(\theta_\sigma(X))$. From this it follows easily that F is continuous and hence an isomorphism between $\mathrm{con}_\sigma^{s+1}(Q)$ and $H^{s+2}(Q,\mathbb{R})$ (note the gain of one derivative). We see thus once again that $\mathrm{Con}_\sigma^{s+1}(Q)$ and $\mathrm{Con}_\varsigma^{s+1}(Q)$ are isomorphic as

ILH-Lie groups for any two global sections $\sigma, \zeta$: $Q \longrightarrow T^*M \setminus \{0\}$, since both are modeled on $H^{s+2}(Q, \mathbb{R})$.

Defining the Hilbert space structure of $\mathcal{S}^{s+2}(T^*M \setminus \{0\})$ as the one induced by the isomorphism j: $H^{s+2}(Q, \mathbb{R}) \longrightarrow \mathcal{S}^{s+2}(T^*M \setminus \{0\})$, it follows that $H^{s+2}(Q, \mathbb{R})$, $\mathcal{S}^{s+2}(T^*M \setminus \{0\})$, $\mathcal{X}_\theta^{s+1}(T^*M \setminus \{0\})$ and $\mathrm{con}_\sigma^{s+1}(Q)$ are all isomorphic as Hilbert spaces. It is desirable to compare the topology of $\mathcal{S}^{s+2}(T^*M \setminus \{0\})$ with the strong $C^1$-Whitney topology. Since all elements of $\mathcal{S}^{s+2}(T^*M \setminus \{0\})$ are $C^2$ by the Sobolev imbedding theorem, we can define a new topology on $\mathcal{S}^{s+2}(T^*M \setminus \{0\})$ in the following way: a neighborhood of zero consists of all those functions $H \in \mathcal{S}^{s+2}(T^*M \setminus \{0\})$ for which $dH$: $(T^*M \setminus \{0\}) \longrightarrow T^*(T^*M \setminus \{0\})$ is $C^1$-close to zero in the strong $C^1$-Whitney topology. Denote by $\mathcal{S}_W^{s+2}(T^*M \setminus \{0\})$ the space $\mathcal{S}^{s+2}(T^*M \setminus \{0\})$ equipped with this new topology. It is then a routine matter to check that j: $H^{s+2}(Q, \mathbb{R}) \longrightarrow \mathcal{S}_W^{s+2}(T^*M \setminus \{0\})$, or equivalently the identity $\mathcal{S}^{s+2}(T^*M \setminus \{0\}) \longrightarrow \mathcal{S}_W^{s+2}(T^*M \setminus \{0\})$, is continuous with discontinuous inverse, i.e. the new topology is strictly coarser than the original one on $\mathcal{S}^{s+2}(T^*M \setminus \{0\})$. This remark is useful in the construction of an explicit chart at e in $\mathcal{D}_\theta^{s+1}(T^*M \setminus \{0\})$.

__Remarks__. (1) The gain of one derivative at the Lie algebra level has a corresponding statement in $\mathcal{D}_\theta^{s+1}(T^*M \setminus \{0\})$: for every $\eta \in \mathcal{D}_\theta^{s+1}(T^*M \setminus \{0\})$, $\tau^* \circ \eta$: $(T^*M \setminus \{0\}) \longrightarrow M$ is of class $H^{s+2}$, where $\tau^*$: $T^*M \longrightarrow M$ is the cotangent bundle projection. Locally, this means that if $\eta(x, \alpha) = (y(x, \alpha), \beta(x, \alpha))$, then y is $H^{s+2}$ jointly in x and $\alpha$. To prove this, note that $\eta^*\theta = \theta$ is equivalent locally to

$$\sum_{i=1}^{n} \beta_i \frac{\partial y^i}{\partial x^k} = \alpha_k, \qquad \sum_{i=1}^{n} \beta_i \frac{\partial y^i}{\partial \alpha_k} = 0, \qquad k = 1, \ldots, n.$$

Since $\eta$ is a diffeomorphism of class $H^{s+1}$, for fixed x, there exists a unique $\alpha$ such that $\beta = (0, \ldots, 1, \ldots, 0)$, the i-th basis vector. For this choice of $\alpha$, the first relation shows that the i-th column of the matrix $\left[\frac{\partial y^i}{\partial x^k}\right]$ is $H^{s+1}$. This says that $y(x, \alpha)$ has all derivatives of order at most s+2 square integrable except the derivatives involving only $\alpha_k$'s. The second relation is an elliptic equation with $H^{s+1}$ coefficients of first order in $y^i$ regarded as a function of $\alpha$ only (its symbol maps $(\xi^i) \in \mathbb{R}^n$, to $(\beta_1 + \dots + \xi^i\beta_i + \dots + \beta_n) \in \mathbb{R}^n$) and thus its solution is of class $H^{s+2}$, i.e. the (s+2)nd derivative of y with respect to $\alpha$ is square integrable and thus y is of class $H^{s+2}$.

(2) Let $\eta \in \mathcal{D}_\theta^{s+1}(T^*M\setminus\{0\})$ be fiber preserving, i.e. $\tau^*\eta(\alpha_x) = \tau^*\eta(\alpha_x')$ for all $\alpha_x, \alpha_x' \in T_x^*M\setminus\{0\}$. Then $\eta$ can be extended $H^{s+1}$-smoothly to the zero section by $\eta(0_x) = 0_y$ for $y = \tau^*\eta(\alpha_x)$, $\alpha_x \in T^*M\setminus\{0\}$. So $\eta$: $T^*M \longrightarrow T^*M$, $\eta^*\theta = \theta$ and hence $\eta = T^*g$ for an $H^{s+2}$ diffeomorphism g: M $\longrightarrow$ M. In particular if $\tau^*\eta(\alpha_x) = x$ for all $\alpha_x \in T^*M\setminus\{0\}$, then $\eta = e$. From this it follows that the effect of $\eta$ on base points uniquely determines $\eta$, i.e. if $\eta, \bar{\eta} \in \mathcal{D}_\theta^{s+1}(T^*M\setminus\{0\})$ satisfy $\tau^*\circ\eta = \tau^*\circ\bar{\eta}$, then $\eta = \bar{\eta}$ (since $\tau^*(\bar{\eta}\circ\eta^{-1})(\alpha_x) = x$).

## 12. Fourier integral operators as an ILH-Lie group and completely integrable PDE's

Fourier integral operators, FIO, generalize pseudodiffeential operators, $\psi$DO, which themselves are a generalization of differential operators, DO.

$$\text{FIO} \supset \psi\text{DO} \supset \text{DO}.$$

(For general background information see e.g. Duistermaat [1973], Hörmander [1971], Kumano-go [1981], Taylor [1980], Treves [1981].) Consider a differential operator P on $\Omega \subset \mathbb{R}^n$, of order m with smooth coefficients $a_\alpha$

$$Pu(x) = \sum_{|\alpha|\leqslant m} a_\alpha(x)\, D_x^\alpha\, u(x), \qquad u \in C^\infty(\Omega).$$

We associate to the operator P the *polynomial*

$$p(x,\xi) = \sum_{|\alpha|\leqslant m} a_\alpha(x)\, \xi^\alpha$$

called the *symbol* of P. Using the Fourier transform $\hat{u}(\xi)$ of u(x) we can write (where $(x \cdot \xi = \sum_{i=1}^{n} x_i \xi_i)$

$$Pu(x) = (2\pi)^{-n} \int e^{ix\cdot\xi} p(x,\xi)\, \hat{u}(\xi) d\xi$$

(12.1)

$$= (2\pi)^{-n} \iint e^{i(x-y)\cdot\xi}\, p(x,\xi)\, u(y) dy d\xi .$$

A pseudodifferential operator is an operator P of the form (12.1) but with a symbol $p(x,\xi)$ of a more general class than polynomials.

A smooth function $p(x,\xi)$ on $\Omega \times \mathbb{R}^n$ belongs to the symbol class $S^m(\Omega)$, if for any compact subset $K \subset \Omega$ and every n-tuples $\alpha$, $\beta$, there is a constant $C_{\alpha,\beta}(K) > 0$ such that

$$|D_x^{\beta} D_{\xi}^{\alpha} p(x,\xi)| \leqslant C_{\alpha,\beta}(K)\ (1 + |\xi|)^{m-|\alpha|}$$

for all $x \in K$, $\xi \in \mathbb{R}^n$. We restrict ourselves to *classical* symbols, i.e. those which have an asymptotic expansion

$$p(x,\xi) \sim \sum_{j=0}^{\infty} p_{m-j}(x,\xi) \qquad (m = \text{order of } p)$$

where each $p_{m-j}(x,\xi) \in C^{\infty}(\Omega \times (\mathbb{R}^n \setminus \{0\}))$ is homogeneous of degree (m-j) in $\xi$, $(p_{m-j}(x,\tau\xi) = \tau^{m-j} \cdot p_{m-j}(x,\xi),\ \tau > 0)$. Then a classical $\psi$DO of order m is of the form

$$Pu(x) = (2\pi)^{-n} \iint e^{i(x-y)\cdot\xi}\, p(x,\xi)\, u(y) dy d\xi \tag{12.2}$$

where $p(x,\xi)$ is a classical symbol of order m. The principal symbol of P is the leading term $p_m(x,\xi)$. These integrals are highly singular but make sense as oscillatory integrals, and the operators are nice in the following sense:

(i) they are invariant under diffeomorphisms, so they can be defined on manifolds M;

(ii) $P: C^{\infty}(M) \longrightarrow C^{\infty}(M)$ extends continuously to distributions;

$P: \mathcal{E}'(M) \longrightarrow \mathcal{E}'(M)$;

(iii) they are closed under multiplication, $P, Q \in \psi DO \Rightarrow P \circ Q \in \psi DO$;

(iv) they have properties which are "close" to those of differential operators:

$$P \in DO \Leftrightarrow P \text{ is local i.e. supp } Pu \subset \text{supp } u;$$

$$P \in \psi DO \Rightarrow P \text{ is pseudolocal i.e. sing supp } Pu \subset \text{sing supp } u.$$

Moreover they preserve the wave front set WF, i.e. $WF(Pu) \subset WF(u)$, ($WF(u) \subset T^*M \setminus \{0\}$ and $\tau^* WF(u) = \text{sing supp } u$).

Fourier integral operators, FIO, generalize pseudodifferential operators; they have properties (i), (ii) and (iii) but generalize (iv) in the following sense. The Fourier integral operator A moves WF by a canonical relation $\Lambda$, i.e.

$$WF(Au) \subset \Lambda \circ WF(u)$$

where $\Lambda \subset (T^*M \setminus \{0\}) \times (T^*M \setminus \{0\})$ is a conic Lagrangian submanifold, locally generated by the phase function $\varphi(x,y,\xi)$, i.e. $\Lambda = \{(x,y,d_{(x,y)}\varphi(x,y,\xi) \mid d_\xi \varphi(x,y,\xi) = 0\}$. Assume $A \in FIO$ is invertible with $A^{-1} \in FIO$. This implies that $\Lambda = \text{graph } (\eta)$ where $\eta: T^*M \setminus \{0\} \longrightarrow T^*M \setminus \{0\}$ is a diffeomorphism. We have:

$$\left.\begin{array}{l} \Lambda \text{ Lagrangian} \Leftrightarrow \eta^* \omega = \omega \\ \Lambda \text{ conic} \Leftrightarrow \eta \text{ homogeneous of degree } + 1 \end{array}\right\} \Leftrightarrow \eta^* \theta = \theta$$

where $\theta$ is the canonical one-form on $T^*M$ and $\omega = d\theta$ the canonical symplectic structure on $T^*M$. Notice that $P \in \psi DO$ implies that $\Lambda$ is the diagonal; i.e. $\Lambda = \text{graph } (e)$, where $e = \text{id}: T^*M \longrightarrow T^*M$ is the identity, so that $\psi DO \subset FIO$. If $\eta$ is near the identity and $\varphi(x,y,\xi)$ is a local generating function for graph $(\eta)$ then a classical Fourier integral operator A of order m can be written in a local chart of M in the form

$$(12.3) \qquad Au(x) = (2\pi)^{-n} \iint e^{i\varphi(x,y,\xi)}\, a(x,\xi)u(y)dyd\xi$$

where $a(x,\xi)$ is a classical symbol of order m. Notice that the phase function $\varphi(x,y,\xi)$ of a Fourier integral operator generalizes the phase function $(x-y)\cdot\xi$ of a pseudodifferential operator in the sense that $\varphi(x,y,\xi)$ is smooth in x, y and $\xi$, $\xi \neq 0$, homogeneous of degree one in $\xi$, and nondegenerate.

Denote by $\mathcal{D}^{\infty}_{\theta}(T^*M\setminus\{0\}) = \{\eta \in \mathcal{D}^{\infty}(T^*M\setminus\{0\}) \mid \eta^*\theta = \theta\}$, the $\theta$-preserving diffeomorphisms on $T^*M\setminus\{0\}$ and by $FIO_*$ the invertible Fourier integral operators on M. We have the surjective map:

$$p\colon FIO_* \longrightarrow \mathcal{D}^{\infty}_{\theta}(T^*M\setminus\{0\})$$

where graph(p(A)) is the canonical relation of A. The kernel of p is $p^{-1}(e) = \psi DO_*$, all invertible $\psi DO$'s. $\mathcal{D}^{\infty}_{\theta}(T^*M\setminus\{0\})$ is a group under composition and $FIO_*$, $\psi DO_*$ are groups under operator multiplication, graded by the order (which is additive). Notice that $p(A\circ B) = p(A) \circ p(B)$, so we get the exact sequence of groups:

$$(12.4) \qquad I \longrightarrow \psi DO_* \overset{j}{\hookrightarrow} FIO_* \overset{p}{\longrightarrow} \mathcal{D}^{\infty}_{\theta}(T^*M\setminus\{0\}) \longrightarrow e$$

(j the inclusion). We want to make this into an exact sequence of *Lie groups*.

Notice that the zero$^{th}$ order operators $(\psi DO_0)_*$ and $(FIO_0)_*$ are groups themselves, and form an exact sequence

$$(12.5) \qquad I \longrightarrow (\psi DO_0)_* \overset{j}{\hookrightarrow} (FIO_0)_* \overset{p}{\longrightarrow} \mathcal{D}^{\infty}_{\theta}(T^*M\setminus\{0\}) \longrightarrow e.$$

Now we are going to give a Lie group structure to this sequence of zero$^{th}$ order operators and then move this structure to all orders by means of a fixed elliptic operator, e.g. $(1 - \Delta)^{m/2}$. For the parameter spaces we look at the corresponding Lie algebras:

$\psi DO_0$, the $\psi DO$'s of order zero

$\psi DO_1$, the $\psi DO$'s of order one

$\mathcal{X}^\infty_\theta(T^*M\setminus\{0\})$, the vector fields X on $T^*M\setminus\{0\}$ such that $L_X\theta = 0$ ($L_X$ the Lie derivative).

Then (12.5) has the corresponding exact sequence of Lie algebras

(12.6) $$0 \longrightarrow \psi DO_0 \overset{i}{\hookrightarrow} \psi DO_1 \overset{\rho}{\longrightarrow} \mathcal{X}^\infty_\theta(T^*M\setminus\{0\}) \longrightarrow 0.$$

<u>**Remark**</u>. Clearly FIO $\neq$ exp($\psi$DO) since $\mathcal{D}_\theta \neq \exp(\mathcal{X}_\theta)$, i.e. we cannot obtain a chart at the identity in FIO by exponentiating the $\psi$DO's.

The idea to construct a manifold structure on $(FIO_0)_*$ is to construct a principal fiber bundle with

base space = $\mathcal{D}_\theta(T^*M\setminus\{0\})$
total space = $(FIO_0)_*$
fiber = $p^{-1}(\eta) = FIO_0(\eta)_* \cong (\psi DO_0)_*$

Then we check that multiplication is differentiable. We do this in several steps

<u>Step 1</u>: Make $\mathcal{D}^\infty_\theta(T^*M\setminus\{0\})$, the group of $\theta$ preserving diffeomorphisms of class $C^\infty$ into an ILH-Lie group with the tower of spaces $\mathcal{D}^s_\theta(T^*M\setminus\{0\})$.

<u>Step 2</u>: Define an $H^s$-norm on $\psi DO_0$ and complete this space. $\psi DO^s_0$ is a Hilbert Lie algebra thus $(\psi DO^s_0)_*$ is a Hilbert Lie group.

<u>Step 3</u>: Piece together $\mathcal{D}^s_\theta(T^*M\setminus\{0\})$ and $(\psi DO^s_0)_*$ by a local section $\sigma$: $U \subset \mathcal{D}^s_\theta(T^*M\setminus\{0\}) \longrightarrow (FIO_0)_*$. This gives $(FIO_0)_*$ the local product structure

$$p^{-1}(U) \cong U \times (\psi DO_0)_*$$

and hence a chart at the identity.

**Step 4**: Move this chart around by the group structure of $\mathcal{D}_\theta$. Compatibility conditions for the group structure and the topology give conditions on $\sigma$ to make $(FIO_0)_*$ a topological group.

**Step 5**: Overlap conditions in local charts give conditions on $\sigma$ to make $(FIO_0)_*$ into a differentiable manifold, and differentiability of group multiplication gives the Lie group structure of $(FIO_0)_*$.

Let us discuss these steps in more detail and state the main theorems for the proofs of which we refer to Adams, Ratiu, Schmid [1984].

**Step 1** has been carried out in the previous section.

**Step 2**: If we want to define a norm on $\psi DO_0$ directly we would end up with a Fréchet space because each $P \in \psi DO_0$ has a symbol of the form $p(x,\xi) = \sum_{j=0}^{-\infty} p_j(x,\xi)$, so we would have to have control over an infinite number of functions and their derivatives, and the infinite product of Hilbert spaces is no longer a Hilbert space. So what we do is cut the symbol at the term $p_{-k}$ for some fixed $k < \infty$. In terms of operators we look at $\psi DO_{m,k} \equiv \psi DO_m/\psi DO_{-k-1}$ and similarly $FIO_{m,k}(\eta) = FIO_m(\eta)/FIO_{-k-1}(\eta)$, $FIO_{m,k} = \bigcup_\eta FIO_{m,k}(\eta)$ (where $FIO_m(\eta) = \{A \in FIO_m \mid p(A) = \eta\}$).

Composition is still well defined in $\psi DO_{0,k}$ and $FIO_{0,k}$ and we denote by $(\psi DO_{0,k})_*$ and $(FIO_{0,k})_*$ the groups of invertible elements in $\psi DO_{0,k}$ and $FIO_{0,k}$. We have the exact sequence of groups:

$$(12.7)\quad I \longrightarrow (\psi DO_{0,k})_* \overset{j}{\hookrightarrow} (FIO_{0,k})_* \overset{p}{\longrightarrow} \mathcal{D}_\theta(T^*M\setminus\{0\}) \longrightarrow e.$$

For $P \in \psi DO_{m,k}$ with symbol $p(x,\xi) = p_m(x,\xi) + \ldots + p_{-k}(x,\xi)$ we define the norm

$$\|P\|^2_{m+k,s} = \|\tilde{p}_m\|^2_{s+k+m} + \|\tilde{p}_{m-1}\|^2_{s+k+m-1} + \dots + \|\tilde{p}_{-k}\|^2_s$$

where $\tilde{p}_{m-j}$ is the restriction of $p_{m-j}$ to the cosphere bundle $S^*M$ and $\|\tilde{p}_{m-j}\|_{s+k+m-j}$ is the $H^{s+k+m-j}$-Sobolev norm on $S^*M$. Let $\psi DO^s_{m,k}$ denote the completion of $\psi DO_{m,k}$ with respect to this norm and for m = 0, $(\psi DO^s_{0,k})_*$ is the group of invertible elements in $\psi DO^s_{0,k}$. Then one proves the following.

$(\psi DO_{0,k})_*$ is an ILH-Lie group with Lie algebra $\psi DO_{0,k}$; in particular, for each $s > n$, $(\psi DO^s_{0,k})_*$ is a Hilbert Lie group with Lie algebra $\psi DO^s_{0,k}$, i.e. $(\psi DO^s_{0,k})_*$ is a smooth ($C^\infty$) Hilbert manifold with smooth group operations and $(\psi DO_{0,k})_* = \varprojlim (\psi DO^s_{0,k})_*$ with the inverse limit topology is a topological group.

<u>**Step 3**</u>: Problem: For $\eta \in \mathcal{D}_\theta(T^*M\setminus\{0\})$ close to the identity we have to construct a *global* generating phase function $\varphi$ for graph($\eta$), so we can write a Fourier integral operator close to the identity with this phase function. This is done by constructing an explicit chart about the identity e in $\mathcal{D}^s_\theta(T^*M\setminus\{0\})$ in the following manner.

(A) Let $H \in \mathcal{S}^{s+1}(T^*M\setminus\{0\})$ be close to zero and define $\varphi_H$: $(T^*M\setminus\{0\}) \times M \longrightarrow \mathbb{R}$ by

$$(*) \qquad \varphi_H(\alpha_x, y) = \langle \alpha_x, (\exp_x^{-1}(y))\rangle + H(\alpha_x).$$

(where exp is defined by a Riemannian metric on M). Then there exists an $\eta \in \mathcal{D}^s_\theta(T^*M\setminus\{0\})$ close to e such that $\varphi_H$ is a global generating function for graph($\eta$).

(B) The map $H \longmapsto \eta$ is a bijection from a neighborhood $V(0) \subset \mathcal{S}^{s+1}(T^*M\setminus\{0\})$ onto a neighborhood $U^s(e) \subset \mathcal{D}^s_\theta(T^*M\setminus\{0\})$. The inverse is given as follows: let $\eta \in \mathcal{D}^s_\theta(T^*M\setminus\{0\})$ be close to e and define H: $T^*M\setminus\{0\} \longrightarrow \mathbb{R}$ by

$$H(\alpha_x) = -\langle \alpha_x, (\exp_x^{-1}(\tau^* \eta^{-1}(\alpha_x))\rangle$$

($\tau^*$: $T^*M \longrightarrow M$ the projection).

Then $\varphi_H$ defined by (*) is a global phase function for graph($\eta$).

Now we define a local section of the sequence (12.5)

$$\sigma \colon U^s \cap \mathcal{D}^\infty_\theta(T^*M \setminus \{0\}) \longrightarrow (FIO_0)_*$$

as follows:

$$(12.8) \quad (\sigma(\eta)u)(x) = (2\pi)^{-n} \int_{T_x^*M} d\xi \int_{\Omega_x} \chi(x,y) e^{i\varphi_H(x,y,\xi)} u(y) \,|\, J(\exp_x) \,|\, dy$$

where $\Omega_x$ is the open neighborhood of x, $\exp_x$ is a local diffeomorphism at x, $J(\exp_x)$ is its Jacobian, and $\chi(x,y)$ is a bump function.

In other words $\sigma(\eta)$ is defined as the Fourier integral operator whose phase function is $\varphi_H$, the global generating function of graph($\eta$), and whose amplitude is 1. Notice that H is smooth if $\eta$ is smooth, in which case $\sigma(\eta)$ is a well defined Fourier integral operator of order zero. Its principal symbol is 1, hence $\sigma(\eta)$ is invertible modulo smoothing, in particular $\sigma(\eta) \in (FIO_{0,k})_*$ for any k. Notice $p(\sigma(\eta)) = \eta$ for all $\eta \in U^s(e)$, therefore $\sigma$ defines a local section of the exact sequence (12.7).

Now we define the topology around the identity in $(FIO_{0,k})_*$ by the bijection $\Phi$: $p^{-1}(U^{2t}) \longrightarrow U^{2t} \times \psi DO^{2(t+k)}_{0,k}$, $\Phi(A) = (p(A), A\circ\sigma(p(A))^{-1})$; $\Phi^{-1}(\eta,P) = P\circ\sigma(\eta)$. To define the topology on $(FIO_{0,k})_*$ we move the open sets in $p^{-1}(U^{2t})$ by right translations. Complete this topological space in the right-uniform structure and denote it by $(FIO^t_{0,k})_*$.

<u>**Step 4**</u>: For each $t > n/2$, $(FIO^t_{0,k})_*$ is a topological group and

$(FIO_{0,k})_* = \bigcap_t (FIO^t_{0,k})_*$, with the inverse limit topology is a topological group.

**Remark.** To prove this statement one has to show that the map $(A,B) \longmapsto AB^{-1}$ is continuous for any $A,B \in (FIO^t_{0,k})_*$. This amounts to showing that the following map in local charts is continuous:

$$(U^{2t} \times \psi DO^{2(t-k)}_{0,k}) \times (U^{2t} \times \psi DO^{2(t-k)}_{0,k}) \longrightarrow U^{2t} \times \psi DO^{2(t-k)}_{0,k}$$

$$((\eta_1,P_1),(\eta_2,P_2)) \longmapsto (\eta_1 \circ \eta_2^{-1}, P_1\sigma(\eta_1)\sigma(\eta_2)^{-1}P_2^{-1}\sigma(\eta_1 \circ \eta_2^{-1})^{-1}),$$

which involves a careful study of products of symbols of Fourier integral operators.

**Step 5:** To prove that the transition maps in local charts are differentiable we have to show that the following map is differentiable:

$$(U^{2t} \cdot \alpha \cap U^{2t} \cdot \beta) \times (\psi DO^{2(t-k)}_{0,k})_* \longrightarrow (\psi DO^{2(t-k)}_{0,k})_*$$

$$(\eta,P) \longmapsto P\sigma(\eta \circ \alpha^{-1})AB^{-1}\,\sigma(\eta \circ \beta^{-1})^{-1},$$

for any $A,B \in (FIO^t_{0,k})_*$, where $\alpha = p(A)$, $\beta = p(B)$. The symbol calculus shows that this map is $C^t$. Then $(FIO^t_{0,k})_*$ is a $C^t$ manifold.

To show that group multiplication in $(FIO^t_{0,k})_*$ is differentiable we have to show that the following map is differentiable:

$$((U^{2(t+p)} \cdot \alpha) \times (\psi DO^{2(t+k+p)}_{0,k})_*) \times ((U^{2t} \cdot \beta) \times (\psi DO^{2(t+k)}_{0,k})_*) \longrightarrow ((U^{2t} \cdot (\alpha\beta) \times (\psi DO^{2(t+k)}_{0,k})_*)$$

$$((\eta_1,P_1),(\eta_2,P_2)) \longmapsto (\eta_1 \circ \eta_2,\ P_1\sigma(\eta_1\alpha^{-1})AP_2\sigma(\eta_2\beta^{-1})A^{-1}\sigma(\eta_1\eta_2\beta^{-1}\alpha^{-1})^{-1})$$

for any $A \in (FIO^{t+p}_{0,k})_*$, $B \in (FIO^{t}_{0,k})_*$, where $\alpha = p(A)$, $\beta = p(B)$.

Summarizing we have:

**Theorem.**

(A) $\{(FIO_{0,k})_*, (FIO^{t}_{0,k})_* \mid t > n/2\}$ is an ILH-Lie group. Explicitly:

(i) For each $t > n/2$, $(FIO^{t}_{0,k})_*$ is a $C^t$ manifold modeled on $\mathcal{S}^{2(t+k+1)}(T^*M \setminus \{0\}) \times \psi DO^{2t}_{0,k}$.

(ii) $(FIO^{t}_{0,k})_*$ is a topological group and $(FIO_{0,k})_* = \varprojlim (FIO^{t}_{0,k})_*$ is a topological group with the inverse limit topology.

(iii) The inclusions $(FIO^{t+1}_{0,k})_* \overset{j}{\hookrightarrow} (FIO^{t}_{0,k})_*$ are $C^t$.

(iv) Group mutliplication $\mu$: $(FIO_{0,k})_* \times (FIO_{0,k})_* \longrightarrow (FIO_{0,k})_*$ extends in a $C^p$ manner to $(FIO^{t+p}_{0,k})_* \times (FIO^{t}_{0,k})_* \longrightarrow (FIO^{t}_{0,k})_*$.

(v) Inversion $\nu$: $(FIO_{0,k})_* \longrightarrow (FIO_{0,k})_*$ extends in a $C^p$ manner to $(FIO^{t+p}_{0,k})_* \longrightarrow (FIO^{t}_{0,k})_*$.

(vi) For $A \in (FIO^{t}_{0,k})_*$, right multiplication $R_A$: $(FIO^{t}_{0,k})_* \longrightarrow (FIO^{t}_{0,k})_*$ is $C^t$.

(B) The Lie algebra of the ILH-Lie group $(FIO_{0,k})_*$ is the ILH-Lie algebra $\{\widetilde{\psi DO}_{1,k}, \widetilde{\psi DO}^{t}_{1,k} \mid t > n/2\}$ where $\widetilde{\psi DO}^{t}_{1,t} = \{P \in \psi DO^{t}_{1,k}$ with purely imaginary principal symbol$\}$. Explicitly:

(i) For each $t > n/2$, $\widetilde{\psi DO}^{t}_{1,k}$ is a Hilbert space.

(ii) $\widetilde{\psi DO}_{1,k} = \varprojlim \widetilde{\psi DO}^{t}_{1,k}$ is a Fréchet space with the inverse limit topology.

(iii) The inclusions $\widetilde{\psi DO}^{t+1}_{1,k} \overset{j}{\hookrightarrow} \widetilde{\psi DO}^{t}_{1,k}$ are continuous and dense.

(iv) The Lie bracket (commutator) is bilinear, continuous, antisymmetric

$$[\ ,\ ]:\ \widetilde{\psi DO}^{s+2}_{1,k} \times \widetilde{\psi DO}^{t+2}_{1,k} \longrightarrow \widetilde{\psi DO}^{\min(s,t)}_{1,k}$$

for all s,t > n and satisfies the Jacobi identity on $\widetilde{\psi DO}^{\min(s,t,r)}_{1,k}$ for elements in $\widetilde{\psi DO}^{s+4}_{1,k} \times \widetilde{\psi DO}^{t+4}_{1,k} \times \widetilde{\psi DO}^{r+4}_{1,k}$.

If we put the exact sequence (12.7) of Lie groups together with the exact sequence of their corresponding Lie algebras we get the following commuting diagram

$$\begin{array}{ccccccccc}
0 & \longrightarrow & \psi DO^{t}_{0,k} & \overset{j}{\hookrightarrow} & \widetilde{\psi DO}^{t}_{1,k} & \overset{\rho}{\longrightarrow} & \mathcal{S}^{t-k+1}(T^*M\setminus\{0\}) & \longrightarrow & 0 \\
 & & \big\downarrow \exp_1 & & \big\downarrow \exp_2 & & \big\downarrow \exp_3 & & \\
I & \longrightarrow & (\psi DO^{t}_{0,k})_* & \overset{j}{\hookrightarrow} & (FIO^{t}_{0,k})_* & \overset{p}{\longrightarrow} & \mathcal{D}^{t-k}_{\theta}(T^*M\setminus\{0\}) & \longrightarrow & 0.
\end{array}$$

The Lie algebra homomorphism $\rho = T_I p$ is just 1/i times the principal symbol. $\exp_1$ is just exponentiation of operators (modulo equivalence classes). $\exp_2$ is also exponentiation of operators; for $P \in \widetilde{\psi DO}^{t}_{1,k}$, $\exp_2 tP$ is the solution of the equation $[\frac{d}{dt} + P]U(t) = 0$ with $U(0) = I$, i.e. a one-parameter family of Fourier integral operators of order zero. $\exp_3$ is given by $\exp_3(tH) = \eta_t$ where $\eta_t$ is the flow of the Hamiltonian vector field $X_H$ on $T^*M\setminus\{0\}$.

The second square commutes by Egorov's theorem, i.e. if U(t) satisfies $[\frac{d}{dt} + P]U(t) = 0$, $U(0) = I$ for $P \in \widetilde{\psi DO}^{t}_{1,k}$, then the canonical transformation of U(t) is the flow of the Hamiltonian vector field $X_{\rho(P)}$ on $T^*M\setminus\{0\}$.

**Remark.** As for any diffeomorphism group, the exponential map $\exp_3$: $\mathcal{S}^{s+1}(T^*M\setminus\{0\}) \longrightarrow \mathcal{D}^{s}_{\theta}(T^*M\setminus\{0\})$ is not onto a neighborhood of the identity, i.e. it cannot be used as a local chart. As a consequence, the same is true for $\exp_2$: $\widetilde{\psi DO}^{t}_{1,k} \longrightarrow (FIO^{t}_{0,k})_*$.

To get the differentiable structure of the full group of

invertible Fourier integral operators, we proceed in the following manner.

We have given $(FIO_{0,k})_*$ an ILH-Lie group structure. Now consider $(FIO_0)_*$. This has the Lie group structure of a direct limit of ILH-Lie groups, i.e. $(FIO_0)_* = \varinjlim (FIO_{0,k})_*$. Next, for any m we can give $FIO_m \cap (FIO)_*$ the structure of a direct limit of ILH-manifolds by using an elliptic operator, e.g. $(1 - \Delta)^{m/2}$, to give an identification of $FIO_m \cap (FIO)_*$ with $(FIO_0)_*$. Multiplication will be smooth between the appropriate spaces. Piecing this together for all m makes $(FIO)_* = \bigcup_m (FIO_m)_*$ a graded direct limit of ILH-Lie groups with Lie algebra $\psi DO$ the space of all pseudodifferential operators.

Let $\psi DO^*$ be the dual of the Lie algebra $\psi DO$ with respect to some pairing <,>. We shall comment later what choices are possible for <,>. The Lie Poisson bracket for any two functions $F,G: \psi DO^* \longrightarrow \mathbb{R}$ is given by $\{F,G\}(A) = \langle A, [\frac{\delta F}{\delta A}, \frac{\delta G}{\delta A}]\rangle$, $A \in \psi DO^*$, $\frac{\delta F}{\delta A}, \frac{\delta G}{\delta A} \in \psi DO$, and [ , ] is the Lie algebra bracket (commutator) in $\psi DO$. Then Hamilton's equations of motion for any energy function H on $\psi DO^*$ are determined by

$$\dot{F} = \{F,H\}.$$

These are equivalent to Hamilton's evolution equation on coadjoint orbits of $(FIO)_*$ with respect to the Kostant-Kirillov symplectic structure.

Suppose we have a vector space decomposition of the Lie algebra $\psi DO$ into a direct sum of two subalgebras, $\psi DO = h \oplus k$. This gives the corresponding decomposition $\psi DO^* = k^o \oplus h^o$ which allows us to identify the duals $h^* \cong k^o$ and $k^* \cong h^o$. Applying the Kostant-Symes theorem (Adler [1979], Kostant [1979], Symes [1980]) we obtain functions in involution as follows. Let $F,G: \psi DO^* \longrightarrow \mathbb{R}$ be two functions that are constant on coadjoint orbits of $(FIO)_*$ in $\psi DO^*$. Then for $A \in h^*$, $\{F_A, G_A\} = 0$, where $F_A$ and $G_A$ are the restrictions of F and G to the coadjoint orbit of A in $h^*$.

As a special case, consider $(FIO)_*$ on $M = S^1$, the circle.

Then each pseudodifferential operator $P \in \psi DO_m$ has the total symbol given by $p(x,\xi) = \sum_{-\infty<j\leq m} p_j(x)\, \xi^j$, where $p_j(x)$ are functions on $S^1$. Let $h = \psi DO_- = \bigcup_{m<0} \psi DO_m$ and $k = \psi DO_+ = \bigcup_{m\geq 0} \psi DO_m$. Note that k is the space of differential operators on $S^1$. Then we have the decomposition $\psi DO = \psi DO_- \oplus \psi DO_+$. For $P \in \psi DO$ with total symbol $p(x,\xi) = \sum p_j(x)\, \xi^j$, define trace $(P) = \int_0^{2\pi} p_{-1}(x)dx$. The commutator of P, Q $\in \psi DO$ can be written in the form $[P,Q] = \partial_\xi A + \partial_x B$, with A,B $\in \psi DO$. From this follows that trace $([P,Q]) = \int_0^{2\pi} ([P,Q])_{-1}dx = 0$, where $([P,Q])_{-1}$ denotes the coefficient of $\xi^{-1}$ in the total symbol of $[P,Q]$. Then the pairing $\langle P,Q\rangle = \text{trace}(P\circ Q) = \int_0^{2\pi} (P\circ Q)_{-1}dx$ identifies $(\psi DO_-)^*$ with $\psi DO_+$ as follows. We represent elements of $\psi DO_-$ in the form $P = \sum_{j=1}^{\infty} c_{-j}(x)\, \xi^{-j} = \sum_{k=0}^{\infty} (\xi + \partial_x)^{-k-1} b_k(x)$, where $b_k = \sum_{\nu=0}^{k} \binom{k}{\nu} \partial_x^\nu c_{-(k+1)+\nu}$. If $A \in \psi DO_+$ is given by the total symbol $a(x,\xi) = \sum_{i=0}^{m} a_i(x)\, \xi^i$, then $A\circ P$ has total symbol $\sum_{i,j} a_i(x)(\xi + \partial_x)^{i-j-1} b_j$ and $\langle A,P\rangle = \int_0^{2\pi}(A\circ P)_{-1}\, dx = \int_0^{2\pi} (a_0 b_0 + \dots + a_m b_m)dx$.

The Lie Poisson bracket of two functions F and G on $\psi DO_-^*$ at a point A is

$$\{F,G\}(A) = \int_0^{2\pi} (A \circ [\frac{\partial F}{\delta A},\frac{\delta G}{\delta A}])_{-1}dx.$$

We assume that F and G are local functions regarded as functions on the symbol space, i.e. they are of the form $F = \int \mathcal{F}dx$ and $G = \int \mathcal{G}dx$ where $\mathcal{F}$ and $\mathcal{G}$ are functions depending on the $a_i$'s and their derivatives up to some finite order. Then for A,B $\in \psi DO_+$, $DF(A)\cdot B = \sum_i \int \frac{\delta \mathcal{F}}{\delta a_i} b_i dx = \langle\frac{\delta F}{\delta A},B\rangle$, so by definition of $\langle,\rangle$ we have

$$\frac{\delta F}{\delta A} = \sum_{k=0}^{m} (\xi + \partial_x)^{-k-1} \frac{\delta \mathcal{F}}{\delta a_k}, \text{ where } \frac{\delta \mathcal{F}}{\delta a_k} = \frac{\partial \mathcal{F}}{\partial a_k} - \partial_x(\frac{\partial \mathcal{F}}{\partial(\partial_x a_k)}) +$$

$\partial_x^2 \left(\frac{\partial \mathcal{F}}{\partial(\partial_k^2 a_k)}\right) - \ldots$ . The Lie Poisson evolution equations $\dot{F} = \{F,H\}$ for a function H on $(\psi DO_-)^*$ are equivalent to the evolution equations

$$\dot{A} = X_H(A) = \mathrm{ad}^*_{\frac{\delta H}{\delta A}} A$$

on $(\psi DO_-)^*$.

Under the identification $(\psi DO_-)^* \cong \psi DO_+$, the infinitesimal coadjoint action of $P \in \psi DO_-$ on $A \in \psi DO_+$ is given by $\mathrm{ad}^*_P A = [P,A]_+$, where the subscript + means taking only the part in $\psi DO_+$. Thus, if A has total symbol $a(x,\xi) = a_0(x) + \ldots + a_m(x)\xi^m$, the terms of degree m and m−1 in $\mathrm{ad}^*_P A$ vanish, so that $a_m$ and $a_{m-1}$ are orbit invariants, and may be set $a_m = 1$, $a_{m-1} = 0$. Hence the space of operators with total symbols of the form $\xi^m + \sum_{i=0}^{m-2} a_i(x)\, \xi^i$ is invariant under the coadjoint action.

As a special case, which is important for KdV, we consider the space of operators A with total symbol $a + \xi^2$, $a \in C^\infty(S^1)$; so A is the time independent Schrödinger operator with potential a. This space is invariant under the coadjoint action. The Lie Poisson bracket of two (local) functions F and G on this space at a point A becomes (with $\frac{\delta F}{\delta A} = (\xi + \partial_x)^{-1}\frac{\delta \mathcal{F}}{\delta a} = \frac{\delta \mathcal{F}}{\delta a}\xi^{-1} - \partial_x \frac{\delta \mathcal{F}}{\delta a}\xi^{-2} + \ldots$)

$$\{F,G\}(A) = \int_0^{2\pi} \frac{\delta \mathcal{F}}{\delta a}\, \partial_x \frac{\delta \mathcal{G}}{\delta a}\, dx. \tag{12.9}$$

This is the bracket first obtained by Gardner [1971]. For the Hamiltonian

$$H = \int_0^{2\pi} (a^3 + a_x^2/2)dx$$

($a_x = \partial_x a$), Hamilton's equations become $\dot{A} = [\frac{\delta H}{\delta A}, A]_+$ which are equivalent to $a_t = 6\, aa_x - a_{xxx}$ which is the KdV equation.

Moreover, for the functionals $H_k(A) = \mathrm{trace}(A^k) = \int_0^{2\pi} (A^k)_{-1}dx$

($k \in \mathbb{N}$), we have $\frac{\delta H_k}{\delta A} = k\, A^{k-1}$ and hence $[A, \frac{\delta H_k}{\delta A}] = [A, kA^{k-1}] = 0$. Thus $H_k$ are constant on coadjoint orbits. Following the Kostant-Symes theorem for these functions, with respect to the decomposition $\psi DO = \psi DO_- \oplus \psi DO_+$ as above, i.e. restricting the $H_k$'s to $(\psi DO_-)^* = \psi DO_+$, gives the Gelfand-Dikii family of commuting integrals for the KdV equation. For example

$$H_0 = \int a dx,\ H_1 = \int (1/2)\, a^2 dx,\ H_2 = \int (a^3 + (1/2)\, a_x^2) dx \ (= H \text{ above}),$$

$$H_3 = \int ((5/8)\, a^4 + (5/4)\, a a_x^2 + (1/8)\, a_{xx}^2) dx.$$

This shows that the KdV equation is a completely integrable Hamiltonian system on a coadjoint orbit of the Lie group $(FIO)_*$. This result was first obtained by Adler [1979] on the Lie algebra of formal pseudodifferential operators; see also Iacob and Sternberg [1980].

**Remarks.**

(1) The Poisson bracket (12.9) is general and can be formed for the dual of any Lie algebra. In Ratiu [1981] it is shown that if $g$ is a Lie algebra with dual $g^*$ and $\epsilon \in g^*$ is fixed, the prescription

$$\{F,G\}(\mu) = \langle \epsilon, [\frac{\delta F}{\delta \mu}, \frac{\delta G}{\delta \mu}] \rangle$$

defines a Poisson bracket induced by a Lie-Poisson structure on the dual of the loop extension of $g$. The same result holds for any real valued Lie algebra 2 cocycle.

(2) Mulase [1984] has studied the geometric structure of the space of formal solutions of the soliton type equations which are equivalent to the hierarchy of the KP (Kadomtsev-Petviashvili) equations. He showed that this space is a Lie group G of the Lie algebra of formal pseudodifferential operators of order -1 on $\mathbb{R}$, explicitly $G = 1 + \psi DO_{-1}$.

Using a representation of the Lie algebra $\psi DO$ he defines an affine coordinate system on G which enables him to calculate explicit formulas for solutions and $\tau$-functions. The KP system is described as a dynamical system on G and it is shown that every orbit of it is locally isomorphic to a certain cohomology group associated with a commutative algebra. Moreover, an orbit is finite dimensional if and only if it is essentially a Jacobian variety of an algebraic curve. Using algebraic geometric methods and soliton equations Shiota [1985] solved the Schottky problem; i.e. the problem of characterization of Jacobians among abelian varieties.

It would be of great interest to understand these results in the framework of the smooth category of infinite dimensional Lie groups.

(3) Due to the pathologies described in Ratiu [1979] regarding the time t-map of the KdV equation, it seems that a more geometric approach should be taken in studying perturbation theory for the KdV equation. We hope that the Lie group theoretical methods of $(FIO)_*$ within the Hamiltonian framework just sketched above might yield such results.

## References


Abarbanel, H.D.I., Holm, D.D., Marsden, J.E. and Ratiu, T.S. [1984], Richardson number criterion for the nonlinear stability of stratified flow, Phys. Rev. Lett. 52, 2352-2355.

Abarbanel, H.D.I., Holm, D.D., Marsden, J.E. and Ratiu, T.S. [1985], Nonlinear stability analysis of stratified ideal fluid equilibria, Transactions Phil. Soc. Cambridge (to appear).

Abraham, R. [1961], Lectures of Smale on differential topology, Mimeographed, Columbia.

Abraham, R. and Marsden, J.E. [1978], Foundations of Mechanics, Second Ed., Addison-Wesley.

Abraham, R., Marsden, J.E. and Ratiu, T.S. [1983], Manifolds, Tensor

Analysis, and Applications, Addison-Wesley, Reading, Mass.

Adams, R. [1975], Sobolev Spaces, Academic Press, New York.

Adams, M., Ratiu, T.S. and Schmid, R. [1984], A Lie group structure for Fourier integral operators, MSRI Preprint 049-84-7, Berkeley, California.

Adler, M. [1979], On a trace functional for formal pseudo-differential operators and the symplectic structure for Korteweg-deVries type equations, Invent. Math. 50, 219-248.

Arms, J.M. [1981], The structure of the solution set for the Yang-Mills equations, Math. Proc. Camb. Phil. Soc. 90, 361-372.

Arms, J.M., Marsden, J.E. and Moncrief, V. [1982], The structure of the space of solutions of Einstein's equations II. Several Killing fields and the Einstein-Yang-Mills equations, Ann. of Phys., 144, 81-106.

Arnold, V.I. [1965], Conditions for nonlinear stability for plane curvilinear flows of an ideal fluid, Doklady Mat. Nauk 162(5), 773-777.

Arnold, V.I. [1966], Sur la géometrie differentielle des groupes de Lie de dimension infinie et ses applications a l'hydrodynamique des fluids parfaits, Ann. Inst. Fourier, Grenoble, 16, 319-361.

Arnold, V.I. [1969], An a priori estimate in the theory of hydrodynamic stability, Am. Math. Soc. Transl. 19, 267-269.

Arnold, V.I. [1978], Mathematical Methods of Classical Mechanics, Springer Graduate Text in Math 60, Springer, New York.

Atiyah, M.F., Hitchin, N.J. and Singer, I.M. [1978], Self-duality in four-dimensional Riemannian geometry, Proc. Royal Soc. London A, 362, 425-461.

Binz, E. [1979], Einstein's evolution equation for the vacuum formulated on the space of differentials of immersions, Springer Lecture Notes in Math. 1037.

Binz, E. [1984], The space of smooth isometric immersions of a compact manifold into an Euclidean space is a Fréchet manifold, C.R. Math. Rep. Acad. Sci. Canada, Vol. VI(5).

Binz, E. and Fischer, H.R. [1978], The manifold of embeddings of a closed manifold, Springer Lecture Notes in Physics 139.

Bokobza-Haggiag, J. [1969], Opérateurs pseudo-différentiels sur une variété différentiable, Ann. Inst. Fourier, Grenoble, 19, 125-177.

Bourbaki, N. [1975], Lie Groups and Lie Algebras, Hermann, Paris.

Brill, D., and Deser, S. [1968], Variational methods and positive energy in relativity, Ann. Phys. 50, 548-570.

Chernoff, P.R. and Marsden, J.E. [1974], Properties of infinite dimensional Hamiltonian systems, Lecture Notes in Math., Vol. 425, Springer, Berlin.

Choquet-Bruhat, Y. and Marsden, J.E. [1976], Solution of the local mass problem in general relativity, C.R. Acad. Sci. (Paris), 282, (1976), 609-612, Comm. Math. Phys. 51, (1976), 283-296.

Choquet-Bruhat, Y., DeWitt-Morette, C. and Dillard-Bleick, M. [1982], Analysis, Manifolds and Physics, 2nd ed., North Holland, Amsterdam.

Coifman, R.R. and Meyer, Y. [1978], Au-delà des Opérateurs Pseudo-Différentiels, Societé math. de France, Astérisque 57.

Duff, G.F.D. [1952], Differential forms in manifolds with boundary, Annals of Math., 56, 115-127.

Duistermaat, J.J. [1973], Fourier Integral Operators, Lecture Notes, Courant Institute of Mathematical Sciences, New York.

Ebin, D.G. [1968], The manifold of Riemannian metrics, Bull. Am. Math. Soc. 74, 1002-1004.

Ebin, D.G. and Marsden, J.E. [1970], Groups of diffeomorphisms and the motion of incompressible fluid, Ann. of Math. 92, 102-163.

Eells, J. [1958], On the geometry of function spaces, Symposium de Topologia Algebrica, UNAM, Mexico City, 303-307.

Fischer, A. [1970], The theory of superspace, in Relativity, ed. M. Carmelli, S. Fickler, and L. Witten, Plenum Press.

Fischer, A. and Marsden, J.E. [1972], The Einstein equations of evolution - a geometric approach, J. Math. Phys. 13, 546-568.

Fischer, A. and Marsden, J.E. [1973], Linearization stability of the Einstein equations, Bull. Am. Math. Soc. 79, 102-163.

Fischer, A. and Marsden, J.E. [1979], The initial value problem and the dynamical formulation of general relativity, General Relativity, An Einstein Centenary Volume, eds. S.W. Hawking and W. Israel, Cambridge, 138-211.

Freed, D.S. and Uhlenbeck, K.K. [1984], Instantons and Four-Manifolds, MSRI Berkeley Publications 1, Springer Verlag.

Gardner, C.F. [1971], Korteweg-deVries equation and generalizations IV. The Korteweg-deVries equation as a Hamiltonian system, J. Math. Phys. 12, 1548-1551.

Garland, H. [1980], The Arithmetic Theory of Loop Groups, Publ. Math. I.H.E.S. 52, 181-312.

Goldin, G.A. and Sharp, D.H. [1983], Particle spin from representations of the diffeomorphism group, Comm. Math. Phys. 92, 217-228.

Goodman, R. and Wallach, N.R. [1984], Structure and unitary cocycle representations of loop groups and the group of diffeomorphisms of the circle, J. fuer reine und angewandte Math. (Crelle J.), 347, 69-133.

Guckenheimer, J. and Holmes, P. [1983], Nonlinear Oscillations, Dynamical Systems, and Bifurcations of Vector Fields, Springer, Applied Math. Sciences, Vol. 42.

Guillemin, V. and Sternberg, S. [1977], Geometric Asymptotics, American Mathematical Society Survey, Vol. 14, AMS, Providence, RI.

Guillemin, V. and Sternberg, S. [1980], The moment map and collective motion, Ann. of Phys. 127, 220-253.

Guillemin, V. and Sternberg, S. [1984], Symplectic Techniques in Physics, Cambridge University Press, Cambridge.

Gutknecht, J. [1977], Die $C^{\infty}_{\Gamma}$-Struktur auf der Diffeomorphismengruppe einer kompakten Mannigfaltigkeit, Ph.D. Thesis, ETH, Zürich.

Hamilton, R. [1982], The inverse function theorem of Nash and Moser, Bull. Am. Math. Soc. 7, 65-222.

Holm, D., Marsden, J.E., Ratiu, T.S. and Weinstein, A. [1985], Nonlinear stability of fluid and plasma systems, Physics Reports.

Holmes, P. and Marsden, J.E. [1983], Horseshoes and Arnold diffusion for Hamiltonian systems on Lie groups, Ind. Univ. Math. J. 32, 273-310.

Hörmander, L. [1971], Fourier integral operators I, Acta Math. 127,

79-183.

Iacob, A. and Sternberg, S. [1980], Coadjoint structures, solitons, and integrability, Lecture Notes in Physics, Vol. 120, Springer.

Isenberg, J. and Marsden, J.E. [1982], A slice theorem for the space of solutions of Einstein's equations, Phys. Rep. 89, 179-222.

Keller, H.H. [1974], Differential Calculus in Locally Convex Spaces, Lecture Notes in Math. 417, Springer, Berlin.

Kostant, B. [1979], The solution to a generalized Toda Lattice and representation theory, Adv. Math. 34, 195-338.

Kuchar, K. [1976], The dynamics of tensor fields in hyperspace, J. Math. Phys. 17: 777-791, 792-800, 801-820; 18: 1589-1597.

Kumano-go, H. [1981], Pseudo-Differential Operators, MIT Press, Cambridge, Mass.

Lang, S. [1972], Differential Manifolds, Addison-Wesley, Reading, Mass.

Lawson, H.B. [1983], The Theory of Gauge Fields in Four Dimensions, CBMS Conference, University of California, Santa Barbara.

Leslie, J. [1967], On a differential structure for the group of diffeomorphisms, Topology 6, 263-271.

Marsden, J.E. [1974], Applications of Global Analysis in Mathematical Physics, Publish or Perish, Berkeley, CA.

Marsden, J.E. [1976], Well-posedness of the equations of a nonhomogeneous perfect fluid, Comm. PDE 1, 215-230.

Marsden, J.E. [1981], Lectures on Geometric Methods in Mathematical

Physics, CBMS, Vol. 37, SIAM, Philadelphia.

Marsden, J.E. [1985], Chaos in dynamical systems by the Poincaré-Melnikov-Arnold method, Proceedings of the ARO Conference on Dynamics, SIAM (to appear).

Marsden, J.E., Ebin, G.D. and Fischer, A. [1972], Diffeomorphism groups, hydrodynamics and relativity, in Proc. 13th Biennial Seminar of Canadian Math. Congress, J.R. Vanstone (ed.), Montreal, 135-279.

Marsden, J.E., Ratiu, T.S. and Weinstein, A. [1984a], Semidirect products and reduction in mechanics, Transact. of AMS, 281, 147-177.

Marsden, J.E., Ratiu, T.S. and Weinstein, A. [1984b], Reduction and Hamiltonian structures on duals of semidirect product Lie algebras, Cont. Math. AMS, Vol. 28, 55-100.

Marsden, J.E. and Weinstein, A. [1982], The Hamiltonian structure of the Maxwell-Vlasov equations, Physica 4D, 394-406.

Marsden, J.E. and Weinstein, A. [1983], Co-adjoint orbits, vortices and Clebsch variables for incompressible fluids, Physica 7D, 305-323.

Marsden, J.E., Weinstein, A., Ratiu, T.S., Schmid, R. and Spencer, R. [1983], Hamiltonian systems with symmetry, coadjoint orbits and plasma physics, Proc. IUTAM-ISIMM Symposium on Modern Developments in Analytical Mechanics, Atti Accad. Sci. Torino, Suppl., Vol. 117, 289-340.

Marsden, J.E., Weinstein, A., Ratiu, T.S., Schmid, R. and Spencer, R. [1985], The Geometry and Dynamics of Fluids and Plasmas (in preparation 1985).

Michor, P. [1980], Manifolds of Differentiable Mappings, Shiva Math. Series, No. 3, Kent, UK.

Milnor, J. [1983], Remarks on infinite dimensional Lie groups, Proc. Summer School on Quantum Gravity, ed. B. DeWitt, Les Houches.

Mitter, P.K. [1980], Geometry of the space of gauge orbits and the Yang-Mills dynamical system, Proc. NATO Advanced Study Institute on Recent Developments in Gauge Theories, Cargese 1979, 265-292, ed. G. t'Hooft et al., Plenum Press, New York.

Moncrief, V. [1975], Spacetime symmetries and linearization stability of the Einstein equations, J. Math. Phys. 16, 493-498.

Moncrief, V. [1977], Gauge symmetries of Yang-Mills fields, Ann. of Phys. 108, 387-400.

Moncrief, V. [1980], Reduction of the Yang-Mills equations, Proc. Differential Geometrical Methods in Mathematical Physics, Salamanca 1979, Lecture Notes in Math. 836, 276-291, Springer, Berlin.

Mulase, M. [1984], Cohomological structure of solutions of soliton equations, isospectral deformations of ordinary differential operators and a characterization of Jacobian varieties, Journal of Diff. Geometry 19(2), 403-430.

Nill, F. [1983], An effective potential for classical Yang-Mills fields as outline for bifurcation on gauge orbit space, Ann. Phys. 149, 179-202.

Omori, H. [1970], On the group of diffeomorphisms on a compact manifold, Proc. Symp. Pure Math. 15, 167-184.

Omori, H. [1974], Infinite Dimensional Lie Transformation Groups, Lecture Notes in Math., 427, Springer, Berlin.

Omori, H., Maeda, Y., Yoshika, A., and Kobayashi, O., [1980-83], On regular Frechet-Lie groups I, II, III, IV, V, VI, Tokyo J. Math. 3, (1980), 353-390; 4 (1981), 221-253; 4 (1981), 255-277; 5 (1982),

365-398; 6 (1983), 39-64; 6 (1983), 217-246.

Palais, R. [1965], Seminar on the Atiyah-Singer Index Theorem, Ann. of Math. Studies, Vol. 57, Princeton University Press, Princeton, NJ.

Palais, R. [1968], Foundations of Global Nonlinear Analysis, Addison-Wesley, Reading, Mass.

Parker, P.E. and Taubes, C.H. [1982], On Witten's proof of the positive energy theorem, Comm. Math. Phys. 84, 223-238.

Pressley, A. and Segal, G. [1984], Loop Groups and Their Representations, Preprint, to appear in Oxford University Press.

Ratiu, T.S. [1979], On the smoothness of the time t-map of the KdV equation and the bifurcation of the eigenvalues of Hill's operator, Lecture Notes in Math. Vol. 755, Springer, Berlin, 1979.

Ratiu, T.S. [1981], Euler-Poisson equations on Lie algebras and the N-dimensional heavy rigid body, Proc. Natl. Acad. Sci. USA 78(3), (1981), 1327-1328, and Amer. Journal of Math. 104(2), (1982), 409-448.

Ratiu, T.S. and Schmid, R. [1981], The differentiable structure of three remarkable diffeomorphism groups, Math. Zeitschr. 177, 81-100.

Schmid, R. [1978], Die Symplectomorphismen-Gruppe als Fréchet-Lie-Gruppe, Thesis, Univ. Zürich.

Schmid, R. [1983], The inverse function theorem of Nash and Moser for the Γ-differentiability, to appear in Proc. Convergence Structures and Applications II, Akademie der Wissenschaften der DDR, Berlin.

Schoen, R. and Yau, S.T. [1979], Positivity of the total mass of a general space-time, Phys. Rev. Lett. 43, 1457-1459.

Shiota, T. [1985], Characterization of Jacobian varieties in terms of soliton equations, preprint.

Singer, I.M. [1978], Some remarks on the Gribov ambiguity, Commun. Math. Phys. 60, 7-12.

Singer, I.M. [1980], The geometry of the orbit space for nonabelian gauge theories, preprint.

Spencer, R. and Schmid, R. [1984], Electrostatic normal modes in an unmagnetized homogeneous Coulomb plasma. A Hamiltonian Approach, Phys. Lett. 101A, 485-490.

Symes, W. [1980], Systems of Toda type, inverse spectral problems and representation theory, Inventiones Math. 59, 13-51.

Symes, W. [1980], Hamiltonian group actions and integrable systems, Physica 1D, 339-374.

Taubes, C.H. [1983], Stability in Yang-Mills theories, Commun. Math. Phys. 91, 235-263.

Taubes, C.H. [1984], Self-dual connections on 4-manifolds with indefinite intersection matrix, Journal of Diff. Geometry 19, 517-560.

Taylor, M. [1981], Pseudodifferential Operators, Princeton University Press, Princeton, NJ.

Trèves, F. [1980], Introduction to Pseudodifferential and Fourier Integral Operators I, II, Plenum Press, New York, NY.

Tromba, A.J. [1976], Almost-Riemannian structures on Banach manifolds, the Morse lemma and the Darboux theorem, Can. J. Math. 28, 640-652.

Uhlenbeck, K.K. [1982], Connections with $L^p$ bounds on curvature, Commun. Math. Phys. 83, 31-42.

Weinstein, A. [1971], Symplectic manifolds and their Lagrangian submanifolds, Advances in Math. 6, 329-346.

Weinstein, A. [1977], Lectures on Symplectic Manifolds, CBMS, Conference Series, Vol. 29, American Mathematical Society, Providence, RI.

Weinstein, A. [1983], The local structure of Poisson manifolds, Journal of Diff. Geometry 18(3), 523-557.

Witten, E. [1981], A new proof of the positive energy theorem, Comm. Math. Phys. 80, 381-402.

# ON LANDAU-LIFSHITZ EQUATION AND INFINITE DIMENSIONAL GROUPS

By

Etsuro Date (Joint Work With Michio Jimbo, Masaki Kashiwara and Tetsuji Miwa)
Department of Mathematics, College of General Education, Kyoto University, Kyoto 606, JAPAN

1. Here we shall briefly discuss relations between the Landau-Lifshitz equation and infinite dimensional groups. These groups will appear in connection with elliptic curves which are closely related to the Landau-Lifshitz equation. For detail, we refer to our paper [1].

The Landau-Lifshitz (L-L) equation is the following non-linear partial differential equation:

$$\mathcal{S}_t = \mathcal{S} \times \mathcal{S}_{xx} + \mathcal{S} \times J\mathcal{S},$$

$$\mathcal{S} = \mathcal{S}(x,t) = (S_1,S_2,S_3),\ S_1^2 + S_2^2 + S_3^2 = 1,$$

$$J = \begin{pmatrix} J_1 & & 0 \\ & J_2 & \\ 0 & & J_3 \end{pmatrix}, \text{ a constant diagonal matrix.}$$

Here $\mathcal{S}_t = \frac{\partial \mathcal{S}}{\partial t}$, etc. and the symbol $\times$ denotes the vector product of 3-dimensional vectors. This equation is a classical equation for non-linear spin waves in a ferromagnet and one of the so-called soliton equations.

As is usual with soliton equations, the L-L equation admits the Lax representation, which was given by Sklyanin [2] and Borovik. Namely, the L-L equation is the integrability condition of the following linear equations:

$$\frac{\partial w}{\partial x_1} = Lw, \; \frac{\partial w}{\partial x_2} = Mw, \; x_1 = x, \; x_2 = -it,$$

$$L = \sum_{\alpha=1}^{3} Z_\alpha S_\alpha \sigma_\alpha,$$

$$M = i \sum_{\alpha,\beta,\gamma=1}^{3} Z_\alpha S_\beta S_{\gamma x} \epsilon^{\alpha\beta\gamma} \sigma_\alpha +$$

$$+ 2\, Z_1 Z_2 Z_3 \sum_{\alpha=1}^{3} Z_\alpha^{-1} S_\alpha \sigma_\alpha,$$

where $\sigma_\alpha$ are the Pauli matrices

$$\sigma_1 = \begin{bmatrix} 0 & 1 \\ 1 & 0 \end{bmatrix}, \; \sigma_2 = \begin{bmatrix} 0 & -i \\ i & 0 \end{bmatrix}, \; \sigma_3 = \begin{bmatrix} 1 & 0 \\ 0 & -1 \end{bmatrix},$$

and

$$\epsilon^{\alpha\beta\gamma} = \begin{cases} 1 \text{ if } (\alpha,\beta,\gamma) \text{ is an even permutation of } (1,2,3) \\ -1 \text{ if } (\alpha,\beta,\gamma) \text{ is an odd permutation of } (1,2,3) \\ 0 \text{ otherwise.} \end{cases}$$

Here the spectral parameters $Z_1$, $Z_2$ and $Z_3$ are connected by the relations

(1) $\quad Z_\alpha^2 - Z_\beta^2 = \frac{1}{4}(J_\alpha - J_\beta), \; \alpha,\beta = 1, 2, 3.$

These equations define an elliptic curve, which we denote by $\tilde{E}$.

Also the L-L equation is rewritten in bilinear differential equations in the sense of Hirota. We introduce new dependent variables f, f*, g and g* by

$$S_1 = \frac{f^*g + fg^*}{f^*f + g^*g}, \; S_2 = -i\,\frac{f^*g - fg^*}{f^*f + g^*g}, \; S_3 = \frac{f^*f - g^*g}{f^*f + g^*g}.$$

Then in these dependent variables the L-L equation takes the following form [3]:

$$D_1(f^*\cdot f + g^*\cdot g) = 0, \quad \left[D_2 - D_1^2\right] (f^*\cdot f - g^*\cdot g) = 0,$$

(2) $$\left[D_2 - D_1^2 + \frac{1}{2}(a^2+b^2)\right] f^*\cdot g + \frac{1}{2}(a^2-b^2)\, g^*\cdot f = 0,$$

$$\left[D_2 - D_1^2 + \frac{1}{2}(a^2+b^2)\right] g^*\cdot f + \frac{1}{2}(a^2-b^2)\, f^*\cdot g = 0,$$

where $a^2 = J_3 - J_1$, $b^2 = J_3 - J_2$. Here the symbol $D_j = D_{x_j}$ denote the Hirota's bilinear differential operators, which is defined as follows: for any $P\in\mathbb{C}[x_1,x_2,\ldots]$, $P(D_x)f\cdot g = P(\partial_{y_1},\partial_{y_2},\ldots)f(x+y)\cdot g(x-y)\big|_{y=0}$.

One of our conclusions is that the solution space of the L-L equation when expressed in terms of f, f*, g and g* is an orbit of an infinite dimensional group whose Lie algebra is a central extension of $\mathfrak{sl}(2,\mathbb{C}[k,k^{-1},\omega])$. Here k and $\omega$ are related by $\omega^2 = (k^2-a^2)(k^2-b^2)$. We denote by E the elliptic curve defined by this equation.

To show the above result we use the language of free fermions, which is the approach we have introduced in the study of transformation groups for soliton equations (see for example [4]). This time we use free fermions on the elliptic curve E. Before going into this case, we think it will be helpful to explain the BKP case first.

2. The BKP hierarchy [5] is the integrability condition of the following linear equations

(3) $$\frac{\partial w(x)}{\partial x_\ell} = \left[\frac{\partial^\ell}{\partial x_1^\ell} + \sum_{m=1}^{\ell-2} b_m(x)\frac{\partial^m}{\partial x_1^m}\right] w(x), \quad x = (x_1,x_3,x_5,\ldots),$$

$\ell = 1, 3, 5, \ldots$ .

Note that operators on the right-hand side of (3) have zero constant terms. This system of linear equations has a formal solution of the following form:

$$w(x,k) = (1 + O(k^{-1}))\, e^{\tilde{\xi}(x,k)},$$

where $\tilde{\xi}(x,k) = \sum_{j>0:\mathrm{odd}} x_j k^j$ and k is a parameter. We call this function the wave function. Then we have the bilinear identity:

$$(4) \qquad \oint \underline{dk}\; w(x,k)w(x',-k) = 1, \quad \underline{dk} = \frac{dk}{2\pi i k}$$

for any x and x'. Here the integration is taken on a small circuit around $k = \infty$. This bilinear identity also characterizes the BKP hierarchy.

The $\tau$-function of the BKP can be introduced by the following formula

$$(5) \qquad w(x,k) = \frac{e^{\tilde{\xi}(x,k)} e^{-2\tilde{\xi}(\tilde{\partial},k^{-1})} \tau(x)}{\tau(x)}$$

where $\tilde{\partial} = (\frac{\partial}{\partial x_1}, \frac{1}{3}\frac{\partial}{\partial x_3}, \frac{1}{5}\frac{\partial}{\partial x_5}, \ldots)$. From the above bilinear identity we obtain bilinear equations for $\tau$. Namely, for any $y = (y_1, y_3, \ldots)$ we have

$$\sum_{j\geq 1} \tilde{p}_j(2y)\tilde{p}_j(-2\tilde{D}_x) \exp\left[\sum_{\ell>0:\mathrm{odd}} y_\ell D_{x_\ell}\right] \tau(x)\cdot\tau(x) = 0$$

where $\tilde{D}_x = (D_{x_1}, \frac{1}{3} D_{x_3}, \frac{1}{5} D_{x_5}, \ldots)$ and $\tilde{p}_j(x)$ are defined by

$$e^{\tilde{\xi}(x,k)} = \sum_{\ell=0}^{\infty} k^\ell \tilde{p}_\ell(x).$$

For example, we have

$$(D_1^6 - 5\, D_1^3 D^3 - 5\, D_3^2 + 9\, D_1 D_5)\tau\cdot\tau = 0.$$

In the BKP case, the totality of $\tau$-functions is an orbit of an infinite dimensional orthogonal group. This can be seen as follows.

Let $\phi_n$ ($n\in\mathbb{Z}$) be generators with the defining relations

$$(6) \qquad [\phi_m, \phi_n]_+ = \phi_m\phi_n + \phi_n\phi_m = (-1)^m \delta_{m+n,0}, \quad m, n\in\mathbb{Z}.$$

Let $W = \bigoplus_{n\in\mathbb{Z}} \phi_n$ be the vector space spanned by $\phi_n$ and denote by $A(w)$ the Clifford algebra generated by W. Put $L_0 = \bigoplus_{n<0} \mathbb{C}\phi_n$, $L_0^* = \bigoplus_{n>0} \mathbb{C}\phi_n$, and define $|vac\rangle$ (resp. $\langle vac|$) to be the residue class of 1 in $\mathcal{F} = A(w)/A(w)L_0$ (resp. $\mathcal{F}^* = A(w)/L_0^*A(w)$). Then, of course, we have

$$(7) \qquad \phi_n|vac\rangle = 0 \quad n < 0, \qquad \langle vac|\phi_n = 0 \quad n > 0.$$

We call a subspace of W Lagrangian if it is maximally totally isotropic. For example, $L_0$ is Lagrangian.

Now, for later purpose, we introduce Fourier transform of $\phi_n$ by $\phi(k) = \sum_{n\in\mathbb{Z}} \phi_n k^n$ where k is a parameter. Then the above commutation relations (6) take the following form

$$[\phi(k), \phi(k')]_+ = \hat{\delta}(-k/k').$$

Here $\hat{\delta}(k) = \sum_{n\in\mathbb{Z}} k^n$ is the delta function with respect to the measure $\underline{dk}$. Namely, for any function f(k), we have

$$\int \underline{dk}\, f(k)\, \hat{\delta}(-k/k') = f(-k').$$

We define the inner product on W by

$$(f.g) = \sum_{n\in\mathbb{Z}} (-1)^n f_n.g_{-n},\ f = \sum_{n\in\mathbb{Z}} (-1)^n f_n \phi_{-n},\ g = \sum_{n\in\mathbb{Z}} (-1)^n g_n \phi_{-n}.$$

Let G(W) be the Clifford group:

$$G(W) = \{g \in A(W) \mid \exists g^{-1},\ \iota(g)Wg^{-1} = W\},$$

where $\iota$: $A(W) \longrightarrow A(W)$ is the algebra automorphism with the property $\iota(w) = -w$, $w\in W$. We denote by O(W) the subgroup of G(W) consisting of elements of G(W) which leave the above inner product invariant. The Lie algebra $\mathfrak{go}(\infty)$ of O(W) is spanned by quadratic elements $\phi_i\phi_{-j}$.

Also we have a bilinear form $\mathcal{F}^* \times \mathcal{F} \longrightarrow \mathbb{C}$ which is induced from the canonical bilinear map $\mathcal{F}^* \times \mathcal{F} \longrightarrow \mathcal{F}^* \otimes_{A(W)} \mathcal{F} \cong \mathbb{C} \oplus \mathbb{C}\phi_0$. We call this bilinear form the vacuum expectation value and denote it by $< >$. For example, we have

$$<\phi(k)\phi(k')> = \frac{1}{2}\,\frac{1-k'/k}{1+k'/k}\,.$$

Other expectation values can be calculated by using Wick's Theorem.

In order to relate these with the BKP, we need one more. We introduce the "Hamiltonian" of the BKP by

$$\text{(8)} \qquad H(x) = \frac{1}{2} \sum_{\substack{\ell>0:\text{odd} \\ n\in\mathbb{Z}}} (-1)^{n+1} x_\ell \phi_n \phi_{-n-\ell},$$

which is an element of $\mathfrak{go}(\infty)$. The action of $H(x)$ on $W$ is diagonalized with respect to the basis $\phi(k)$:

$$\phi(x,k) = e^{H(x)}\phi(k)e^{-H(x)} = e^{\tilde{\xi}(x,k)}\phi(k).$$

After these preliminaries, our conclusion immediately follows from the following expression of $\tau(x)$: For any element $g$ of even Clifford group (= elements of $G(w)$ invariant under $\iota$), the function

$$\text{(9)} \qquad \begin{aligned} \tau(x,g) &= <e^{H(x)}ge^{-H(x)}> \\ &= <e^{H(x)}g> \end{aligned}$$

is a $\tau$-function of the BKP. The second equality follows from the fact $H(x)\,|\,\text{vac}> = 0$, which is a consequence of (7) and (8). This statement can be proved by showing that the corresponding wave function defined by (5) satisfies the bilinear identity (4). By using this expression, we can also show that $\tau$-functions of the BKP are parameterized by the totality of Lagrangians (= $SO(W)/I$, $I = \{T\in SO(W) \mid TL_0 = L_0\}$).

We give an example of $\tau$-functions. If we take $g = \exp\left[\sum_{j=1}^{N} a_j \frac{p_j+q_j}{p_j-q_j} \phi(p_j)\phi(q_j)\right]$ with arbitrary N, $a_j$, $p_j$ and $q_j$ then the corresponding $\tau$-function is given by

$$\tau(x) = \sum_{n=0}^{N} \sum_{1\leq j_1<\dots<j_n\leq N} a_{j_1}\dots a_{j_n} \prod_{\ell<\ell'} c_{j_\ell j_{\ell'}} \times$$

$$\times \exp\left[\sum_{\ell=1}^{n} (\tilde{\xi}(x,p_{j_\ell}) + \tilde{\xi}(x,q_{j_\ell}))\right]$$

where

$$c_{ij} = \frac{(p_i-p_j)(p_i-q_j)(q_i-p_j)(q_i-q_j)}{(p_i+p_j)(p_i+q_j)(q_i+p_j)(q_i+q_j)} .$$

This is called an N-soliton solution of the BKP.

We note that the above $\tau$-function can be expressed by using the vertex operator:

$$\tau(x) = \exp\left[\sum_{j=1}^{N} a_j \frac{p_j+q_j}{p_j-q_j} Z(p_j,q_j)\right] \cdot 1 .$$

Here Z(p,q) is the vertex operator for the BKP:

$$\text{(10)} \qquad Z(p,q) = \frac{1}{2}\,\frac{p-q}{p+q}\, e^{\tilde{\xi}(x,p)+\tilde{\xi}(x,q)} e^{-2(\tilde{\xi}(\tilde{\partial},p^{-1})+\tilde{\xi}(\tilde{\partial},q^{-1}))} .$$

This vertex operator gives a realization of $\mathfrak{go}(\infty)$. Also if we impose additional condition on the BKP (reduction), for example if we require $\frac{\partial\tau}{\partial x_{\ell j}} = 0$ $j=1,3,\dots$ for an odd $\ell$, then the corresponding vertex operator is given by putting $p^\ell = -q^\ell$ in (10). This vertex operator coincides with that of the affine Lie algebra $A^{(2)}_{\ell-1}$.

3. Now we turn to the L-L equation.

Our starting point was the form of the 2-soliton solution of the L-L equation given by Bogdan-Kovalev [6], which has the following form:

$$f = 1 + e^{\xi_1+\xi_2},\ f^* = 1 + c_{12}h_1h_2\, e^{\xi_1+\xi_2},$$

$$g = e^{\xi_1} + e^{\xi_2},\ g^* = h_1e^{\xi_1} + h_2e^{\xi_2},$$

where

$$\xi_j = \xi_j^0 + k_jx_1 + \omega_jx_2,\ \omega_j^2 = (k_j^2 - a^2)(k^2 - b^2),$$

(11)

$$h_j = \frac{\omega_j+k_j^2-a^2}{\omega_j-k_j^2+a^2},\ c_{12} = \frac{k_1-k_2}{k_1+k_2}\,\frac{\omega_1(k_2^2-a^2)-\omega_2(k_1^2-a^2)}{\omega_1(k_2^2-a^2)+\omega_2(k_1^2-a^2)},$$

and $k_j$, $\xi_j^0$ are arbitrary constants. Here again the elliptic curve E: $\omega^2 = (k^2-a^2)(k^2-b^2)$ appears. The elliptic curve $\tilde{E}$ (1) is an unramified double cover of E:

$$\pi:\ \tilde{E} \longrightarrow E$$

$$\tilde{P} = (Z_1,Z_2,Z_3) \longmapsto P = (k,\omega),\ \ k = 2\,Z_3,\ \omega = 4\,Z_1Z_2.$$

Our first task is to identify the factor $c_{12}$ in (11). For that purpose we reexamine the BKP. If we regard the parameter k in the BKP case as the coordinate on $\mathbb{P}^1$, then the vacuum expectation value of $\phi(k)\phi(k')$

$$\langle\phi(k)\phi(k')\rangle = \frac{1}{2}\,\frac{k-k'}{k+k'}$$

can be interpreted as the Cauchy kernel on $\mathbb{P}^1$. Also the factor $c_{ij}$ in the BKP case is expressed in terms of this kernel.

We seek an analogy. Let #: $E \longrightarrow E$ be the involution on E defined by $P^\# = (k,\omega)^\# = (-k,-\omega)$ and denote by $\infty_\pm$ the points on E with $(k,\omega) = (\infty,\infty)$ and $\omega/k^2 = \pm 1$. We introduce the Cauchy kernel K(P,P') on E given by

$$K(P,P') = \frac{1}{2}\,\frac{\omega+k^2-\omega'-k'^2}{k+k'}, \quad P = (k,\omega),\ P' = (k',\omega'),$$

which has the following properties: the kernel $K(P,P')$ has its poles in $P$ at $P = P'^{\#}$ and $P = \infty_+$ and its zeros in $P$ at $P = P'$ and $P = \infty_-$. Also it satisfies

$$\int dPK(P,P') = 1$$

where $dP = \frac{dk}{2\pi i\omega}$ and the integration is taken around $P = P'^{\#}$. Then the factor $c_{12}$ in (11) is expressed as

$$c_{12} = \frac{4}{a^2-b^2}\,K(P_1,P_2)K(P_2^{\#},P_1^{\#}),\ P_j = (k_j,\omega_j)\ j=1,2.$$

We note that the function $K(P,P') + K(P',P)$ can be regarded as the delta function with respect to the measure $dP$, provided the integration contour is suitably chosen (cf. [1]).

After knowing these, we can follow the way which we have taken for the BKP case.

We prepair free fermions $\phi(P)$ on the elliptic curve E:

$$[\phi(P),\phi(P')]_+ = K(P,P') + K(P',P).$$

Define the vacuum expectation value of $\phi(P)\phi(P')$ by

$$\langle\phi(P)\phi(P')\rangle = K(P,P').$$

We introduce time evolutions

$$(12)\qquad \xi_n(x,P) = n\log h + \sum_{\ell>0:\mathrm{odd}} x_\ell k^\ell + \sum_{\ell>0:\mathrm{even}} x_\ell (c)^{1/2}\left(h^{\frac{\ell}{2}} - h^{-\frac{\ell}{2}}\right)$$

where $c = \frac{a^2-b^2}{4}$ and $h(P) = \frac{\omega+k^2-a^2}{\omega-k^2+a^2}$. We note that coefficients of $x_\ell$ in (12) are functions on E which have poles only at $P = \infty_\pm$ and are odd with respect to the involution #. This is also an analogue of the BKP case, provided we consider the involution $\mathbb{P}' \ni k \longmapsto -k \in \mathbb{P}'$ in the BKP case. We can find $H_n(x)$ which is quadratic in $\phi(P)$ and has the properties

$$e^{H_n(x)}\phi(P)e^{-H_n(x)} = e^{\xi_n(x,P)}\phi(P),$$

and

$$H_n(x)\,|vac\rangle = 0 \text{ (see [1])}.$$

Since the L-L equation is a "subholonomic" equation, we need further procedure.

Let $\phi^{(i)}(P)$, i=1,2 be two copies of free fermions on E and let g be an element of the Clifford group generated by $\phi^{(i)}(P)$, i=1,2. We put

$$\tau_{n_1,n_2}(x^{(1)},x^{(2)}) = \langle e^H g\rangle,$$

$$\sigma_{n_1,n_2}(x^{(1)},x^{(2)}) = \langle \phi_0^{(1)}\phi_0^{(2)}e^H g\rangle,$$

where $H = H_{n_1}^{(1)}(x^{(1)}) + H_{n_2}^{(2)}(x^{(1)})$ and $\phi_0^{(i)} = \int_{\infty_+\cup\infty_-} dP\ \phi^{(i)}(P)$, i=1,2. In order to obtain the L-L equation, we impose restriction on the choice of g. We require that g satisfies the relation

$$(g\phi^{(1)}(P),\ g\phi^{(2)}(P)) = (\phi^{(1)}(P)g,\ \phi^{(2)}(P)g)T_g(P),$$

where $T_g(P)$ is a 2×2 matrix with $T_g(P)\,{}^tT_g(P^\#) = 1$ and $\det T_g(P) = 1$. The set of such g forms a subgroup $G_L$ of the Clifford group. For an element $g\in G_L$, we can show that $e^H g e^{-H}$ is invariant under the change $(n_1,n_2) \longmapsto (n_1+n,n_2+n)$, $(x^{(1)},x^{(2)}) \longmapsto (x^{(1)}+x,x^{(2)}+x)$. Accordingly $\tau_{n_1,n_2}$ and $\sigma_{n_1,n_2}$ depend only on $n = n_1 - n_2$, $x = x^{(1)} - x^{(2)}$. Then our conclusion follows from the following assertion:

Functions $f = \tau_{n_1,n_2}$, $f^* = \tau_{n_1+1,n_2}$, $g = (c)^{1/2}\sigma_{n_1,n_2}$, $g^* = (c)^{1/2}\sigma_{n_1+1,n_2}$ solve the L-L equation in the bilinear form (2). This statement can be shown by using the bilinear identity of the L-L equation (see [1]). The Lie algebra of $G_L$ is given by $\mathfrak{sl}(2,\mathbb{C}[k,k^{-1},\omega])\oplus\mathbb{C}1$ where the rule of the central extension is given by

$$c(A(P),A'(P)) = \underset{k=\infty_+}{\mathrm{Res}}\ \mathrm{Tr}(A(P)dA'(P)).$$

## References


1. E. Date, M. Jimbo, M. Kashiwara and T. Miwa. J. Phys. A 16(1983), 221-236.

2. E. K. Sklyanin. On complete integrability of the Landau-Lifshitz equation; LOMI Preprint E-3-1979.

3. R. Hirota. J. Phys. Soc. Japan 51(1982), 323-331.

4. E. Date, M. Jimbo, M. Kashiwara and T. Miwa. In Proceedings of RIMS symposium (ed. M. Jimbo, T. Miwa), 39-120, World Scientific, 1983.

5. E. Date, M. Jimbo, M. Kashiwara and T. Miwa. Physica 4D(1982), 343-365.

6. M. M. Bogdan and A. S. Kovalev. JETP Lett. 31(1980), 424-427.

# FLAG MANIFOLDS AND INFINITE DIMENSIONAL KÄHLER GEOMETRY

by

D. S. Freed

Let G be a connected, compact Lie group, A (generalized) *flag manifold* for G is the quotient of G by the centralizer of a torus. Hermitian symmetric spaces (e.g. complex projective spaces and Grassmannians), which are of the form $G/C(\mathbb{T})$ for a circle $\mathbb{T} \subset G$, are flag manifolds. The generic examples, though, are flag manifolds of the form G/T for T a maximal torus. The name derives from the manifold U(n)/diagonals of flags $\{0 = V_0 \subset V_1 \subset V_2 \subset \dots \subset V_n = \mathbb{C}^n\}$ in complex n-space. Flag manifolds enjoy many favorable geometric properties. They can be realized as coadjoint orbits of G, and thus carry an invariant symplectic form. There is also a complex description of flag manifolds as quotients of the complex group $G_{\mathbb{C}}$. The symplectic and complex structures merge nicely: flag manifolds are homogeneous Kähler. Using these two properties--homogeneity and the Kähler condition--we easily compute curvature formulas. For special metrics flag manifolds are Kähler-Einstein. In any metric the Ricci curvature is positive, from which we deduce a vanishing theorem in cohomology. The full flag manifold G/T plays an important role in the representation theory of G à la Borel-Weil-Bott, and in that story the vanishing theorem plays a crucial part.

Starting with a compact Lie group G we can also form the space $\Omega G$ of smooth *based* loops in G. This is the set of smooth maps $S^1 \longrightarrow G$ which carry a fixed point on $S^1$ to the identity element of G. Now $\Omega G$ and flag manifolds are known to be similar--witness Bott's application of Morse Theory to compute their cohomology [B1]. In fact, the analogy runs far deeper than cohomology: $\Omega G$ is a flag manifold for an infinite dimensional group. Let LG denote the group of *all* smooth loops on G, and $\mathbb{T} \ltimes LG$ the semi-direct product by the circle action which rotates the loops.

Then $\Omega G$ is the quotient of $\mathbb{T} \ltimes LG$ by the centralizer $\mathbb{T} \times G$ of the circle $\mathbb{T}$. (Here $G \subset LG$ are the point loops.) The geometry of $\Omega G$ can be explored as for finite dimensional flag manifolds. Hence $\Omega G$ is homogeneous Kähler, and its connection and curvature have simple expressions in terms of Toeplitz operators. What is most significant is that the curvature is trace class (with a modified trace), and so can be used to calculate the "first Chern class" of $\Omega G$. In particular, $\Omega G$ is Kähler-Einstein.

At this point we should interject a few remarks about infinite dimensional manifolds. Those which arise in geometric problems are usually of the form $\mathfrak{M}$ = Map(M,N) for M,N finite dimensional. Roughly speaking, the study of $\mathfrak{M}$ involves analysis on M and geometry on N. Since the manifold with the simplest analysis is $M = S^1$, and that with the simplest geometry is N = G, it is natural to study loop groups before tackling the general case. Now $\mathfrak{M}$ has various completions as a Hilbert manifold, but always has a trivial tangent bundle. This theorem, due to Kuiper, states that the group of invertible transformations on a Hilbert space is contractible; consequently, any Hilbert manifold is parallelizable. In particular, $\mathfrak{M}$ has no characteristic classes. What we have done on $\Omega G$, in the Kähler category, is to reduce the structure group from unitaries on a Hilbert space, which is contractible, to the group of unitaries of the form (identity + trace class), which has the homotopy type of $U(\infty)$. This reduction should be given by the holonomy bundle of the Kähler metric, and then the first Chern class can be computed as the trace of the curvature by a generalization of Chern-Weil Theory. Unfortunately, we were unable to surmount some technical difficulties in constructing the holonomy bundle. As a temporary measure, we introduce an honest reduction of the bundle of frames via a family of Fredholm operators parametrized by LG as a substitute for the holonomy bundle. The characteristic classes of this family are calculated using a new index theorem.

The group $\mathbb{T} \ltimes LG$ has a "maximal torus" $\mathbb{T} \times T$, where $T \subset G$ is a maximal torus of the compact group. The quotient $\mathfrak{F}$ is the full flag manifold, analogous to G/T in the finite dimensional case, and so plays an important role in the representation theory of $\mathbb{T} \ltimes$

LG. As $\mathcal{F}$ fibers holomorphically over $\Omega G$ with fiber G/T, we can combine our results about G/T and $\Omega G$ to draw conclusions about the geometry and representation theory of loop groups. Our arguments at this stage are purely formal, since we will unabashedly apply finite dimensional principles to the infinite dimensional situation, and we have self-consciously branded our metatheorems with quotation marks. One feature not seen in finite dimensions is the central extension $\mathbb{T} \ltimes LG$. It is the group most honestly called the affine Kac-Moody group, and is the one for which the positive energy representations are honest representations.

The geometric approach to the representation theory of loop groups is due to Segal. Our account stresses the *intrinsic* geometry of $\Omega G$. As our primary interest is infinite dimensional geometry, not representation theory, the latter half of this paper should be read as a case study advocating the holonomy bundle in infinite dimensions as the appropriate geometric analogue of the frame bundle in finite dimensions. Representation theory is brought in to verify the simplest topological invariant of the holonomy bundle, its first Chern class. Surprisingly, the theory of instantons on $S^4$ provides additional favorable evidence, as we describe in §5.

In this paper we only sketch proofs and omit many details. A complete account of this material, with emphasis on definite dimensional geometry, can be found in the author's thesis [F]. There the connection between Toeplitz generators and Bott Periodicity is also explored, from which the higher Chern classes of $\Omega G$ can be computed.

My knowledge of loop groups derives mainly from the writings of Pressley and Segal [P1], [P2], [PS], [S1], [S2]. In addition, I received patient personal instruction from Andrew Pressley and Martin Guest. I take this opportunity to thank them, as well as Bert Kostant, Shrawan Kumar, and Iz Singer for informative discussions on various aspects of this work.

## §1. Finite Dimensional Flag Manifolds

From now on G is a fixed connected compact Lie group, which for simplicity we take to be simple and simply connected. Consider the coadjoint action of G on the dual of its Lie algebra $\mathfrak{g}^*$. The stabilizer of $\mu \in \mathfrak{g}^*$ is the centralizer of the torus generated in G by $e^{tX_\mu}$, where $X_\mu$ is the element of $\mathfrak{g}$ identified with $\mu$ via the Killing form. In other words, the coadjoint orbits of G are flag manifolds. Conversely, every flag manifold appears as a coadjoint orbit. Now by a well-known prescription of Kostant and Kirillov, coadjoint orbits carry natural symplectic forms. On the other hand, coadjoint orbits of G have a complex description $G_{\mathbb{C}}/P$ for some complex (parabolic) subgroup P of the complex group $G_{\mathbb{C}}$ corresponding to G. The complex and symplectic structures together give a nondegenerate symmetric bilinear form at each point of the coadjoint orbit. If the structures are compatible, this form is positive definite. Therefore, coadjoint orbits are Kähler manifolds.

Fix a maximal torus $T \subset G$ with Lie algebra $\mathfrak{t}$. We regard $\mathfrak{t}^* \subset \mathfrak{g}^*$ using the Killing form; then coadjoint orbits in $\mathfrak{g}^*$ restrict to Weyl group orbits in $\mathfrak{t}^*$. The type of the coadjoint orbit is determined by the intersection of the Weyl group orbit with the infinitesimal diagram (i.e. hyperplanes in $\mathfrak{t}^*$ consisting of singular elements). The restriction of the generic orbit lies strictly in the interior of the chambers and is diffeomorphic to the full flag manifold G/T. We restrict our attention to these generic orbits, although our geometric results (except for Corollary 1.12) extend to the general case. For intermediate flag manifolds the set of *complementary roots* [BH] replace the set of all roots in our formulas.

Under the action of T the Lie algebra of G decomposes as

$$\mathfrak{g} = \mathfrak{t} \oplus \mathfrak{m},$$

with $\mathfrak{m}$ the sum of the root spaces relative to T. Complexifying,

$$\mathfrak{g}_{\mathbb{C}} = \mathfrak{t}_{\mathbb{C}} \oplus \mathfrak{m}_+ \oplus \mathfrak{m}_-, \tag{1.1}$$

where $\mathfrak{m}_{\mathbb{C}} = \mathfrak{m}_+ \oplus \mathfrak{m}_-$ is the splitting into positive and negative roots relative to a fixed Weyl chamber. We identify the complexified tangent space to G/T at the basepoint p with $\mathfrak{m}_{\mathbb{C}}$; then the splitting into $\mathfrak{m}_+ \oplus \mathfrak{m}_-$ defines a homogeneous almost complex structure on G/T via the G action. As $\mathfrak{m}_+$ is closed under brackets, this almost complex structure is integrable. Coadjoint orbits diffeomorphic to G/T are parametrized by the interior of the Weyl chamber, which also parametrizes the positive homogeneous symplectic forms, hence homogeneous Kähler structures, on G/T. Now the interior of the Weyl chamber is a subset of $\mathfrak{t}^* = H^1(T;\mathbb{R}) \simeq H^2(G/T;\mathbb{R})$. This last identification via transgression can be realized canonically on the form level in view of the following simple

**Lemma 1.2.** *Each element of* $H^2(G/T;\mathbb{R})$ *has a unique representative as a closed invariant 2-form on* G/T.

Explicitly, for each root $\alpha$ let $\theta^\alpha \in \mathfrak{g}^*$ be a (suitably normalized) left invariant 1-form on G dual to the $\alpha$ root space. Then for $\mu \in \mathfrak{t}^*$,

(1.3) $$d\mu = \frac{1}{2\pi i} \sum_{\alpha>0} (\mu,\alpha)\, \theta^\alpha \wedge \theta^{-\alpha},$$

where $(\cdot,\cdot)$ is the Killing form, transferred to $\mathfrak{t}^*$, and the summation runs over the positive roots. The right-hand side of (1.3) is the pullback of the sought after 2-form $\omega_\mu$ on G/T. This form is of type (1,1) and is positive if and only if $\mu$ is in the interior of the Weyl chamber. It is then the Kähler form corresponding to $\mu$.

For any of these metrics G/T is a compact homogeneous Kähler manifold. Wang [W] classified compact simply connected homogeneous complex manifolds, and proved that the ones which are Kähler are necessarily flag manifolds. We exploit this combination of homogeneity and Kähler to compute the connection and curvature.

Fix a Kähler metric $\omega_\mu$ on G/T. The left action of G on G/T has an infinitesimal version which assigns a vector field $\xi_X$ to each $X \in \mathfrak{g}$. Since the Kähler structure is invariant, $\xi_X$ is both a Killing vector field and an infinitesimal automorphism of the complex structure. Recall that the map $X \longmapsto \xi_X$ identifies $\mathfrak{m} \simeq T_p(G/T)$ at the basepoint p, so by complexification identifies $\mathfrak{m}_{\mathbb{C}} \simeq T_p(G/T)_{\mathbb{C}}$. Let $\nabla$ denote the Kähler covariant derivative and $\mathcal{L}$ the Lie derivative. Then for any $X \in \mathfrak{g}$, the difference $\nabla_{\xi_X} - \mathcal{L}_{\xi_X}$ is tensorial, and so defines a linear transformation on $T_p(G/T) \simeq \mathfrak{m}$. Since $\xi_X$ preserves the complex structure, under complexification this transformation separately preserves $\mathfrak{m}_+$ and $\mathfrak{m}_-$. Thus we obtain a map

$$\varphi\colon \mathfrak{g}_{\mathbb{C}} \longrightarrow \mathfrak{gl}(\mathfrak{m}_+) \tag{1.4}$$

as the $\mathbb{C}$-linear extension of

$$X \longmapsto (\nabla_{\xi_X} - \mathcal{L}_{\xi_X})\big|_{\mathfrak{m}_+}.$$

There is a simple formula for $\varphi$ in terms of the projection $\pi_+\colon \mathfrak{g}_{\mathbb{C}} \longrightarrow \mathfrak{m}_+$ determined by the decomposition (1.1).

**Theorem 1.5.** (a) $\varphi(H) = \operatorname{ad} H$ *for* $H \in \mathfrak{t}_{\mathbb{C}}$;

(b) $\varphi(\bar{X}) = \pi_+ \circ \operatorname{ad} \bar{X}$ *for* $\bar{X} \in \mathfrak{m}_-$;

(c) $\varphi(X) = -\varphi(\bar{X})^*$ *for* $X \in \mathfrak{m}_+$.

In (c) the adjoint is taken with respect to the Kähler form $\omega_\mu$, which restricts to a Hermitian metric on $\mathfrak{m}_+$. We can express $\varphi(X)$ in terms of $H_\mu \in \mathfrak{t}$, the dual to $\mu \in \mathfrak{t}^*$ via the Killing form, as

$$\varphi(X) = (\operatorname{ad} H_\mu)^{-1}(\operatorname{ad} X)(\operatorname{ad} H_\mu), \qquad X \in \mathfrak{m}_+. \tag{1.6}$$

It is worth pointing out that (b) is independent of the homogeneous Kähler metric. This formula becomes particularly nice when we identify $\mathfrak{m}_+ \simeq (\mathfrak{m}_-)^*$ via the Killing form; then $\varphi(\bar{X})$ is the coadjoint action of $\bar{X} \in \mathfrak{m}_-$.

Before giving the proof of (1.5) we recall some basic facts about the Kähler connection. A Kähler manifold is first of all a

Riemannian manifold, and so has a unique torsionfree metric connection--the Levi-Civita connection. On the other hand, the tangent bundle is holomorphic, and any holomorphic bundle with a Hermitian metric has a unique metric connection--the Hermitian connection--which agrees with the $\bar{\partial}$ operator in antiholomorphic directions. In particular, it vanishes on holomorphic vector fields in antiholomorphic directions. Note that in these directions the connection is independent of the metric. Finally, the Kähler condition is satisfied precisely when the Levi-Civita connection and the Hermitian connection coincide.

**Proof of (1.5).**

(a) This is simply the isotropy representation which defines the holomorphic tangent bundle. More explicitly, $\xi_H(p) = 0$ so that $\nabla_{\xi_H} = 0$ at p. Also, the map $X \longmapsto \xi_X$ is an antihomomorphism of Lie algebras ($[\xi_X, \xi_Y] = -\xi_{[X,Y]}$), from which $\varphi(H) = \mathrm{ad}\, H$ is immediate.

(b) For $Y \in \mathfrak{m}_+$ let $(\xi_Y)_+$ denote the (1,0) component of the vector field $\xi_Y$. This is a holomorphic vector field ($\xi_Y$ preserves the complex structure), and we use it to compute $\varphi(\bar{X})Y$. By the remarks preceeding the proof,

$$\nabla_{\xi_{\bar{X}}}(\xi_Y)_+ = \bar{\partial}((\xi_Y)_+)(\xi_{\bar{X}}) = 0,$$

since $\xi_{\bar{X}}(p)$ is of type (0,1) and $(\xi_Y)_+$ is holomorphic. As noted in the previous paragraph, $\mathcal{L}_{\xi_{\bar{X}}}\xi_Y = -\xi_{[\bar{X},Y]}$, and an easy argument now shows that $\mathcal{L}_{\xi_{\bar{X}}}(\xi_Y)_+(p) = -\xi_{\pi_+([\bar{X},Y])}$ as desired.

(c) Both the Kähler connection and Lie derivative preserve the metric and complex structure on real vectors $\xi_X$. Therefore, $\varphi$ maps the real Lie algebra $\mathfrak{g}$ into skew-Hermitian transformations, whence (c).

The curvature of the Kähler metric is

(1.7) $$R(X,Y) = [\varphi(X),\varphi(Y)] - \varphi([X,Y]).$$

R is an invariant 2-form which, on real vectors, is a skew-Hermitian transformation of the holomorphic tangent space at each point; (1.7) is the expression for R at the basepoint p. It is a quite general property of Kähler metrics that the curvature is of type (1,1), a fact which also follows immediately in this case from (1.5). That said,

**Theorem 1.8.** *For* $X \in \mathfrak{m}_+$ *and* $\bar{Y} \in \mathfrak{m}_-$,

$$R(X,\bar{Y}) = [(\mathrm{ad}\ H_\mu)^{-1}(\mathrm{ad}\ X)(\mathrm{ad}\ H_\mu),\ \pi_+ \circ \mathrm{ad}\ \bar{Y}]$$

$$- \pi_+(\mathrm{ad}\ [X,\bar{Y}]_{\mathfrak{t}_{\mathbb{C}} \oplus \mathfrak{m}_-}) - (\mathrm{ad}\ H_\mu)^{-1}(\mathrm{ad}\ [X,\bar{Y}]_{\mathfrak{m}_+})(\mathrm{ad}\ H_\mu).$$

On a Kähler manifold the Ricci tensor, which in Riemannian geometry is the symmetric bilinear form

(1.9) $$\mathrm{Ric}(X,Y) = \mathrm{Trace}\ \{Z \longmapsto R(Z,X)Y\},$$

can be realized as the (1,1) form

(1.10) $$\sigma(X,\bar{Y}) = \mathrm{Trace}\ R(X,\bar{Y}).$$

Furthermore, by Chern-Weil theory $\frac{i}{2\pi}\sigma$ has cohomological significance: it represents the first Chern class of the holomorphic tangent bundle. For G/T direct calculation from (1.8) proves

**Proposition 1.11.** *The Ricci form* $\frac{i}{2\pi}\sigma$ *for* G/T *is the invariant* (1,1) *form* $\omega_{2\rho}$ *(cf.* (1.3)) *corresponding to* $2\rho = \sum_{\alpha>0} \alpha \in \mathfrak{t}^*$, *the sum of the positive roots of* G.

We can also give an *a priori* argument. The holomorphic tangent bundle of G/T splits into a direct sum of line bundles, one for each positive root $\alpha$. In fact, $\alpha$ exponentiates to a character of T, and the line bundle $L_\alpha$ corresponding to $\alpha$ is the bundle associated to

$G \longrightarrow G/T$ via this character. Its first Chern class is the transgression of $\alpha$, and so $c_1(G/T)$ is the transgression of

$\sum_{\alpha>0} \alpha = 2\rho$. Since the Kähler metric is invariant, the Ricci form $\frac{i}{2\pi}\sigma$ is an invariant representative of the cohomology class $2\rho$. By (1.2) there is a unique such representative, and it is given by $\omega_{2\rho}$. Remarkably, the Ricci curvature is independent of the homogeneous Kähler metric. In particular, G/T is homogeneous Kähler-Einstein for the metrics $\omega_{r\rho}$, $r>0$.

The weight $\rho$ plays an important role in representation theory, for example in the Weyl Character Formula. Recall that $\rho$, defined above as half the sum of the positive roots, is also the sum of the fundamental weights. In the infinite dimensional setting of loop groups, the sum of the positive roots diverges, whereas the sum of the fundamental weights still makes sense. Our curvature computation in §2 "regularizes" the divergent sum of positive roots, and the answer agrees with the sum of the fundamental weights.

We now outline several consequences of (1.11) which are quite standard for finite dimensional flag manifolds.

**Corollary 1.12.** G/T *admits a spin structure.*

For a complex manifold the second Stiefel-Whitney class $w_2$, which is the obstruction to a spin structure, is the reduction mod 2 of the first Chern class. Since $c_1(G/T)$ is even, $w_2(G/T) = 0$. The complex and spin points of view are easily reconciled on a *Kähler* manifold: The Dirac operator is realized as $\bar{\partial} + \bar{\partial}^*$ on the complex $\Omega^{0,*}(L)$, with L a holomorphic square root of the canonical bundle $\Lambda^n T^*$. The choice of L corresponds to a choice of spin structure. Since G/T is simply connected, there is a unique spin structure (cf. Corollary 1.17b), and $L = L_{-\rho}$ which again exhibits the importance of the weight $\rho$. We adhere to the complex viewpoint, although what follows could be expressed in terms of Dirac. Corollary 1.12 also follows from the observation that the real tangent bundle of G/T is stably trivial (since the normal bundle to $G/T \hookrightarrow \mathfrak{g}^*$ is trivial), whence its Stiefel-Whitney and Pontrjagin classes vanish.

The computation of Ricci curvature leads to an easy vanishing theorem.

**Theorem 1.13.** *The* $\bar\partial$*-cohomology* $H^{0,q}_{\bar\partial}(G/T)$ *vanishes for* $q>0$. *For* $q = 0$, $H^{0,0}_{\bar\partial}(G/T) = \mathbb{C}$.

The proof uses the Weitzenböck formula

$$\bar\partial\bar\partial^* + \bar\partial^*\bar\partial = \bar\nabla^*\bar\nabla + \mathrm{Ric} \qquad \text{on } \Omega^{0,q} \tag{1.14}$$

for Kähler manifolds [MK]. In this formula $\bar\nabla\colon \Omega^{0,q} \longrightarrow \Omega^{0,q} \otimes \Omega^{0,1}$ is the antiholomorphic piece of the covariant derivative. Ric is viewed first as an endomorphism on (0,1) forms (by "raising an index"), and then extended in the obvious way to (0,q) forms. Ric acts trivially on (0,0) forms. The operator $\bar\nabla^*\bar\nabla$ is nonnegative, so if Ric is positive, as it is on G/T, then the kernel of the Hodge Laplacian $\Delta_{\bar\partial}$ on the left-hand side is trivial in positive degrees. More explicitly, let $d\mu$ denote the Kähler measure on G/T. Then for $\alpha\in\Omega^{0,q}$,

$$\int_{G/T} (\Delta_{\bar\partial}\alpha,\alpha)d\mu = \int_{G/T} \{|\bar\nabla\alpha|^2 + (\mathrm{Ric}(\alpha),\alpha)\}\, d\mu. \tag{1.15}$$

If $\Delta_{\bar\partial}\alpha = 0$, then either $\alpha = 0$ or $q = 0$ and $\bar\nabla\alpha = 0$. In the latter case, $\alpha$ is an ordinary function, and $\bar\nabla\alpha = 0$ means that $\alpha$ is constant. But the Hodge Theorem identifies $\ker(\Delta_{\bar\partial}) \simeq H^{0,q}_{\bar\partial}$, whence (1.13). Furthermore, the Dolbeault Theorem identifies $\bar\partial$-cohomology with holomorphic cohomology, that is cohomology with coefficients in the sheaf of germs of holomorphic functions $\mathcal{O}$.

**Corollary 1.16.**

$$H^q(G/T;\mathcal{O}) = \begin{cases} \mathbb{C} & \text{for } q=0; \\ 0 & \text{for } q>0. \end{cases}$$

For $q = 0$ and $q = 1,2$ there is a nice geometric interpretation.

**Corollary 1.17.** (a) *Every global holomorphic function on* G/T *is constant.*

(b) *Every line bundle over* G/T *carries a unique holomorphic structure.*

Part (b) follows from the long exact sequence associated to

$$0 \longrightarrow \mathbb{Z} \longrightarrow \mathcal{O} \xrightarrow{\exp} \mathcal{O}^* \longrightarrow 0,$$

where $\mathcal{O}^*$ consists of germs of nonzero holomorphic functions. We emphasize the importance of a finite measure for defining the Hilbert space of $L^2$ forms implicit in the proof of the vanishing theorem.

Line bundles over G/T correspond to elements of $H^2(G/T;\mathbb{Z})$ which, by transgression, is identified with the weight lattice in $\mathfrak{t}^*$. As noted above, the bundle $L_\lambda$ corresponding to $\lambda$ is an associated bundle to $G \longrightarrow G/T$, and so is homogeneous. By (1.17) $L_\lambda$ has a unique holomorphic structure. Hence the space of holomorphic forms with values in $L_\lambda$ is a representation space for G. The Borel-Weil-Bott Theorem [B2] determines this representation. In the simplest case,

**Theorem 1.18.** *If* $\lambda+\rho$ *is in the interior of the positive Weyl chamber, then* $H^q(G/T;L_\lambda) = 0$ *for* $q>0$ *and* $H^0(G/T;L_\lambda)$ *is the irreducible representation* $V_\lambda$ *of* G *with highest weight* $\lambda$.

**Proof.** The vanishing statement follows from a refinement of (1.14) for line bundles. Namely, if L is a holomorphic line bundle over a Kähler manifold, and F is the curvature of some Hermitian metric on

L, viewed as an endomorphism of $\Omega^{0,q}(L)$, then

$$\bar{\partial}\bar{\partial}^* + \bar{\partial}^*\bar{\partial} = \bar{\nabla}^*\bar{\nabla} + \mathrm{Ric} + F \qquad \text{on } \Omega^{0,q}(L). \tag{1.19}$$

For $L_\lambda$ we choose an invariant metric. Since the curvature form is then invariant and represents $c_1(L_\lambda)$, it must be given by $\omega_\lambda$. So Ric + F is the endomorphism corresponding to the positive form $\omega_{\lambda+2\rho}$. The vanishing of higher cohomology is now immediate (cf. (1.15)). The identification of $H^0(G/T;L_\lambda)$ proceeds via the Peter-Weyl Theorem. Briefly, $L^2$ sections of $L_\lambda$ are $L^2$ complex-valued functions on G which transform correctly under right translation by T. If f: $G \longrightarrow \mathbb{C}$ is a holomorphic section of $L_\lambda$, then for any $\bar{X} \in \mathfrak{m}_-$, extended to a *left* invariant complex vector field on G, we have $\bar{X} \cdot f = 0$. Furthermore, $H \cdot f = -\lambda(H) f$ for any $H \in \mathfrak{t}_{\mathbb{C}}$, since the left invariant vector field H generates *right* translation on G. (This is the infinitesimal version of the transformation law alluded to above.) So under the action of G on $L^2(G)$ by *right* translation, f is a lowest weight vector for the representation of highest weight $\lambda$. By Peter-Weyl the decomposition of $L^2(G)$ under left and right translation is $L^2(G) = \bigoplus_{\nu \in \hat{G}} V_\nu \otimes V_\nu^*$, and so the *left* G-module

$H^0(G/T;L_\lambda) = V_\lambda$. (See [GS] for details.)

Finally, the Weyl Character Formula can be derived in this context from the Holomorphic Lefschetz Fixed Point Theorem. Left translation on G/T by a generic element of T lifts to an action on $\Omega^{0,*}(L_\lambda)$. The Lefshetz formula computes the alternating sum of the trace of this action on the cohomology $H^*(G/T;L_\lambda)$ which, by (1.18), is the character of the representation $V_\lambda$. The computation, carried out in [AB], yields the character formula.

## §2. Based Loops

We turn now to the group $\Omega G$ of *smooth* based loops on G. (Let us agree to defer the technicalities of infinite dimensional manifolds to §3.) It will be advantageous to regard $\Omega G$ not as a group, but as a homogeneous space for the semidirect product $\mathbb{T} \ltimes LG$ of the circle group $\mathbb{T}$ acting on the group LG of *all* smooth loops by rotation. The centralizer of $\mathbb{T}$ consists of $\mathbb{T}$ together with point loops $G \subset LG$, so that

$$\Omega G = (\mathbb{T} \ltimes LG)/C(\mathbb{T}) = (\mathbb{T} \ltimes LG)/(\mathbb{T} \times G)$$

is an infinite dimensional (intermediate) flag manifold. As in the finite dimensional case it can be realized by certain coadjoint orbits in $\mathrm{Lie}(\mathbb{T} \ltimes LG)^* = (\mathbb{R} \ltimes L\mathfrak{g})^*$. In fact, the orbits diffeomorphic to $\Omega G$ are those passing through a point $\langle r,0\rangle \in (\mathbb{R} \ltimes L\mathfrak{g})^*$ with $r>0$, and so are parametrized by the positive reals. We describe the Kähler structure on $\Omega G$ in terms of the decomposition

$$(2.1) \quad (\mathbb{R} \ltimes L\mathfrak{g})_{\mathbb{C}} = (\mathbb{R} \oplus \mathfrak{g})_{\mathbb{C}} \oplus (\bigoplus_{n>0} z^n \mathfrak{g}_{\mathbb{C}}) \oplus (\bigoplus_{n>0} z^{-n} \mathfrak{g}_{\mathbb{C}})$$

$$= \quad \mathfrak{h}_{\mathbb{C}} \quad \oplus \quad \mathfrak{m}_+ \quad \oplus \quad \mathfrak{m}_- \quad .$$

Thus $\mathfrak{h}_{\mathbb{C}}$ is the complexified Lie algebra of the isotropy group $\mathbb{T} \times G$, and the complexified tangent space to $\Omega G$ at the basepoint p is identified with $\mathfrak{m}_+ \oplus \mathfrak{m}_-$, the holomorphic part $\mathfrak{m}_+$ consisting of loops on $\mathfrak{g}_{\mathbb{C}}$ of strictly positive Fourier series, and the antiholomorphic part $\mathfrak{m}_-$ consisting of loops on $\mathfrak{g}_{\mathbb{C}}$ of strictly negative Fourier series. We write $z^n = e^{in\theta}$. As before, the resulting invariant complex structure is integrable since $\mathfrak{m}_+$ is closed under brackets. Now there is only a one parameter family of positive symplectic forms, corresponding to the fact that $\Omega G$ is not the generic coadjoint orbit, but is rather a "minimal" flag manifold. The (normalized) positive symplectic form is

$$(2.2) \qquad \omega(X,Y) = \frac{1}{2\pi} \int_0^{2\pi} (\dot{X}(\theta), Y(\theta))_{\mathfrak{g}_{\mathbb{C}}} \, d\theta.$$

There are several points to explain. First, $(\cdot,\cdot)_{\mathfrak{g}_{\mathbb{C}}}$ is minus the Killing form on $\mathfrak{g}_{\mathbb{C}}$. Next, we have implicitly extended $\omega$ by complex linearity. In (2.2) X and Y should be interpreted as elements of $\mathfrak{m}_{\mathbb{C}} \simeq \mathfrak{m}_+ \oplus \mathfrak{m}_- \simeq T_p(\Omega G)_{\mathbb{C}}$. This defines a 2-form at p which is invariant under $\mathbb{T} \times G$, and so extends to a 2-form on $\Omega G$. Alternatively, view $\Omega G$ as a group and interpret X and Y as elements of the Lie algebra $\Omega\mathfrak{g}_{\mathbb{C}}$, the tangent space at the identity. Then extend the 2-form to be left invariant. A simple calculation shows that the two points of view yield the same 2-form on $\Omega G$. The symplectic form $\omega$ is the Kähler form for the Kähler metric

$$(2.3) \qquad \langle X,Y\rangle = \frac{1}{2\pi} \int_0^{2\pi} \left(\left|\frac{d}{d\theta}\right| X(\theta), Y(\theta)\right)_{\mathfrak{g}_{\mathbb{C}}} d\theta.$$

Again the two interpretations of X and Y are valid. The symmetric form (2.3) is the *Sobolev* $H_{1/2}$ *metric* ($\left|\frac{d}{d\theta}\right| = \Delta^{1/2}$ for $\Delta = \frac{-d^2}{d\theta^2}$ the Laplacian on $S^1$), as will be explained in §3. Writing $a,b \in \mathfrak{g}_{\mathbb{C}}$,

$$\omega(z^n a, z^m b) = \begin{cases} in(a,b)_{\mathfrak{g}_{\mathbb{C}}} & \text{if } n+m=0; \\ 0 & \text{otherwise,} \end{cases}$$

$$\langle z^n a, z^m b\rangle = \begin{cases} |n|(a,b)_{\mathfrak{g}_{\mathbb{C}}} & \text{if } n+m=0; \\ 0 & \text{otherwise.} \end{cases}$$

We emphasize that $\Omega G$ is only a partial flag manifold. The full flag manifold (generic coadjoint orbit) $\mathcal{F} = (\mathbb{T} \ltimes LG)/(\mathbb{T} \times T)$ is topologically a product $\Omega G \times G/T$. It is not surprising, then, that $\Omega G$ behaves in some respects like finite dimensional intermediate flag manifolds (say projective spaces and Grassmannians) and in other respects like the full flag manifold G/T. We remark that in finite dimensions many flag manifolds of the form $G/C(\mathbb{T})$ are Hermitian symmetric spaces. Unfortunately, though, the Kähler metric on $\Omega G$ is not symmetric, since a symmetric left invariant metric on a group is necessarily bi-invariant. As with finite dimensional intermediate flag manifolds, $\mathfrak{m}_+$ is the sum of the root spaces for the positive complementary roots

$$\{<n,\alpha>: n>0 \text{ and } \alpha \text{ is a root of } G \text{ or } 0, \text{ with } 0 \text{ occurring } \dim T \text{ times}\}.$$

(The other positive roots of $\mathbb{T} \ltimes LG$ are $\{<0,\alpha>: \alpha$ is a positive root of $G\}$.) The sum of the positive complementary roots diverges to $<\infty,0>$, so that the first Chern class $c_1(\Omega G)$ is not computable directly from the roots as in finite dimensions. Instead, we turn to curvature.

An infinite dimensional Riemannian manifold (modeled on a Hilbert space) has a unique torsionfree connection. In other words, the Levi-Civita Theorem holds in infinite dimensions. Similarly, the Hermitian connection exists in the infinite dimensional holomorphic category. Therefore, an infinite dimensional Kähler manifold has a Kähler connection whose basic properties carry over from finite dimensions. Theorem 1.5, then, has a direct translation to $\Omega G$. Thus

$$\varphi: (\mathbb{R} \ltimes L\mathfrak{g})_{\mathbb{C}} \longrightarrow \mathfrak{gl}(\mathfrak{m}_+)$$

is given by the same equations as before, but with $\mathfrak{h}_{\mathbb{C}}$ replacing $\mathfrak{t}_{\mathbb{C}}$. Now the resulting formulas have nice expressions in terms of *Toeplitz operators.*

Toeplitz operators are defined classically as follows. Let $L^2(S^1;\mathbb{C})_{hol}$ denote the Hardy space of $L^2$ holomorphic functions on the circle. For each smooth function f on the circle, define

$$T_f: L^2(S^1;\mathbb{C})_{hol} \longrightarrow L^2(S^1;\mathbb{C})_{hol}$$

$$\phi \longmapsto (f\cdot\phi)_+,$$

where $(f\cdot\phi)_+$ is the holomorphic function on $S^1$ which is the holomorphic part of the product $f\cdot\phi$. In terms of Fourier series, we expand $f \cdot \phi = \sum_{n \in \mathbb{Z}} c_n z^n$; then $(f \cdot \phi)_+ = \sum_{n \geq 0} c_n z^n$. The Toeplitz operators we need are a super up version. Our basic Hilbert space $H_s(S^1;\mathfrak{g}_{\mathbb{C}})_{hol}$ is the completion of $\mathfrak{m}_+$ in the $H_s$ metric for $s = 1/2 + \epsilon$, $\epsilon>0$. In coordinates,

$$H_s(S^1;\mathfrak{g}_{\mathbb{C}})_{hol} = \{\phi = \sum_{n>0} a_n z^n : a_n \in \mathfrak{g}_{\mathbb{C}}, \sum_{n>0} n^{2s} |a_n|^2 < \infty\}.$$

We will explain more in §3, and for now revert to the notation $\mathfrak{m}_+$, even where we mean the Hilbert space completion. Then for $X: S^1 \longrightarrow \mathfrak{g}_{\mathbb{C}}$ we define the Toeplitz operator

$$\begin{aligned} T_X: \mathfrak{m}_+ &\longrightarrow \mathfrak{m}_+ \\ Y &\longmapsto [X,Y]_+. \end{aligned} \tag{2.4}$$

The bracket is computed pointwise, and now "+" denotes projection onto the strictly positive components of the Fourier series (omitting the constant term). In terms of Toeplitz operators the connection formulas (1.5) are

**Theorem 2.5.** (a) $\varphi(H) = T_H$ *for* $H \in \mathfrak{h}_{\mathbb{C}}$;

(b) $\varphi(\bar{X}) = T_{\bar{X}}$ *for* $\bar{X} \in \mathfrak{m}_-$;

(c) $\varphi(X) = -T^*_{\bar{X}}$ *for* $X \in \mathfrak{m}_+$.

The Toeplitz operator in (a) is simply a multiplication operator--there is no projection. In (b) the identification $\mathfrak{m}_+ \simeq (\mathfrak{m}_-)^*$ displays Toeplitz operators as the coadjoint representation of $\mathfrak{m}_-$. Here the "Killing form" used to make the identification is the $L^2$ metric

$$(X,Y) = \frac{1}{2\pi} \int_0^{2\pi} (X(\theta),Y(\theta))_{\mathfrak{g}_{\mathbb{C}}} \, d\theta.$$

(The honest Killing form diverges.) Following (1.6) we can write (c) more explicitly. Now ad $H_\mu = \frac{d}{d\theta} = iz\frac{d}{dz} = D$ is the generator of $\mathbb{R}$ in $\mathbb{R} \ltimes Lg$, and

$$\varphi(X) = D^{-1}T_X D \qquad \text{for } X \in \mathfrak{m}_+. \tag{2.6}$$

$T_X$ is a multiplication operator since X is holomorphic.

We reiterate: Theorem 2.5 is a theorem, not an analogy! The proof for the finite dimensional case goes over verbatim to the loop group.

The curvature is expressed in terms of Toeplitz operators too.

**Theorem 2.7.** *For* $X \in \mathfrak{m}_+$ *and* $\bar{Y} \in \mathfrak{m}_-$,

$$R(X,\bar{Y}) = [D^{-1}T_X D, T_{\bar{Y}}] - T_{[X,\bar{Y}]_{\mathfrak{h}_{\mathbb{C}} \oplus \mathfrak{m}_-}} - D^{-1}T_{[X,\bar{Y}]_{\mathfrak{m}_+}} D.$$

At this point, following the agenda of §1, we will compute Trace $R(X,\bar{Y})$. But as $R(X,\bar{Y})$ is an operator on an infinite dimensional Hilbert space, there is no guarantee that such a trace exists. In fact, there is a rigorous notion of what it means for a bounded operator on a Hilbert space to be *trace class*; regrettably, $R(X,\bar{Y})$ does not fall into that class. For smooth $X,\bar{Y}$ the curvature $R(X,\bar{Y})$ is a pseudodifferential operator of order -1 on the circle, and therefore its trace norm diverges logarithmically. Yet we can write

$$H_s(D^2;\mathfrak{g}_{\mathbb{C}})_{hol} = H_s(D^2;\mathbb{C})_{hol} \otimes \mathfrak{g}_{\mathbb{C}}$$

and define an operator $\text{Trace}_{\mathfrak{g}_{\mathbb{C}}}(R(\bar{X},Y))$ on $H_s(D^2;\mathbb{C})_{hol}$ by taking the (finite) trace on $\mathfrak{g}_{\mathbb{C}}$. This new operator is pseudodifferential of order -2, therefore is trace class, and taking its trace we obtain

**Proposition 2.8.** $\widetilde{\text{Trace}}\, R(X,\bar{Y}) = \frac{1}{2\pi i}\int_0^{2\pi} (\dot{X}(\theta),\bar{Y}(\theta))_{\mathfrak{g}_{\mathbb{C}}}\, d\theta = -i\omega(X,\bar{Y}).$

We write a tilde over "Trace" as a reminder of our modified notion of trace. Proposition 2.8 implies that $\Omega G$ is Kähler-Einstein. We remark that after removing the Lie algebra by taking $\mathrm{Trace}_{\mathfrak{g}_{\mathbb{C}}}$, what is left is the trace of the commutator of a Toeplitz operator and its adjoint. Such a trace is well known in the Operator Algebra literature (see [HH], for example) when the adjoint is taken with respect to $L^2$. What we find is that the $H_{1/2}$ adjoint gives the same answer.

**Proof.** We verify (2.8) on the basis elements $X = z^n a$, $\bar{Y} = z^{-m}b$. One sees easily that $\widetilde{\mathrm{Trace}}$ vanishes unless $n = m$. In that case the third term of (2.7) is zero, and the second term

$$T_{[z^n a, z^{-n} b]_{\mathfrak{h}_{\mathbb{C}} \oplus \mathfrak{m}_-}} = \mathrm{ad}\ [a,b]$$

vanishes after taking $\mathrm{Trace}_{\mathfrak{g}_{\mathbb{C}}}$. The first term

$$[D^{-1}T_{z^n a}D, T_{z^{-n}b}](z^{\ell}c) = \delta_{\ell>n}\left(\frac{\ell-n}{\ell}\right) z^{\ell}[a[bc]] - \left(\frac{\ell}{\ell+n}\right) z^{\ell}[b[ac]],$$

where $\delta_{\ell>n}$ indicates that the term appears only if $\ell>n$. Performing $\mathrm{Trace}_{\mathfrak{g}_{\mathbb{C}}}$ we obtain

$$-\left\{\delta_{\ell>n}\left(\frac{\ell-n}{\ell}\right) - \left(\frac{\ell}{\ell+n}\right)\right\} z^{\ell}\ (a,b)_{\mathfrak{g}_{\mathbb{C}}},$$

since $\mathrm{Trace}\ \{c \longmapsto [a[bc]]\} = \mathrm{Trace}\ \{c \longmapsto [b[ac]]\} = -(a,b)_{\mathfrak{g}_{\mathbb{C}}}$ is the Killing form. Finally, sum over $\ell>0$ to compute the Hilbert space trace:

$$\begin{aligned}\widetilde{\mathrm{Trace}}\ R(z^n a, z^{-n}b) &= -\sum_{\ell>0}\left\{\delta_{\ell>n}\left(\frac{\ell-n}{\ell}\right) - \left(\frac{\ell}{\ell+n}\right)\right\}(a,b)_{\mathfrak{g}_{\mathbb{C}}}\\ &= n\ (a,b)_{\mathfrak{g}_{\mathbb{C}}}\\ &= -i\omega(z^n a, z^{-n}b).\end{aligned}$$

As in the finite dimensional case the element of cohomology represented by the Ricci form $\frac{i}{2\pi}\,\widetilde{\text{Trace}}\;R(X,\bar{Y}) = \frac{1}{2\pi}\omega$ has topological meaning, although the interpretation as a first Chern class is more involved (§3). In any case, to determine the cohomology class represented by $\frac{1}{2\pi}\omega$ we need to understand $H^2(\Omega G;\mathbb{Z})$. Since $\pi_1(G) = \pi_2(G) = 0$ and $\pi_3(G) = \mathbb{Z}$, the based loop group $\Omega G$ is connected, simply connected, and $\pi_2(\Omega G) = \mathbb{Z}$. Hence $H_2(\Omega G) = \mathbb{Z}$ by the Hurewicz Theorem, from which $H^2(\Omega G;\mathbb{Z}) = \mathbb{Z}$. Our task is to construct an explicit generator for this group in de Rham cohomology. There is an isomorphism $H^2(\Omega G;\mathbb{Z}) \simeq H^3(G;\mathbb{Z})$ by transgression, so we must first produce a generator for $H^3(G;\mathbb{Z})$. Fortunately, Bott and Samelson [BS] solved this problem by Morse Theory (cf. the discussion in [AHS, p. 453]). We write a representative $\beta_G$ for the generator as an invariant form, i.e. a form on the Lie algebra, and for $G = SU(2)$ it is easily determined as

$$\beta_{SU(2)}(X,Y,Z) = \frac{1}{32\pi^2}\,([X,Y],Z)_{\mathfrak{su}(2)}, \qquad X,Y,Z\in\mathfrak{su}(2),$$

where $(\cdot,\cdot)_{\mathfrak{su}(2)}$ is the Killing form. For any simple G the *highest* root space of $\mathfrak{g}$ determines an inclusion $\mathfrak{su}(2) \hookrightarrow \mathfrak{g}$ and hence a homomorphism $SU(2) \longrightarrow G$ which, according to Bott and Samelson, is a generator for $\pi_3(G)$. By comparing the Killing form of $\mathfrak{g}$ to that of $\mathfrak{su}(2)$, and denoting by $n_G$ the reciprocal square length of the highest root (relative to the Killing form transferred to $\mathfrak{g}^*$), we conclude that

$$\beta_G(X,Y,Z) = \frac{1}{16\pi^2 n_G}\,([X,Y],Z)_{\mathfrak{g}}, \qquad X,Y,Z\in\mathfrak{g}$$

represents a generator of $H^3(G;\mathbb{Z})$. The integer $n_G$ is sometimes termed the *dual Coxeter number* of G, and is given in the following table.

| $G$ | $SU(n)$ | $Spin(n)$, $n\geq 5$ | $Sp(n)$ | $G_2$ | $F_4$ | $E_6$ | $E_7$ | $E_8$ |
|---|---|---|---|---|---|---|---|---|
| $n_G$ | $n$ | $n-2$ | $n+1$ | 4 | 9 | 12 | 18 | 30 |

The transgression of $\beta_G$ to the based loop group is calculated from the evaluation map $\epsilon$: $S^1 \times \Omega G \longrightarrow G$. Pull $\beta_G$ back via $\epsilon$ and then integrate over $S^1$. The resulting form is not invariant, but is cohomologous to the invariant form [S1, p. 328]

$$(2.9) \quad \gamma_{\Omega G}(X,Y) = \frac{1}{8\pi^2 n_G} \int_0^{2\pi} (\dot{X}(\theta),Y(\theta))_{\mathfrak{g}_{\mathbb{C}}} \, d\theta, \quad X,Y \in \Omega\mathfrak{g}.$$

(The remarks after (2.3) apply to the interpretation of (2.9).) The form $\gamma_{\Omega G}$ represents the desired positive generator of $H^2(\Omega G;\mathbb{Z})$.

With this topology behind us we compare $\frac{1}{2\pi}\omega$ and $\gamma_{\Omega G}$ to conclude

**Proposition 2.10.** *The "first Chern class" of* $\Omega G$, *as defined by the curvature, is* $2n_G$ *times the generator of* $H^2(\Omega G;\mathbb{Z})$.

In §4 and §5 we will show how this answer fits in with the theory of Kac-Moody algebras and the theory of instantons. But first we must discuss the geometry of infinite dimensional manifolds to explain what we mean by "first Chern class."

## §3. Infinite Dimensional Manifolds

Finite dimensional manifolds are open patches of $\mathbb{R}^n$ sewn together smoothly. In a similar manner, infinite dimensional manifolds are quilted from open sets in an infinite dimensional topological vector space. Whereas the linear structure of $\mathbb{R}^n$ uniquely determines its topological structure, if the structures are to be compatible, infinite dimensional topological vector spaces exhibit many new phenomena. It is best to avoid most of them, particularly for calculus, and for this reason manifolds modeled on Banach spaces, especially Hilbert spaces, are preferable, though not always available. The differential topology of infinite dimensional manifolds was worked out in the 1960's by Palais, Eells, and others. Those infinite manifolds which arise in geometry and physics are often of the form Map(M,N) for finite dimensional M,N (or are a twisted version of this), and these spaces have nice Hilbert completions [P1]. We briefly recall the construction for $LG = \mathrm{Map}(S^1,G)$. Let $\Delta = \frac{-d^2}{d\theta^2}$ denote the Laplace operator on $S^1$. For $s \geq 0$ the Sobolev $H_s$ metric on the tangent space $Lg = \mathrm{Map}(S^1,g)$ is defined by

$$(3.1)\qquad \langle X,Y\rangle_{H_s} = \frac{1}{2\pi}\int_0^{2\pi} ((1+\Delta)^s X(\theta), Y(\theta))_g \, d\theta, \qquad X,Y \in Lg.$$

This formula makes sense for smooth loops, and the Hilbert space completion with respect to (3.1) is denoted $H_s(S^1,g)$. For $s>1/2$ the $H_s$ loops are continuous (Sobolev Embedding Theorem). There are corresponding completions $H_s(S^1,G)$ in the continuous range $(s>1/2)$ which are Hilbert manifolds modeled on $H_s(S^1,g)$. To construct coordinate charts we note that $\exp\colon g \longrightarrow G$ defines $\mathrm{Exp}\colon H_s(S^1,g) \longrightarrow H_s(S^1,G)$ pointwise, which gives a local chart near the identity. Left translation by *smooth* loops provide charts about every point, and the Sobolev Composition Lemma ensures that the transition functions are smooth. Furthermore, the (pointwise) group multiplication and inversion operation are smooth (Sobolev

Multiplication Theorem). Altogether, then, $H_s(S^1,G)$ is a Hilbert Lie Group for $s>1/2$. (Compare [FU, Appendix A].)

There is a slight simplification when we come to the based loop space $\Omega G$. Here the Laplacian $\Delta$ has no kernel, and we take

$$(3.2)\qquad \langle X,Y\rangle_{H_s} = \frac{1}{2\pi}\int_0^{2\pi} (\Delta^s X(\theta),Y(\theta))_{\mathfrak{g}}\, d\theta, \qquad X,Y\in\Omega\mathfrak{g}$$

as the Sobolev $H_s$ metric. For $s = 1/2$ this agrees with (2.3), which explains our terminology for the Kähler metric. Incidentally, the *Riemannian* curvature of the other $H_s$ metrics is quite tractable, although the formulas are not nearly so pretty as (2.5) and (2.7). We hope to record these results elsewhere.

Our use of the $H_{1/2}$ metric requires some clarification, as the $H_{1/2}$ loops do not form a smooth manifold. But for any $\epsilon>0$ the $H_{1/2+\epsilon}$ loops do form a smooth manifold, and we choose to regard the $H_{1/2}$ metric as an (incomplete) metric on $H_{1/2+\epsilon}$ loops. Note that $H_{1/2+\epsilon}$ is closed under multiplication, so the Toeplitz operators (2.4) are well defined. The $H_{1/2+\epsilon}$ loops have the attractive feature that we can let $\epsilon \longrightarrow 0$ and watch the loop space acquire singularities. Pressley and Segal [PS] study the limiting *based* $H_{1/2}$ loop space as an infinite dimensional singular projective variety.

The technicalities of the differentiable structure temporarily disposed of, we turn to Differential Geometry. On a finite dimensional smooth n-manifold M there is a canonical principal $GL(n;\mathbb{R})$ bundle $GL(M) \longrightarrow M$, the *frame bundle* of M. Encoded in GL(M) are the characteristic classes (Pontrjagin and Steifel-Whitney) of M, its first order intrinsic geometry, and relationships between the topology and geometry of M. More refined geometry involves a reduction of the structure group. For example, if $n = 2m$ is even, a reduction to $GL(m,\mathbb{C})$ is an almost complex structure, and then there are Chern classes corresponding to the primitive cohomology of $GL(m,\mathbb{C})$. A further reduction to U(m) is equivalent to a Hermitian metric. In this framework the integrability of a structure (e.g. the Kähler condition) amounts to the existence of a torsionfree connection on the reduced bundle.

There is an immediate surprise when we construct the frame bundle $GL(\mathcal{M})$ of an infinite dimensional smooth Hilbert manifold $\mathcal{M}$. Now the structure group is $GL(\mathcal{H}_{\mathbb{R}})$, the group of bounded invertible linear transformations on a real Hilbert space $\mathcal{H}_{\mathbb{R}}$. A theorem of Kuiper [K] asserts that $GL(\mathcal{H}_{\mathbb{R}})$ is contractible. As a result, the bundle $GL(\mathcal{M})$ is trivial; in other words, every infinite Hilbert manifold is parallelizable. In particular, there are no (nontrivial) characteristic classes. The complex situation is no better--$GL(\mathcal{H}_{\mathbb{C}})$ is also contractible. Fortunately, though, many subgroups of operators in $GL(\mathcal{H})$ have highly nontrivial topology. The group $GL(\infty)$ of invertible operators of the form (identity + finite rank) is the first example. Its homotopy groups in both the real case and the complex case are determined by the Bott Periodicity Theorem. Various completions of $GL(\infty)$ have the same homotopy type [P2]. The nicest topologicaly is the closure of $GL(\infty)$ in $GL(\mathcal{H})$, the group $GL_{cpt}(\mathcal{H})$ of invertible operators of the form (identity + compact). Lying between $GL(\infty)$ and $GL_{cpt}(\mathcal{H})$ are the groups $GL_p(\mathcal{H})$ consisting of invertibles of the form (identity + A), with A in the $p^{th}$ Schatten ideal $L^p(\mathcal{H})$, $1 \leqslant p < \infty$. (See [Si] for a discussion of $L^p(\mathcal{H})$.) For $p = 1$, A is trace class and for $p = 2$, A is Hilbert-Schmidt. The $L^p$ norm topologizes $GL_p(\mathcal{H})$. In this topology $GL_p(\mathcal{H})$ is a Banach Lie Group and is homotopy equivalent to $GL(\infty)$. Summarizing, there is a hierarchy of groups

$$(3.3) \qquad GL(\infty) \subset GL_1(\mathcal{H}) \subset GL_2(\mathcal{H}) \subset GL_3(\mathcal{H}) \subset \dots \subset GL_{cpt}(\mathcal{H})$$

in which each inclusion is a homotopy equivalence. (From now on we drop "$\mathcal{H}$" from the notation.) A principal bundle over $\mathcal{M}$ with one of these groups as structure group has characteristic classes corresponding to the nontrivial cohomology of $BGL(\infty)$. In the real case there are Pontrjagin and Stiefel-Whitney classes of arbitrarily high dimension, in the complex case Chern classes in every even dimension.

With this background it is clear how to proceed with the frame bundle $GL(\mathcal{M})$. We want a reduction $GL_{cpt}(\mathcal{M})$ to a $GL_{cpt}$ bundle.

(Such a reduction is called a *Fredholm structure* in the literature.) With luck we will instead reduce to $GL_p$ for some p, and this extra refinement will facilitate the computation of characteristic classes. Let us recall a general principle, which holds in finite and infinite dimensions: If $P \longrightarrow M$ is a principal G-bundle, then the reductions to $H \subseteq G$ are classified up to topological equivalence by homotopy classes of sections of the G/H bundle $P/H = P \times_G G/H \longrightarrow M$. For the case at hand, the fiber of $GL(\mathcal{M})/GL_{cpt}$ is $GL/GL_{cpt}$ which is homotopy equivalent to $BGL(\infty)$. This follows since GL is contractible and $GL_{cpt} \sim GL(\infty)$. So reductions of the frame bundle are classified topologically by homotopy classes of maps $\mathcal{M} \longrightarrow BGL(\infty)$. As there may be several inequivalent maps, hence several inequivalent reductions, our problem is this: How do we choose a *geometrically meaningful* reduction of the frame bundle? We interpret our curvature computation in §2 as leading to a natural solution for the complex frame bundle of $\mathcal{M} = \Omega G$.

Before explaining further, we wish to make the space $GL/GL_{cpt} = \mathfrak{B}$ more concrete. Notice that $GL_{cpt}$ is a *closed* normal subgroup of GL, since the compact operators form a closed two-sided ideal inside the algebra of all bounded operators, so that the quotient $\mathfrak{B}$ is a bonafide topological group. Let $\mathfrak{gl}(\mathcal{H})$ denote the algebra of bounded operators on $\mathcal{H}$ and $cpt(\mathcal{H})$ the ideal of compact operators. The quotient $\mathfrak{gl}/cpt$ is called the *Calkin algebra*, and the inclusion $GL \hookrightarrow \mathfrak{gl}$ induces an isomorphism

$$\mathfrak{B} = GL/GL_{cpt} \simeq Inv(\mathfrak{gl}/cpt)_0 \tag{3.4}$$

onto the identity component of the invertible elements in the Calkin algebra. It is a standard result in Functional Analysis that the inverse image of these invertibles under the projection $\mathfrak{gl} \longrightarrow \mathfrak{gl}/cpt$ is the space of *Fredholm operators* $Fred_0(\mathcal{H})$ of index zero. Recall that $A \in \mathfrak{gl}$ is Fredholm if A has closed range and both its kernel and cokernel are finite dimensional; the difference of these dimensions is the index of A. The fiber of the projection

$\mathrm{Fred}_0 \longrightarrow \mathfrak{V}$ is the contractible space of compact operators, and a general theorem of Bartle and Graves [BG] implies that $\mathfrak{V}$ is homotopy equivalent to $\mathrm{Fred}_0$ (cf. [AS, §2]). This proves

**Proposition 3.5.** *Reductions of the frame bundle* $GL(\mathfrak{M})$ *to* $GL_{cpt}$ *up to homotopy equivalence correspond to homotopy classes of maps* $\mathfrak{M} \longrightarrow \mathrm{Fred}_0$, *i.e. to families of Fredholm operators of index zero parametrized by* $\mathfrak{M}$.

A more precise statement is preferable: A reduction of the frame bundle to $GL_{cpt}$ corresponds to a choice at each point $x \in \mathfrak{M}$ of an equivalence class $[A_x]$ of Fredholm operators on the tangent space $T_x(\mathfrak{M})$, the equivalence being up to compact operators. Then one can lift (via Bartle-Graves) to a family of Fredholms $\{A_x\}$. In this proposition we have implicitly assumed a trivialization $x \longmapsto \mathcal{e}_x$ of $GL(\mathfrak{M})$: $\mathcal{e}_x$ is a distinguished basis of $T_x(\mathfrak{M})$. Then the desired reduction consists of all frames $\mathcal{e}_x'$ such that the matrix of $A_x$ with respect to $\mathcal{e}_x$ in the domain and $\mathcal{e}_x'$ in the range is of the form (identity + compact).

In the preceding paragraph we can replace $GL_{cpt}$ with $GL_p$, and (3.4) becomes

$$\mathfrak{V}_p = GL/GL_p = \mathrm{Inv}(\mathfrak{gl}/L^p)_0. \tag{3.6}$$

Unfortunately, $L^p$ is not a closed subspace of $\mathfrak{gl}$, so it is not clear how to make $\mathfrak{V}_p$ into a *topological* group. (As a consequence, the subscript "0" in (3.6) refers not to the identity component, but to cosets of index zero operators.) There is a sequence of group homomorphisms

$$\mathfrak{V}_1 \longrightarrow \mathfrak{V}_2 \longrightarrow \mathfrak{V}_3 \longrightarrow \cdots \longrightarrow \mathfrak{V} \tag{3.7}$$

quite analogous to (3.3). The inverse image of each $\mathfrak{V}_p$ inside $\mathfrak{gl}$ is $\mathrm{Fred}_0$. Disregarding the topology, reductions of $GL(\mathfrak{M})$ to $GL_p$

correspond (formally) to sections of the associated $\mathcal{B}_p$ bundle $GL(\mathcal{M})/GL_p \longrightarrow M$, that is, to families of index zero Fredholm operators $[A_x]$ up to equivalence by $L^p$. Of course, an actual index zero Fredholm family $\{A_x\}$ gives an honest reduction to $GL_p$ by choosing only those bases $\mathcal{Q}_x$ such that the matrix of $A_x$ takes the form (identity + $L^p$).

There is a version of Chern-Weil Theory in our present infinite dimensional context. For definiteness take $\mathcal{H} = \mathcal{H}_{\mathbb{C}}$ complex. Then any of the groups (3.3) has real cohomology $H^*(GL(\infty)) = \Lambda(\omega_1,\omega_3,\omega_5,\ldots)$, an exterior algebra on primitive generators in every odd dimension. The $\omega_{2k-1}$ transgress in the universal fibration to the universal Chern classes $c_k \in H^{2k}(BGL(\infty))$. Suppose that $\pi\colon \mathcal{P} \longrightarrow \mathcal{M}$ is a *smooth* principal $GL_p$-bundle with connection A. Thus A is a smooth one-form on $\mathcal{P}$ with values in $\mathrm{Lie}(GL_p) = L^p$. Consider first the case $p = 1$. Then $\alpha_1 = \frac{i}{2\pi}$ Trace A is a well-defined one-form on $\mathcal{P}$, since A is trace class. Furthermore, by the definition of a connection form $\alpha_1$ restricts on each fiber to the *invariant* representative of $\omega_1 \in H^1(GL_1) \simeq H^1(GL(\infty))$. So $\alpha_1$ is an explicit form extending the cohomology class $\omega_1$, and also $d\alpha_1 = \frac{i}{2\pi}$ Trace $\pi^*(F)$, where $F \in \Omega^2(\mathcal{M})$ is the curvature of A. By definition the form $\frac{i}{2\pi}$ Trace F in cohomology is the transgression of $\omega_1$ in the bundle $\mathcal{P} \longrightarrow \mathcal{M}$. Since transgression is functorial, this cohomology class is the pullback of the universal first Chern class $c_1 \in H^1(BGL(\infty))$ via the classifying map $\mathcal{M} \longrightarrow BGL_1 \sim BGL(\infty)$, and so is the first Chern class of $\mathcal{P}$. A similar argument works for $\omega_{2k-1}$, but the expression for the extension $\alpha_{2k-1}$ is more complicated [CS]. For p>1 only the primitive generators $\omega_{2k-1} \in H^*(GL_p)$ with $k \geqslant \frac{p+1}{2}$ have invariant representatives, and only the generators with $k \geqslant p$ can be transgressed in this straightforward manner using a connection.

As a final preliminary to interpreting our curvature results, we recall a construction for principal G-bundles $P \longrightarrow M$ in *finite dimensions*. Suppose P is endowed with a connection, which we now think of as a distribution of horizontal planes, and fix a basepoint $p \in P$. Then the union of all horizontal curves originating at p forms a new, possibly smaller, principal bundle $Q \longrightarrow M$ called the *holonomy bundle*. The holonomy bundle depends on the connection, but is independent (up to isomorphism) of the basepoint. A theorem of Ambrose and Singer [AmS] identifies the Lie algebra of its structure group, the holonomy group, in terms of curvature.

**Theorem 3.8.** *For a finite dimensional principal G-bundle* $P \longrightarrow M$ *with connection, the holonomy algebra of the holonomy bundle* Q *is the subspace of* $\mathfrak{g} = \text{Lie}(G)$ *spanned by the curvature* $R_q(X,Y)$ *as* $q$ *ranges over* Q *and* X,Y *over* $T_q(Q)$. *Furthermore, the original connection on* P *restricts to a connection on the reduced bundle* Q.

Notice that if the holonomy algebra is an *ideal* in $\mathfrak{g}$, then we can let $q$ range over all of P, since curvature changes by conjugation as we move in a fiber.

Return now to the based loop group $\Omega G$. The Kähler structure reduces its $GL(\mathcal{H}_{\mathbb{R}})$ frame bundle to the bundle $U(\Omega G)$ of unitary frames, whose structure group $U(\mathcal{H}_{\mathbb{C}})$ of unitary operators on $\mathcal{H}_{\mathbb{C}}$ is still contractible. In §2 we explicitly calculated the curvature of the Kähler connection, and found that it is always "trace class," but for a modified trace. Ignoring this modification momentarily, and assuming that the Ambrose-Singer Theorem works for $U(\Omega G)$, we would conclude that the holonomy bundle $\mathfrak{Q}$ of $U(\Omega G)$ has structure group contained in the group $U_1(\mathcal{H}_{\mathbb{C}})$ of unitaries of the form (identity + trace class), and that the Kähler connection reduces to a connection on $\mathfrak{Q}$. The Chern-Weil Theory described above would then imply that the Ricci form (2.8) represents the first Chern class of $\mathfrak{Q}$. The bundle $\mathfrak{Q}$ would be a *geometrically constructed* reduction of the frame bundle, and we would define $\mathfrak{Q}$, together with its connection, as the *geometric frame bundle* of $\Omega G$. In particular, we would define the Chern classes of $\Omega G$ to be the Chern classes of $\mathfrak{Q}$. Indeed, this interpretation fits with representation theory and instantons, as will be described in §4 and §5. But as our use of the conditional tense sadly reminds us, there are technical difficulties to be overcome before the holonomy bundle can be properly baptized. First, our modified trace, which is quite natural and arises in other geometric contexts, seems awkward from an Operator Algebras point of view. Our major stumbling block is the extension of Ambrose-Singer to infinite dimensions. The usual proof in finite dimensions invokes the Frobenius Theorem. However, Frobenius in infinite dimensions is quite delicate; in particular, the fact that trace class operators do not form a *closed* subspace of all operators causes considerable trouble. Regrettably, then, we have reached the first of our "theorems."

"Theorem" 3.9. *The holonomy bundle* $\mathfrak{Q}$ *of the Kähler connection on the unitary frame bundle* $U(\Omega G)$ *exists, has structure group contained in* $U_1$, *and has a connection induced from the Kähler connection. The Ricci form* (2.8) *represents* $c_1(\mathfrak{Q})$.

We have immersed the reader in these infinite dimensional technicalities to show precisely where the difficulties lie in our interpretation of the "holonomy bundle" as providing a natural geometric frame bundle. In §4 and §5 we abandon all rigor and blithely apply finite dimensional theorems to $\Omega G$. Before indulging ourselves, however, we pause to explain the relationship of our putative bundle $\mathfrak{Q}$ with a construction of Segal [S1], [PS]. At the same time we will introduce a *rigorously constructed* alternative to the holonomy bundle.

To display a nontrivial central extension of the loop group, Segal considers the family of Toeplitz operators

$$T^{(0)}: f \longmapsto T_f$$

on LG. Here $T_f$: $\mathfrak{m}_+ \longrightarrow \mathfrak{m}_+$ operates pointwise via the adjoint representation of G on $\mathfrak{g}$; thus $T_f(\phi) = (\mathrm{Ad}\ f \cdot \phi)_+$. Since $T_f$ is Fredholm of index zero, by Proposition 3.5 the Toeplitz family $T^{(0)}$ defines a reduction of the unitary frame bundle $U(\Omega G)$ to a $U_{cpt}$

bundle $\mathfrak{Q}_0$. Conversely, the "holonomy bundle" $\mathfrak{Q}$, if it exists, has a $U_{cpt}$ extension which would be given by a family of index zero Fredholm operators. Our strategy is to produce a Fredholm family $T^{(1)}$ from which we can define an honest $U_{cpt}$ bundle $\mathfrak{Q}_1$. We prove that $\mathfrak{Q}_1$ is isomorphic to $\mathfrak{Q}_0$ and give plausibility arguments to assert that $\mathfrak{Q}_1$ is isomorphic to $\mathfrak{Q}$.

Our plausibility arguments are based on a new index theorem for very special families of Fredholm operators.

**Proposition 3.10.** *Suppose* $\mathcal{G}$ *is a smooth group and* $T\colon \mathcal{G} \longrightarrow \mathrm{Fred}_0$ *a smooth family of operators such that the composition* $\mathcal{G} \xrightarrow{T} \mathrm{Fred}_0 \longrightarrow \mathfrak{P}_1$ *is a homomorphism of groups. Then the pullback of the integral generator of* $H^2(\mathrm{Fred}_0;\mathbb{Z})$ *via* $T$ *has an invariant two-form representative on* $\mathcal{G}$ *given by the formula*

$$c(X,Y) = \frac{i}{2\pi}\,\mathrm{Trace}\,\{[\dot{T}(X),\dot{T}(Y)] - \dot{T}([X,Y])\}, \qquad X,Y \in \mathrm{Lie}\,(\mathcal{G}).$$

The differential $\dot{T}\colon \mathrm{Lie}\,(\mathcal{G}) \longrightarrow \mathfrak{gl}$ at the identity is a Lie algebra homomorphism modulo trace class operators, so $c(X,Y)$ is well-defined. The proof proceeds by constructing a *smooth* group extension $GL_1 \longrightarrow \tilde{\mathcal{G}} \longrightarrow \mathcal{G}$ corresponding to $GL_1 \longrightarrow GL \longrightarrow \mathfrak{P}_1$ and proving that $c$ is an explicit transgression of the generator in $H^1(GL_1;\mathbb{Z})$. There is an analogous result for other $\mathfrak{P}_p$ and higher Chern classes. There is a corollary to Proposition 3.10 for homogeneous spaces $\mathcal{G}/H$, which applies when the family $T$ restricts to a homomorphism on $H$. In that case there is an induced $GL_1$ bundle over $\mathcal{G}/H$ whose Chern classes are given by formulas as above. (See [F] for details.)

Ignoring again the difference between Trace and $\widetilde{\mathrm{Trace}}$, suppose that we can find a family $T^{(1)}\colon LG \longrightarrow \mathrm{Fred}_0$ satisfying the hypotheses of (3.10) and with $\dot{T}^{(1)} = \varphi$ for $\varphi$ defined in (2.5) and (2.6). Assume further that $T^{(1)}$ is a homomorphism when restricted to point loops. Let $\mathfrak{Q}_1$ denote the $U_{cpt}$ bundle over $\Omega G$ defined by $T^{(1)}$. Then not only will $c_1(\mathfrak{Q}_1)$ = "$c_1(\mathfrak{Q})$," but all $c_k(\mathfrak{Q}_1)$ = "$c_k(\mathfrak{Q})$" by the higher order analogues of Proposition 3.10 and of Chern-Weil Theory. (To see that $c_1(\mathfrak{Q}_1)$ = "$c_1(\mathfrak{Q})$," compare (3.10) with (1.7) and (1.10).) The family

$$(3.11)\qquad T^{(1)}: f \longmapsto T_{f_-} + D^{-1}T_{f_+}D$$

has the desired properties. To make sense of (3.11) we use the adjoint representation $G \longrightarrow \mathrm{Ad}G \subset GL(\mathfrak{g})$ and expand $f \in LG$ as

$$f(e^{i\theta}) = \sum_{n\in\mathbb{Z}} f_n e^{in\theta}$$

with $f_n \in \mathfrak{gl}(\mathfrak{g}_{\mathbb{C}})$. Then set

$$f_-(e^{i\theta}) = \sum_{n\leqslant 0} f_n e^{in\theta},$$

$$f_+(e^{i\theta}) = \sum_{n>0} f_n e^{in\theta}.$$

It is easy to prove that $T^{(1)}$: $LG \longrightarrow \mathfrak{gl}(\mathcal{H}_{\mathbb{C}})$ is continuous, and for finite series

$$f(e^{i\theta}) = \sum_{-N\leqslant n\leqslant N} f_n e^{in\theta},$$

$$g(e^{i\theta}) = \sum_{-M\leqslant m\leqslant M} g_m e^{im\theta},$$

simple estimates show that $T_f^{(1)}T_g^{(1)} - T_{fg}^{(1)}$ is compact. Thus $T_f^{(1)}T_g^{(1)} = T_{fg}^{(1)}$ modulo compact operators for *all* f,g, and setting $g = f^{-1}$ it follows that $T_f^{(1)}$ is Fredholm of index zero. $T^{(1)}$ gives rise to $\mathfrak{Q}_1$, and the relation to Segal's bundle is

**Theorem 3.12.** *The* $U_{cpt}$ *bundle* $\mathfrak{Q}_0$ *given by the Toeplitz family* $T^{(0)}$ *is topologically equivalent to the* $U_{cpt}$ *bundle* $\mathfrak{Q}_1$ *given by* $T^{(1)}$.

**Proof.** The Fredholm families

$$T^{(s)}: f \longmapsto T_{f_-} + D^{-s}T_{f_+}D^s, \qquad 0 \leqslant s \leqslant 1$$

provide an explicit homotopy between $T^{(0)}$ and $T^{(1)}$. Now apply Proposition 3.5.

**Corollary 3.13.** *The Chern classes of* $\mathfrak{L}_1$ *coincide with those of* $\mathfrak{L}$.

The Chern classes of $\mathfrak{L}_0$ can be computed directly from Proposition 3.10 once we observe that the Toeplitz family defines a *homomorphism* $LG \xrightarrow{T^{(0)}} \mathrm{Fred}_0 \longrightarrow \mathfrak{B}_1$. Direct calculation using (3.10) yields

**Proposition 3.14.** *The first Chern class of* $\mathfrak{L}_0$ *is* $2n_G$ *times the generator of* $H^2(\Omega G;\mathbb{Z})$.

This is the calculation known to Operator Algebraists, and the precise version we need can be found in Segal [S1, Proposition 7.4]. (It is instructive to compare this calculation to our calculation (2.8).) There he interprets this class as the cocycle for a Lie algebra extension. An immediate consequence of (3.14) and (3.13) is

**Corollary 3.15.** *The first Chern class of* $\mathfrak{L}_1$ *is* $2n_G$ *times the generator of* $H^2(\Omega G;\mathbb{Z})$.

Let's summarize our point of view on these results. We want to make sense of a holonomy bundle $\mathfrak{L}$ for the Kähler connection as a reduction of $U(\Omega G)$ to $U_1$. Failing this, we constructed instead a $U_{cpt}$ bundle $\mathfrak{L}_1$ from a family of Fredholm operators such that if the holonomy bundle $\mathfrak{L}$ exists, its $U_{cpt}$ extension is topologically equivalent to $\mathfrak{L}_1$. This Fredholm family also gives rise to a $U_1$ bundle (cf. the discussion following (3.7)) which we view as a temporary substitute for $\mathfrak{L}$. The bundle $\mathfrak{L}_1$ is isomorphic to the $U_{cpt}$ bundle $\mathfrak{L}_0$ obtained from the Toeplitz family, and the first Chern class $c_1(\mathfrak{L}_0)$ agrees with the "first Chern class" of our fictional bundle $\mathfrak{L}$. The Toeplitz bundle $\mathfrak{L}_0$ is related to Bott Periodicity and allows us to compute the higher Chern classes of $\Omega G$, which we could not do from $\mathfrak{L}$, even if it were rigorously constructed (cf. [F]).

## §4. Future Applications to Group Theory

At this stage we have nominated a candidate for the geometric frame bundle of $\Omega G$. We now add credibility to our choice by verifying the value of $c_1(\Omega G)$ in two distinct geometric situations. Our arguments here are formal analogies with finite dimensional theorems, and we try to point out where the technology is lacking to make them rigorous in infinite dimensions. In the present section we outline a development of the representation theory of loop groups by analogy with §1, a prograeme due to Graeme Segal. The first Chern class enters in the character formula via its interpretation as a sum of positive roots. In §5 we relate $c_1(\Omega G)$ to the moduli space of instantons on the 4-sphere.

The based loop group $\Omega G = (\mathbb{T} \ltimes LG)/(\mathbb{T} \times G)$ is an intermediate flag manifold for $\mathbb{T} \ltimes LG$, just as $\mathbb{CP}^{n-1}$ is an intermediate flag manifold for $U(n)$. The full flag manifold $\mathcal{F} = (\mathbb{T} \ltimes LG)/(\mathbb{T} \times T)$ fibers holomorphically over $\Omega G$ with fiber $G/T$. Proposition 1.5 holds for any flag manifold, and we use it to calculate the curvature of $\mathcal{F}$. Under $\mathbb{T} \times T$ there is a decomposition

$$(4.1) \qquad (\mathbb{R} \ltimes L\mathfrak{g})_{\mathbb{C}} = \mathfrak{h}_{\mathbb{C}} \oplus \mathfrak{m}_+ \oplus \mathfrak{m}_-,$$

where

$$\mathfrak{h}_{\mathbb{C}} = (\mathbb{R} \oplus \mathfrak{t})_{\mathbb{C}},$$

$$\mathfrak{m}_+ = [\bigoplus_{\alpha>0} \mathfrak{g}_{\alpha}] \oplus [\bigoplus_{n>0} z^n \mathfrak{g}_{\mathbb{C}}],$$

$$\mathfrak{m}_- = [\bigoplus_{\alpha>0} \mathfrak{g}_{-\alpha}] \oplus [\bigoplus_{n>0} z^{-n} \mathfrak{g}_{\mathbb{C}}].$$

Then writing $\mathfrak{m}_{\pm} = \mathfrak{m}_{\pm}^{(0)} \oplus \mathfrak{m}_{\pm}^{(1)}$ and combining (1.5) with (2.5), we see that the operators $\varphi(H)$, $\varphi(\bar{X})$, $\varphi(X)$ for $\mathcal{F}$ are block triangular. For example, if $\bar{X} = \bar{X}^{(0)} + \bar{X}^{(1)} \in \mathfrak{m}_-^{(0)} \oplus \mathfrak{m}_-^{(1)}$,

$$\varphi(\bar{X}) = \left[\begin{array}{c|c} \pi_+ \circ \mathrm{ad}\ \bar{X}^{(0)} & \pi_0 \circ \mathrm{ad}\ \bar{X}^{(1)} \\ \hline 0 & \mathrm{ad}\ \bar{X}^{(0)} + T_{\bar{X}^{(1)}} \end{array}\right].$$

The curvature can be computed, and its $\widetilde{\mathrm{Trace}}$ turns out to be the sum of (1.11) and (2.8).

**Proposition 4.2.** *The "first Chern class"* $c_1(\mathcal{F}) \in H^2(\mathcal{F})$ *is the transgression of* $2\langle n_G, \rho\rangle \in H^1(\mathbb{T} \times T)$.

Recall that the first Chern class of G/T can be computed by summing the positive roots of G. For the affine Kac-Moody algebra (4.1), though, the sum of the positive roots diverges to $\langle\infty, 2\rho\rangle$. Now in finite dimensions the sum of the positive roots is also twice the sum of the fundamental weights. The Kac-Moody algebra has well-defined fundamental weights, and

**Proposition 4.3.** *The sum of the fundamental weights of the affine Kac-Moody Lie algebra* $\mathbb{R} \ltimes Lg$ *is* $\langle n_G, \rho\rangle$.

This is [Ka, Exercise 7.16]. Therefore, our "first Chern class," which we interpret as a regulated sum of the positive roots, agrees with twice the sum of the fundamental weights. Proposition 4.3 provides some confirmation that our reduction of the frame bundle, one of many possible reductions, is geometrically correct.

An immediate consequence of (2.10) and (4.2) is (cf. (1.10))

**Proposition 4.4.** *The manifolds* $\Omega G$ *and* $\mathcal{F}$ *are "spin."*

This means that the second Stiefel-Whitney class of the reduced frame bundle vanishes. This "holonomy bundle" has structure group $U_1$ = unitaries on $\mathcal{H}_{\mathbb{C}}$ of the form (identity + trace class). There is a natural extension to $SO_1$ = orthogonals on $\mathcal{H}_{\mathbb{R}}$ of the form (identity + trace class) which preserve orientation. Then $\pi_1(SO_1) = \mathbb{Z}_2$, since $SO_1 \sim SO(\infty)$, and we denote the double cover group by $Spin_1$. Proposition 4.4 implies that the $SO_1$ frame bundle lifts to a $Spin_1$ bundle. Now there is the possibility of spinors, a Dirac operator, etc.

Following §1 we next want a vanishing theorem in cohomology.

**"Theorem" 4.5.** *The $\bar\partial$-cohomology $H^{0,q}_{\bar\partial}(\Omega G)$ and $H^{0,q}_{\bar\partial}(\mathfrak{F})$ vanish for $q>0$. Also, $H^{0,0}_{\bar\partial}(\Omega G) = H^{0,0}_{\bar\partial}(\mathfrak{F}) = \mathbb{C}$.*

On a complex Hilbert manifold there is a well-defined $\bar\partial$ operator. But $\bar\partial^*$ is not immediate. In finite dimensions $\bar\partial^*$ can be defined either as the adjoint $\bar\partial$ via integration or directly by a local formula. On an arbitrary Hilbert manifold there is no obvious measure, hence no integration. On $\Omega G$, however, there are nice *Wiener measures*, and one could try to define $\bar\partial^*$ using them. The local formula goes awry in general since it involves a trace, which typically diverges. It is tempting to think that our bundle $\mathfrak{L}$, whose Lie algebra would consist of trace class operators, could be useful here. If we succeed in defining $\bar\partial^*$, hence $\Delta_{\bar\partial}$, and the same time define $\bar\nabla^*\bar\nabla$, then the Weitzenböck formula (1.14) would hold. Now we definitely need a measure, along with integration by parts, to carry out (1.15). Then our result that Ricci is positive would imply that $\Delta_{\bar\partial}$ has no kernel for $q>0$. Finally, to identify $\ker(\Delta_{\bar\partial}) \simeq H^{0,q}_{\bar\partial}$ we would need a Hodge Theorem on $\Omega G$. It seems that much analysis on $\Omega G$ must be developed before (4.5) can be proved by these methods.

On an infinite-dimensional Hilbert manifold the de Rham Theorem presents no difficulties. However, the local $\bar\partial$ Poincaré Lemma, which is the crux of the Dolbeault Theorem, remains to be proved in the infinite case. So as of now there is no

**"Corollary" 4.6.** $H^q(\Omega G;\mathcal{O}) = H^q(\mathfrak{F};\mathcal{O}) = \begin{cases} \mathbb{C} & \text{for } q=0; \\ 0 & \text{for } q>0. \end{cases}$

There is little doubt that (4.5) and (4.6) are true. Kumar [Ku] has proved a version of these results for Lie algebra cohomology. Segal [S2] outlined a proof of (4.6) exploiting the Birkhoff decomposition of flag manifolds.

Once (4.6) is proved rigorously, there are no barriers to the next step.

**"Corollary" 4.7.** (a) *Every global holomorphic function on* $\Omega G$ *(or* $\mathcal{F}$*) is constant.*

(b) *Every line bundle over* $\Omega G$ *(or* $\mathcal{F}$*) carries a unique holomorphic structure.*

In fact, (a) can be proved [PS] from the fact that a dense subset of $\Omega G$ (or $\mathcal{F}$) can be expressed as the union of finite dimensional compact projective varieties (Bruhat decomposition).

Our next construction--the central extension of LG (to be distinguished from the noncentral extension $\mathbb{T} \ltimes LG$)--has no finite dimensional analogue. (There is also a central extension $\mathbb{T} \ltimes \widetilde{LG}$ of $\mathbb{T} \ltimes LG$. This group is most properly called the affine Kac-Moody group.) It is a pleasant exercise to verify that on the Lie algebra level, $H^2(Lg_{\mathbb{C}},\mathbb{C}) = \mathbb{C}$, the nontrivial cocycle being given by (2.2), which corresponds to the existence of central extensions for $Lg_{\mathbb{C}}$. Corresponding group extensions do not exist automatically; in fact, they exist only for integral cocycles. The central extensions arise naturally in representation theory since positive energy representations of $\mathbb{T} \ltimes LG$ are projective [S2]. We will first construct central extensions of the complex group $LG_{\mathbb{C}}$ of maps $S^1 \longrightarrow G_{\mathbb{C}}$. Since $\Omega G$ has realization $\Omega G = LG_{\mathbb{C}}/\mathcal{P}$ as a complex quotient, $LG_{\mathbb{C}}$ acts holomorphically on $\Omega G$ (cf. [A] for this complex description). Now the central extensions correspond to elements of $H^2(\Omega G;\mathbb{Z})$, hence to live bundles $\mathcal{L} \longrightarrow \Omega G$. Fix one. Then by (4.7)(b) $\mathcal{L}$ carries a *unique* holomorphic structure. It follows that the action of any $f \in LG_{\mathbb{C}}$ is covered by a holomorphic automorphism of $\mathcal{L}$. Furthermore, by (4.7)(a) the holomorphic gauge transformations of $\mathcal{L}$ (holomorphic automorphisms of $\mathcal{L}$ covering the identity) are constant. Therefore, the group $\widetilde{LG}_{\mathbb{C}}$ of holomorphic automorphisms of $\mathcal{L}$ covering the action of $LG_{\mathbb{C}}$ is an extension of $LG_{\mathbb{C}}$ by the nonzero complex numbers $\mathbb{C}^{\times}$. Then $\widetilde{LG}_{\mathbb{C}} \longrightarrow LG_{\mathbb{C}}$ is a $\mathbb{C}^{\times}$ bundle whose first Chern class is the invariant class coming from the element of $H^2(L\mathfrak{g}_{\mathbb{C}};\mathbb{C})$ which defines the corresponding Lie algebra extension.

Although line bundles over $\Omega G$ correspond to elements of $H^2(\Omega G;\mathbb{Z})$ by abstract topology, a concrete realization is preferable. Our holonomy bundle $\mathfrak{Q}$ would provide such a realization. For $\mathfrak{Q}$ has structure group $U_1$, and the determinant homomorphism det: $U_1 \longrightarrow \mathbb{T}$ would give a *circle* bundle over $\Omega G$. Associated is the line bundle $\mathcal{L}$ with Chern class $2n_G$. Since $\mathcal{L}$ comes from a circle bundle, it carries a natural metric. On the other hand, the holonomy bundle $\mathfrak{Q} \subset U(\Omega G)$ has a *holomorphic* extension $\mathfrak{Q}_{\mathbb{C}} \subset GL_{\mathbb{C}}(\Omega G)$ with structure group $GL_1(\mathcal{H}_{\mathbb{C}})$. The holomorphic determinant det: $GL_1 \longrightarrow \mathbb{C}^{\times}$ displays $\mathcal{L}$ as an associated bundle to $\mathfrak{Q}_{\mathbb{C}}$ and exhibits its holomorphic structure explicitly. We construct a *real* central extension $\mathbb{T} \longrightarrow \widetilde{LG} \longrightarrow LG$ by considering all holomorphic automorphisms of $\mathcal{L}$ covering the action of LG and preserving the metric. It follows that $\pi_1(\widetilde{LG}) = \mathbb{Z}_{2n_G}$, and the universal central extension is the simply connected cover of LG. We remark that Segal [S1] uses the $GL_1$ bundle constructed from the Toeplitz family to produce a central extension of LG.

The earlier remarks on the vanishing theorem apply to the vanishing part of the Borel-Weil Theorem for loop groups. Our argument in finite dimensions to recognize the representation in $H^0$ relied on the Peter-Weyl Theorem. There is a version of Peter-Weyl for algebraic loop groups [KP], and Borel-Weil holds in that context. In our geometric context we would like to study $L^2$ functions on LG, and this would involve Wiener measure in a crucial way. Segal states the Borel-Weil Theorem for loop groups in [S2], presumably with a somewhat different proof in mind.

Finally, a version of the Lefschetz Fixed Point Theorem for loop groups would imply the Kac Character Formula [Ka]. In the derivation the sum of the positive roots must be regulated, most likely using the bundle $\mathfrak{Q}$. Thus $\bar{\rho} = \langle n_G, \rho \rangle$ enters the final result. Again we find confirmation for our geometric frame bundle: $\bar{\rho}$ is the exact expression entering in the character formula.

## §5. Relation with Instantons

Recently, Atiyah [A] and Donaldson [D] proved that the moduli space of based instantons on $S^4$ for a *classical* compact Lie group G is diffeomorphic to the moduli space of based holomorphic maps $\mathbb{CP}^1 \longrightarrow \Omega G$. There is a degree k indexing the components of the moduli space in each instance, and these degrees are identified under the diffeomorphism. It would take us too far afield to elaborate on this theorem, which is the state of the art of a long development relating instantons to algebraic geometry. Rather, we use it to provide additional geometric confirmation, coming now from left field, so to speak, for our value of $c_1(\Omega G)$.

We need the following easy Riemann-Roch Theorem, which is valid for finite dimensional X.

**Proposition 5.1.** *Suppose* $f: \mathbb{CP}^1 \longrightarrow X$ *is a based holomorphic map into a finite dimensional complex manifold* X, *and assume that* f *is a manifold point in the moduli space of all based holomorphic maps* $\mathbb{CP}^1 \longrightarrow X$. *Then the complex dimension of that moduli space at* f *is* $f^*(c_1(X))[\mathbb{CP}^1]$.

**Proof.** The infinitesimal deformations of f are holomorphic sections of $f^*(TX) \longrightarrow \mathbb{CP}^1$. By Grothendieck's Theorem $f^*(TX)$ splits into a direct sum $\oplus\mathcal{O}(a_i)$ of line bundles, $\mathcal{O}(a)$ the $a^{th}$ power of the hyperplane bundle, and since the dimension of based holomorphic sections of $\mathcal{O}(a_i)$ is $a_i$, the desired dimension is $\Sigma a_i = c_1(f^*(TX))[\mathbb{CP}^1] = f^*(c_1(X))[\mathbb{CP}^1]$.

Now we simply close our eyes and apply Proposition 5.1 to the infinite dimensional based loop group.

**"Corollary" 5.2.** *The complex dimension of the space of based holomorphic maps* $\mathbb{CP}^1 \longrightarrow \Omega G$ *of degree* k *is* $2kn_G$.

By the Atiyah–Donaldson Theorem this should be the complex dimension of the space of based k-instantons. Thus the real dimension of the moduli space of unbased k-instantons should be $4kn_G$ - dim G. This is indeed the correct formula [AHS, Table 8.1]. Note, incidentally, that it is valid for *all* groups G and provides a small bit of evidence than the Atiyah–Donaldson result extends to exceptional groups. From our point of view, this calculation strengthens our assertion that $\mathfrak{Q}$ is the correct geometric frame bundle of $\Omega G$.

## Bibliography

[AmS] W. Ambrose, I.M. Singer, A theorem on holonomy, *Trans. Amer. Math. Soc.*, **75** (1953), 428–443.

[A] M.F. Atiyah, Instantons in two and four dimensions, *Comm. Math. Phys.*, **93** (1984), 437–451.

[AB] M.F. Atiyah, R. Bott, A Lefschetz fixed point formula for elliptic complexes: II. Applications, *Ann. of Math.*, **88** (1968), 451–491.

[AHS] M.F. Atiyah, N. Hitchin, I.M. Singer, Self-duality in four-dimensional Riemannian geometry, *Proc. R. Soc. London* A, **362** (1978), 425–461.

[AS] M.F. Atiyah, I.M. Singer, Index theory for skew-adjoint Fredholm operators, *Publ. Math. Inst. Hautes Etudes Sci. (Paris)*, **37** (1969), 305–326.

[BG] R.G. Bartle and L.M. Graves, Mappings between function spaces, *Trans. Amer. Math. Soc.*, **72** (1952), 400–413.

[BH] A. Borel, F. Hirzebrech, Characteristic classes and homogeneous spaces I, *Amer. J. Math.*, **80** (1958), 458–538.

[B1] R. Bott, An application of the Morse theory to the topology of Lie groups, *Bull. Soc. Math. France*, **84** (1956), 251–281.

[B2] R. Bott, Homogeneous vector bundles, *Ann. of Math.*, **66** (1957), 203–248.

[BS] R. Bott and H. Samelson, Applications of the Theory of Morse to symmetric spaces, *Amer. J. Math.*, **80** (1958), 964–1029.

[CS] S.S. Chern and J. Simons, Characteristic forms and geometric invariants, *Ann. of Math.*, **99** (1974), 48–69.

[D] S.K. Donaldson, Instantons and geometric invariant theory, *Comm. Math. Phys.*, **93** (1984), 453–460.

[F] D.S. Freed, Geometry of loop groups, P.h.D. Thesis, University of California, Berkeley, 1985.

[FU] D.S. Freed, K.K. Uhlenbeck, *Instantons and Four-Manifolds*, Mathematical Sciences Research Institute Publications, Volume 1, Springer-Verlag, New York, 1984.

[GS] P. Griffiths, W. Schmid, Locally homogeneous complex manifolds, *Acta. Math.*, 123 (1969), 253–302.

[HH] J.W. Helton, R.E. Howe, Integral operators: traces, index, and homology, in *Proceedings of a Conference on Operator Theory*, Lecture Notes in Mathematics, Volume 345, Springer-Verlag, 1973, pp. 141–209.

[Ka] V.G. Kac, *Infinite Dimensional Lie Algebras*, Birkhäuser, Boston, 1983.

[KP] V.G. Kac, D.H. Peterson, Regular functions on certain infinite dimensional groups, in *Arithmetic and Geometry*, ed. M. Artin and J. Tate, Birkhäuser, Boston, 1983, pp. 141–166.

[K] N.H. Kuiper, The homotopy type of the unitary group of Hilbert space, *Topology*, **3** (1965), 19–30.

[Ku] S. Kumar, Rational homotopy theory of flag varieties associated to Kac–Moody groups, this volume.

[MK] J. Morrow, K. Kedaira, *Complex Manifolds*, Holt, Rinehart, and Winston, Inc., New York, 1971.

[P1] R.S. Palais, *Foundations of Global Non-Linear Analysis*, W.A. Benjamin, Inc., New York, 1968.

[P2] R.S. Palais, On the homotopy type of certain groups of operators, *Topology*, **3** (1965), 271-279.

[Pr1] A.N. Pressley, Decompositions of the space of loops on a Lie group, *Topology*, **19** (1980), 65-79.

[Pr2] A.N. Pressley, The energy flow on the loop space of a compact Lie group, *J. London Math. Soc.* (2), **26** (1982), 557-566.

[PS] A.N. Pressley, G.B. Segal, *Loop Groups*, Oxford University Press, to appear.

[S1] G.B. Segal, Unitary representations of some infinite dimensional groups, *Comm. Math. Phys.*, **80** (1981), 301-342.

[S2] G.B. Segal, Notes from talk at Arbeitstagung, Bonn, 1984.

[Si] B. Simon, *Trace Ideals and Their Applications*, London Mathematical Society Lecture Note Series, Volume 35, Cambridge University Press, Cambridge, 1979.

[W] H.C. Wang, Closed manifolds with homogeneous complex structures, *Am. J. Math.*, **76** (1954), 1-32.

ors, and a self-adjointness criterion of E. Nelson [Ne].) This does not imply even local integrability to $\mathcal{D}$, however, since ubgroup of $\mathcal{D}$ generated by the one-parameter subgroups is re dense. The standard methods of "integrating" a Lie algebra sentation for finite-dimensional Lie algebras and groups simply do pply in this situation.

We should point out that for $c = 1$ and $h = -m^2/4$, m an er, the integrability conjecture was proved by G. Segal [S] using pletely different methods. For $c \in \mathbb{Z}$, $c \geqslant 1$ and certain es of h, it was also proved in [G-W1], in connection with unitary esentations of the loop algebras $\tilde{g}$ and their central extensions (the affine algebras), where $g$ is a finite-dimensional simple Lie ebra.

**The Category** $\mathcal{U}$

We shall denote by $\mathcal{U}$ the category of all *unitarizable* ghest weight $\hat{d}$ modules $(\pi, V)$. That is, V satisfies conditions 1(i)-(iii), and has a positive-definite contravariant form $<\cdot,\cdot>$. If $W \subset V$ is a submodule, then so is $W^{\perp}$ (orthogonal complement elative to $<\cdot,\cdot>$), and both submodules are the direct sum of inite-dimensional $d_0$-eigenspaces. From this the following properties of the category $\mathcal{U}$ are easily verified ($U(g)$ denotes the universal enveloping algebra of a Lie algebra $g$):

(a) If $(\pi, V) \in \mathcal{U}$ has highest weight (h,c) and highest weight vector $v_0$, then the cyclic submodule $U(\hat{d})\cdot v_0$ is irreducible and isomorphic to L(h,c). The contravariant form on V restricts to a positive multiple of the contravariant form on L(h,c).

(b) If $(\pi, V) \in \mathcal{U}$ has highest weight (h,c) and p is a positive integer, then the p-fold tensor product $(\pi^{\otimes p}, V^{\otimes p}) \in \mathcal{U}$ and has highest weight (ph,pc).

(c) If $(\pi, V) \in \mathcal{U}$, then for n = 1, 2, ..., V is the orthogonal direct sum of irreducible highest weight modules for the $sl_2(\mathbb{C})$ subalgebras



# POSITIVE-ENERGY REPRESENTATIONS OF THE GROUP OF DIFFEOMORPHISMS OF THE CIRCLE

By

Roe Goodman (Joint Work With Nolan R. Wallach)
Rutgers University, New Brunswick, NJ 08903

**Abstract**

Let $\mathcal{D}$ be the group of orientation-preserving diffeomorphisms of the circle $S^1$. Then $\mathcal{D}$ is Fréchet Lie group with Lie algebra $(d_\infty)_{\mathbb{R}}$ the smooth real vector fields on $S^1$. Let $d_{\mathbb{R}}$ be the subalgebra of real vector fields with finite Fourier series. This lecture outlines a proof that every infinitesimally unitary projective positive-energy representation of $d_{\mathbb{R}}$ integrates to a continuous projective unitary representation of $\mathcal{D}$. This result was conjectured by V. Kac.

## 1. Projective Representations of $d$

Let $\mathcal{D}$ denote the group of orientation-preserving diffeomorphisms of the unit circle $S^1$, and let $d_\infty$ denote the Lie algebra of smooth, complex vector fields on $S^1$. Then $\mathcal{D}$ can be looked upon as a Fréchet Lie group with Lie algebra $(d_\infty)_{\mathbb{R}}$, the smooth real vector fields on the circle. We denote by $d$ the Lie algebra of complex vector fields on the circle with finite Fourier series, and by $d_{\mathbb{R}}$ the real vector fields in $d$. The vector fields

$$d_n = e^{in\theta} \frac{d}{i\,d\theta},$$

for $n \in \mathbb{Z}$, are a basis for $d$. Let $\hat{d}$ be the central extension of $d$ defined by the cocycle

Research partially supported by NSF Grant MCS83-01582.

$$\omega(d_n,d_m) = \delta_{n,-m} \frac{n(n^2-1)}{12} .$$

Up to a scalar multiple and coboundary, $\omega$ is the unique two-cocycle on $d$ such that $\omega$ is zero on the $sl_2$ subalgebra $\mathbb{C}d_0 \oplus \mathbb{C}d_1 \oplus \mathbb{C}d_{-1}$. Also $\omega$ extends by continuity to a two-cocycle on $d_\infty$ which is explicitly given by the integral formula

$$\omega(X,Y) = \frac{i}{24\pi} \int_0^{2\pi} \left\{\frac{d^2 f}{d\theta^2}(e^{i\theta}) + f(e^{i\theta})\right\} \frac{dg}{d\theta}(e^{i\theta})\, d\theta,$$

when $X = f\frac{d}{d\theta}$ and $Y = g\frac{d}{d\theta}$.

Let $\kappa$ be the central element for the extension. Then $\hat{d}$ has basis $\{d_n\}_{n\in\mathbb{Z}}$ and $\kappa$. The commutation relations are

$$[d_n,d_m] = (m-n)d_{m+n} + \delta_{n,-m} \frac{n(n^2-1)}{12} \kappa, \qquad [d_n,\kappa] = 0.$$

The Lie algebra $\hat{d}$ is also referred to as the *Virasoro* algebra in the physics literature.

Let $(\pi,V)$ be a representation of $\hat{d}$ with $\pi(\kappa) = cI$, where $c \in \mathbb{C}$. Then we shall look upon $(\pi,V)$ as a projective representation of $d$ with commutation relations

$$[\pi(X),\pi(Y)] = \pi([X,Y]) + c\omega(X,Y)I,$$

for $X,Y \in d$. We will call $(\pi,V)$ a *highest weight* representation or a *positive energy* representation if there is a complex number h such that

(i) the operator $\pi(d_0)$ diagonalizes on V with eigenvalues of the form $h-n$, $n \in \mathbb{N}$;

(ii) if $h-n$ is an eigenvalue, then the eigenspace $V_{h-n}$ is finite dimensional;



(iii) $\dim (V_h) = 1$.

We shall call the pair (h,c) the *hi*

representation. A non-zero vector $v_0$

*highest weight vector*. Note that we

highest weight vector is cyclic.

On $d_\infty$ we define the anti-involuti

$d_n^* = d_{-n}$. That is,

$$(f(e^{i\theta}) \frac{d}{d\theta})^* = - \overline{f(e^{i\theta})} \frac{d}{d\theta}$$

Thus X is a real vector field if and only if X*

a projective $d$-module. Suppose $\langle\cdot,\cdot\rangle$ is a H

Then this form is called *contravariant* if

$$\langle\pi(X)v,w\rangle = \langle v,\pi(X^*)w\rangle$$

for $X \in d$ and $v,w \in V$.

For every pair of complex numbers (h,c) there

highest weight module $(\pi_{h,c}, L(h,c))$ for $\hat{d}$, where h

weight of $d_0$ and c is the scalar by which $\kappa$ acts.

these modules was initiated by V. Kac [Ka1], who ga

formula for the determinant of the contravariant form.

he proved in [Ka2], using his determinant formu

contravariant form is positive-definite in the range o

and he made the natural conjecture that whenever this fo

definite, then L(h,c) "integrates" to a projective represen

In this talk I shall outline a proof of Kac's conjecture.

does not use the Kac determinant formula, nor does it

detailed classification of the values of (h,c) such

contravariant form is positive definite. The details of the

in [G-W1] and [G-W2].

It is relatively straightforward to prove that the repre

$\pi_{h,c}$, if unitarizable, can be integrated along every one-p

subgroup of $\mathcal{D}$. (This uses a priori estimates obtained from

with canonical basis $\{E_n, F_n, H_n\}$. Here

$$E_n = \frac{i}{n} d_n, \qquad F_n = \frac{i}{n} d_{-n}, \qquad H_n = \frac{2}{n} d_0 - \frac{(n^2-1)}{12n} c.$$

Note that $H^* = H$ and $E^* = -F$. It is easy to show (without using the Kac determinant formula) that the highest weights of modules in $\mathcal{U}$ satisfy

$$h \leqslant 0 \text{ and } c \geqslant 0.$$

Furthermore, if $c = 0$, then $h = 0$, i.e. a non-trivial irreducible unitarizable highest weight representation of $\mathfrak{d}$ always has a non-trivial cocycle.

Fix a module $(\pi, V) \in \mathcal{U}$, and denote by $H_0$ the Hilbert space completion of V relative to $\langle\cdot,\cdot\rangle$. Let A denote the operator on $H_0$ that is the closure of $I - \pi(d_0)$ on V. Since $d_0^* = d_0$, we see by the contravariance of $\langle\cdot,\cdot\rangle$ that A is a positive self-adjoint operator with discrete spectrum contained in the set $\{1 - h + n \mid n \in \mathbb{N}\}$, and the eigenspaces of A are finite-dimensional.

Let $H_t$ for $t \in \mathbb{R}$ denote the Hilbert space completion of V relative to the inner product

$$\langle v,w\rangle_t = \langle A^t v, A^t w\rangle.$$

The form $\langle\cdot,\cdot\rangle$ then extends continuously to a sesquilinear pairing of $H_t$ and $H_{-t}$. We set

$$H_\infty(\pi) = \bigcap_{t \geqslant 0} H_t \quad \text{and} \quad H_{-\infty}(\pi) = \bigcup_{t \leqslant 0} H_t.$$

Give $H_\infty(\pi)$ the usual Fréchet topology as the intersection of the Hilbert spaces $H_t$, and give $H_{-\infty}(\pi)$ the inductive limit topology. The form $\langle\cdot,\cdot\rangle$ extends by continuity to a non-singular pairing between these spaces. Using *a priori* estimates obtained from Casimir operators for the $\mathfrak{sl}_2$ subalgebras spanned by $\{E_n, F_n, H_n\}$, one proves the following:

**Proposition.** $(\pi,V)$ extends to continuous projective representations of $\mathfrak{d}_\infty$ on $H_\infty(\pi)$ and $H_{-\infty}(\pi)$, such that

$$\langle\pi(X)v,w\rangle = \langle v,\pi(X^*)w\rangle$$

for $v \in H_\infty(\pi)$ and $w \in H_{-\infty}(\pi)$. If $X \in \mathfrak{d}_\infty$ and $X^* = X$, then $\pi(X)$, as an unbounded operator on $H_0$, is essentially self-adjoint on $H_\infty(\pi)$.

Denote the adjoint action of $\mathfrak{D}$ on $\mathfrak{d}_\infty$ by $X \to X^\phi$, for $X \in \mathfrak{d}_\infty$ and $\phi \in \mathfrak{D}$. Thus

$$X^\phi f = (X(f^{\phi^{-1}}))^\phi$$

for $f \in C^\infty(S^1)$. Define the $\phi$-twisted cocycle $\omega^\phi$ by

$$\omega^\phi(X,Y) = \omega(X^\phi,Y^\phi),$$

for $\phi \in \mathfrak{D}$ and $X,Y \in \mathfrak{d}_\infty$. Then $\omega^\phi \neq \omega$ in general. However, the cocycles $\omega$ and $\omega^\phi$ are cohomologous:

**Lemma.** If $\phi \in \mathfrak{D}$ then there exists a unique linear functional $\alpha_\phi \in \mathfrak{d}_\infty'$ (continuous dual) such that $\alpha_\phi(d)$ is a jointly continuous function of $\phi \in \mathfrak{D}$ and $d \in \mathfrak{d}_\infty$, and

$$\omega(X^\phi,Y^\phi) = \omega(X,Y) - \alpha_\phi([X,Y])$$

for $X,Y \in \mathfrak{d}_\infty$ and $\phi \in \mathfrak{D}$.

I can now make more precise the notion of *integrability* of a module $(\pi,V)$ in the category $\mathcal{U}$. Given $\phi \in \mathfrak{D}$ and $X \in \mathfrak{d}_\infty$, define

$$\pi^\phi(X) = \pi(X^\phi) + c\ \alpha_\phi(X),$$

as an operator on $H_\infty(\pi)$. From the defining relation for $\alpha_\phi$, it is a simple calculation to verify that the map $X \to \pi^{\phi}(X)$ is a projective representation of $\mathcal{d}_{\infty}$ on $H_{\infty}(\pi)$ with cocycle $c\omega$. Furthermore, the contravariant form for $(\pi,V)$ is also contravariant for $\pi^{\phi}$.

**Definition.** The module $(\pi,V)$ is *integrable* if there exists a unitary cocycle representation $\sigma$ of $\mathcal{D}$ on $H_0$ such that

(2.1) $\sigma(\phi)$ leaves $H_{\infty}$ invariant, and the map $\mathcal{D} \times H_{\infty} \to H_{\infty}$ given by $\phi$, $v \to \sigma(\phi)v$ is continuous;

(2.2) For all $\phi \in \mathcal{D}$ and $X \in \mathcal{d}_{\infty}$, one has

$$\sigma(\phi)\pi^{\phi}(X) = \pi(X)\sigma(\phi).$$

Thus integrability means that the twisted representations $\pi^{\phi}$ of $\mathcal{d}_{\infty}$ are all unitarily equivalent, and the equivalence can be smoothly implemented by a global cocycle representation of $\mathcal{D}$. The main result of this talk is then the following:

**Theorem.** Suppose $h \leqslant 0$ and $c > 0$ is such that the irreducible module $L(h,c)$ is unitarizable. Then it is integrable. Furthermore the representation $\sigma$ is uniquely determined by (2.1) and (2.2), up to multiplication by a continuous function from $\mathcal{D}$ to $S^1$.

## 3. Construction of $\sigma$

The proof of the theorem is rather involved. The basic idea is to use the uniqueness of the contravariant form on the modules $L(h,c)$. Here is an outline of some of the principal steps, with many analytical technicalities surpressed:

(1) Perturb the energy operator and the vacuum space by the adjoint action of $\mathcal{D}$.

Let $(\pi,V) \in \mathcal{U}$ have highest weight $(h,c)$, and let $\phi \in \mathcal{D}$. By the proposition of §2, the operator $\pi^{\phi}(d_0)$ is defined on $H_\infty(\pi)$. We show that for $\phi$ sufficiently near 1 in $\mathcal{D}$, this operator is bounded above, with discrete spectrum of finite multiplicity. Let $\mu_{h,c}(\phi)$ be the highest eigenvalue. The corresponding eigenspace is one-dimensional, and we can pick a unit vector ("vacuum state") $v_0(\phi)$ in this space which depends smoothly on $\phi$. Clearly, if $(\pi,V)$ is integrable, then $h = \mu_{h,c}(\phi)$. In general, call $h - \mu_{h,c}(\phi)$ the *anomalous phase shift* produced by $\phi$. We shall say that $(\pi,V)$ satisfies *condition* $(\psi)$ if the anomalous phase shift is zero for all $\phi$ near 1 in $\mathcal{D}$.

(2) If L(h,c) is unitarizable and satisfies condition $(\psi)$, then it is integrable.

(Local version) Write $\pi$ for $\pi_{h,c}$. Let $\phi$ be near 1 in $\mathcal{D}$. Let $V_\phi$ be the cyclic subspace generated by $v_0(\phi)$ under the $\phi$-twisted action of $\mathfrak{d}$ on $H_\infty(\pi)$. Then $(\pi^{\phi},V_\phi)$ is a highest weight module with highest weight (h,c), because there is no anomalous phase shift, and $V_\phi$ has a positive-definite contravariant form. By the uniqueness of L(h,c) and its contravariant form, together with a density argument, we obtain a unitary operator $\sigma_0(\phi)$ with the desired intertwining property. From the estimates in §2, one shows that $\sigma_0(\phi)$ acts continuously on $H_\infty(\pi)$ and depends smoothly on $\phi$. The uniqueness property above implies that $\sigma_0$ is a local cocycle representation of $\mathcal{D}$.

(Global version) Let $\mathcal{D}_1$ be the subgroup of $\mathcal{D}$ fixing the point $1 \in S^1$. We pass from a local to a global cocycle representation of $\mathcal{D}$ by a straightforward topological argument, using the factorization $\mathcal{D} = T^1 \times \mathcal{D}_1$ and the contractibility of $\mathcal{D}_1$, since we already have $T^1$ represented by the one-parameter group generated by $\pi^{\phi}(d_0)$. By the semi-boundedness of this operator, $\sigma$ is a positive-energy representation.

(3) Construction of Fock models.

We construct a family of "Fock models" $(\Phi_{h,c}, V)$ in the category $\mathcal{U}$, with V a fixed pre-Hilbert space and highest weights (h,c) comprising the convex set

$$\Sigma = \{c \geq 1, \quad h \leq (1 - c)/24\}.$$

The action of $\hat{\mathfrak{d}}$ is by quadratic sums of the creation and annihilation operators (Virasoro operators). This is analogous to the imbedding of the symplectic Lie algebra as quadratic elements in the Weyl algebra. The creation and annihilation operators themselves give a cocycle representation of the commutative Lie algebra $\tilde{\mathfrak{g}}_\infty = C^\infty(S^1)$. The natural action of $\mathcal{D}$ on $\tilde{\mathfrak{g}}_\infty$ preserves the cocycle. When c = 1 we can use the uniqueness theorem for Fock representations and estimates similar to those in (1) above to construct a unitary cocycle representation of $\mathcal{D}$ on V which implements the action of $\mathcal{D}$ on $\tilde{\mathfrak{g}}_\infty$. We can then use Schur's lemma to show that condition $(\psi)$ is satisfied by L(h,1).

(4) Verification of condition $(\psi)$ for all L(h,c) $\in \mathcal{U}$.

From (3) we know that L(h,1) satisfies $(\psi)$ for all $h \leq 0$. If $(\pi,V) \in \mathcal{U}$ and n is a positive integer, then it is immediate that $(\pi,V)$ satisfies $(\psi)$ if and only if $(\pi^{\otimes n}, V^{\otimes n})$ satisfies $(\psi)$. This implies that if $r > 0$ is rational and both L(h,c) and L(rh,rc) are in $\mathcal{U}$, then condition $(\psi)$ for L(h,c) is equivalent to condition $(\psi)$ for L(rh,rc). Let $\Sigma$ be as in (3). Using the Fock models and a continuity argument, we show that if $(h,c) \in \Sigma$, then L(h,c) satisfies $(\psi)$. Thus from all these special cases we see that if L(h,c) $\in \mathcal{U}$ and $h < 0$, then L(h,c) satisfies $(\psi)$. Indeed, there is a rational number $r > 0$ so that $(rh,rc) \in \Sigma$; see Figure 1:

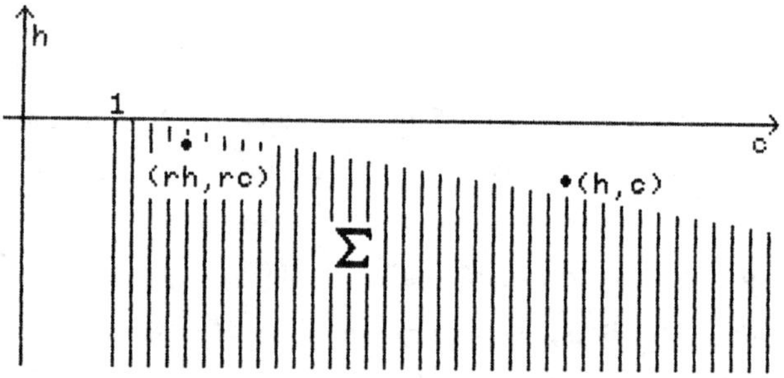


Fig. 1. Region $\Sigma$ of unitarity from Fock models (phase shift at (h,c) outside $\Sigma$ determined by phase shift at point (rh,rc) in $\Sigma$.)

It now only remains to consider the modules $L(0,c) \in \mathcal{U}$, with $c > 0$. Since these modules are on the boundary of the region of unitarizability, a non-zero anomalous phase shift produced by $\phi \in \mathcal{D}$ would move the highest weight of the $\phi$-twisted representation into the region $h < 0$, where we already know there are no anomalous phase shifts. Composing the action of $\phi$ with the action of $\phi^{-1}$, we then find that $\phi$ cannot give an anomalous phase shift on L(0,c). This contradiction completes the proof.

## References

[G-W1] Goodman, R., and Wallach, N. R., Structure and unitary cocycle representations of loop groups and the group of diffeomorphisms of the circle, J. fur reine und angewandte Math. (Crelles J.), Vol. 347(1984), 69-133.

[G-W2] Goodman, R., and Wallach, N. R., Projective Unitary Positive-Energy Representations of Diff($S^1$), J. Functional Analysis (to appear).

[Ka1] Kac, V. G., Highest weight representations of infinite dimensional Lie algebras. Proceedings of ICM, Helsinki(1978), 299-304.

[Ka2] Kac, V. G., Some problems on infinite-dimensional Lie algebras and their representations, in "Lie Algebras and Related Topics." Lecture Notes in Mathematics, Vol. 933, Berlin, Heidelberg, New York: Springer 1982.

[Ne] Nelson, E., Time-ordered operator products of sharp-time quadratic forms, J. Functional Analysis 11(1972), 211-219.

[Se] Segal, G., Unitary representations of some infinite dimensional groups, Commun. Math. Phys. 80(1981), 301-342.

For further discussion of the group $\mathfrak{D}$ and representations of infinite-dimensional Lie groups, see R. S. Hamilton, The inverse function theorem of Nash and Moser, Bull. Amer. Math. Soc. (New Series) 7(1982), 65-222, and A. A. Kirillov's article in Springer Math Lecture Notes #970 "Twistor Geometry and Non-Linear Systems".

# INSTANTONS AND HARMONIC MAPS

By

M. A. Guest

Differential equations arising in Physics often appear as the Euler-Lagrange equations for a certain functional on a space of maps; in this lecture we shall be concerned with two recent examples, namely Yang-Mills theory and the theory of $\sigma$-models. Two striking features here are (a) the existence of special solutions known as instantons, and (b) the possibility of a topological relation between the parameter space or moduli space of instantons and the space on which the functional is defined. Technical problems with the Yang-Mills functional have prompted comparison with a general situation where some results already exist, i.e. the study of the energy functional

$$E: f \longmapsto \frac{1}{2} \int_M |df|^2$$

defined on the space of smooth maps Map(M,N) between (compact) Riemannian manifolds M,N. The critical points for E are, by definition, harmonic maps. This actually includes the usual $\sigma$-model example [10,11,20] where $M = \mathbb{C}P^1$ and $N = \mathbb{C}P^n$. Recent work of M. F. Atiyah and S. K. Donaldson (see [2]) indicates that this is much more than a useful analogy, however: Yang-Mills instantons for a G-bundle over $S^4$ may be identified with "$\sigma$-model instantons" if $M = \mathbb{C}P^1$ and N is replaced by the infinite dimensional Lie group $\Omega G$ consisting of loops on G.

Evidence suggests [29] that $\Omega G$ (for a compact Lie group G) behaves very much as a compact homogeneous Kähler manifold, and it is consequently of some interest to examine harmonic maps between such spaces. In view of the increasingly diverse literature on the general theory of harmonic maps, we shall give a brief introduction to those aspects relevent here, in order to reveal certain simple but important features. In particular, we describe a new construction of harmonic maps using representations (of Lie groups), in the hope that this may be palatable to Lie theorists. The first section considers the relation between harmonic maps and holomorphic maps, the latter being the "instantons" for the energy functional. In the next section, representation theory is introduced. In the last section, we summarize some known results relating the topology of the moduli space of instantons to the topology of the appropriate space of maps, and discuss the role of harmonic maps.

## Harmonic Maps and Holomorphic Maps

Let $f\colon M \longrightarrow N$ be a (smooth) map of (compact) Riemannian manifolds M,N whose metrics will be denoted g,h respectively. It is a critical point for the energy functional

$$E\colon f \longmapsto \frac{1}{2}\int_M |df|^2 = \frac{1}{2}\int_M \langle g,f^*h\rangle = \frac{1}{2}\int_M \mathrm{tr}(f^*h)$$

(i.e. it is harmonic) if and only if it satisfies the second order differential equation

$$\mathrm{tr}\ \nabla^V(df) = 0$$

where df is considered as a section of the vector bundle $V = T^*M \otimes f^*TN$ and $\nabla^V$ is the connection induced in V from the Levi-Civitȧ connections $\nabla^{TM}$, $\nabla^{TN}$ of M,N. (The trace is taken with respect to the metric g.) This can be seen by the usual method of taking a variation of f. The $f^*TN$-valued symmetric 2-form $\nabla^V(df)$ is given explicitly by the formula:

$$(\nabla^V_X df)(Y) = \nabla^{f^*TN}_X df(Y) - df(\nabla^{TM}_X Y), \quad X,Y \in \Gamma(TM)$$

In future, we shall usually simplify notation by omitting superscripts from the symbol $\nabla$. To make the discussion reasonably self-contained, we shall sketch the proofs of some basic facts concerning harmonic maps, referring to [14] and the surveys [12,13] for more information.

A central theme will be the relation between harmonic and holomorphic maps, so we assume now that M and N are Kähler manifolds. The complexified derivative $df \otimes \mathbb{C}$ decomposes to give the maps

$$\partial f: T_{1,0}M \longrightarrow T_{1,0}N, \quad \bar{\partial} f: T_{0,1}M \longrightarrow T_{1,0}N$$

and their complex conjugates, and the energy functional may be rewritten

$$\begin{aligned} E(f) &= \frac{1}{2}\int_M |df|^2 = \frac{1}{4}\int_M |df \otimes \mathbb{C}|^2 \\ &= \frac{1}{4}\int_M (|\partial f|^2 + |\bar{\partial} f|^2 + |\bar{\partial}\bar{f}|^2 + |\partial\bar{f}|^2) \\ &= \frac{1}{2}\int_M (|\partial f|^2 + |\bar{\partial} f|^2) \\ &= \int_M \|\partial f\|^2 + \int_M \|\bar{\partial} f\|^2 \end{aligned}$$

where $|\ |$ refers to the Riemannian norm and $\|\ \|$ to the associated Hermitian norm. Hence we may write

$$E(f) = E'(f) + E''(f)$$

where $E'(f) = \int_M \|\partial f\|^2$ and $E''(f) = \int_M \|\bar{\partial} f\|^2$. Note that f is holomorphic if and only if $E''(f) = 0$, and antiholomorphic if and only if $E'(f) = 0$. This is significant because of the following result of A.

Lichnerowicz [26].

**Proposition 1.** $E'(f) - E''(f)$ depends only on the (smooth) homotopy class of f.

**Proof.** A straightforward calculation shows that $E'(f) - E''(f)$

$= \int_M \langle \omega^M, f^*\omega^N \rangle$ where $\omega^M, \omega^N$ are the Kähler forms of M,N respectively. Let $\{f_t\}$ be a smooth variation of $f = f_0$, with $t \in [0,1]$. Then $f_1^*\omega^N - f_0^*\omega^N = \int_0^1 \frac{\partial}{\partial t} f_t^*\omega^N \, dt = d\alpha$ for some 1-form $\alpha$. Writing $\langle \omega^M, f^*\omega^N \rangle = \pm f^*\omega^N \wedge ({*}\omega^M)$ where $*$ is the Hodge operator, one finds $\langle \omega^M, f_1^*\omega^N \rangle - \langle \omega^M, f_0^*\omega^N \rangle = \pm(f_1^*\omega^N - f_0^*\omega^N) \wedge ({*}\omega^M) = \pm d\alpha \wedge ({*}\omega^M) = \pm d(\alpha \wedge ({*}\omega^M))$, and the result follows. (This is a sketch of the proof given in [13], §8.)

The role of holomorphic maps (between Kähler manifolds) is now easy to clarify.

**Proposition 2.** (a) If f is holomorphic, then f is an absolute minimum for $E: \mathrm{Map}_*(M,N) \longrightarrow \mathbb{R}$ (and hence harmonic), where $\mathrm{Map}_*(M,N)$ is the component of Map(M,N) containing f. (b) Conversely, if f is an absolute minimum for $E: \mathrm{Map}_*(M,N) \longrightarrow \mathbb{R}$, and if $\mathrm{Map}_*(M,N)$ contains some holomorphic map f', then f is holomorphic.

**Proof.** (a) If $\{f_t\}$ is a (smooth) variation of f, then $E(f) = E'(f) = E'(f) - E''(f) = E'(f_t) - E''(f_t)$ (by proposition 1) $\leq E(f_t)$. (b) From proposition 1, $E(f) - E(f') = 2(E''(f) - E''(f'))$, hence f is an absolute minimum of $E'': \mathrm{Map}_*(M,N) \longrightarrow \mathbb{R}$. Thus $E''(f) \leq E''(f') = 0$, and f is holomorphic.

**Remarks.**

(1) In [26], harmonicity of holomorphic maps between more general manifolds was proved. See [30] for a recent exposition.

(2) Even if it is not known whether $Map_*(M,N)$ contains a holomorphic map, it may still be possible to show that an absolute minimum of E is holomorphic. This could be expected to yield non-trivial results, e.g. existence of holomorphic maps or non-existence of harmonic maps. For example, see [16,32].

(3) Proposition 2 is valid if "holomorphic" is replaced by "antiholomorphic." It is worth emphasizing that f is holomorphic if it satisfies the <u>first</u> <u>order</u> equation $\bar{\partial} f = 0$, and that for given Riemannian metrics g,h, many different Kähler structures may exist.

As an example of the kind of situation we shall meet, consider the case $M = \mathbb{C}P^1$, $N = G/T$ where G is a compact simple Lie group and T is a maximal torus (if $G = SU_2$, $M \cong N$). We shall only consider <u>invariant</u> complex structures and metrics, namely those which respect the adjoint action of T on the tangent space $T_0(G/T)$. There is the usual Lie algebra decomposition $L(G) = L(T) \oplus \sum_{\alpha\in\Delta^+} V_\alpha$ where $\Delta^+$ denotes the set of positive roots of G with respect to some choice of fundamental Weyl chamber $D \subseteq L(T)$, and $V_\alpha$ is the "root space" of $\alpha$. Complexifying, $L(G) \otimes \mathbb{C} = L(T) \otimes \mathbb{C} \oplus \sum_{\alpha\in\Delta^+} E_\alpha \oplus \sum_{\alpha\in\Delta^+} E_{-\alpha}$ where $V_\alpha \otimes \mathbb{C} = E_\alpha \oplus E_{-\alpha}$ is the decomposition into eigenspaces of $Ad \otimes \mathbb{C}$. An invariant almost complex structure on G/T is specified by a choice of sign $\epsilon_\alpha$ for each $\alpha \in \Delta^+$, since this determines an invariant complex structure on each space $V_\alpha$. The endomorphism J of $E_\alpha \oplus E_{-\alpha}$ thus defined is given by $Je_{\epsilon_\alpha \alpha} = ie_{\epsilon_\alpha \alpha}$, $Je_{-\epsilon_\alpha \alpha} = -ie_{-\epsilon_\alpha \alpha}$ for $0 \neq e_\alpha \in E_\alpha$, $0 \neq e_{-\alpha} \in E_{-\alpha}$. This identifies $T_{1,0}(G/T)$ with $\sum_{\alpha\in\Delta^+} E_{\epsilon_\alpha \alpha}$, and the almost complex structure is integrable if this defines an integrable distribution in $T(G/T) \otimes \mathbb{C}$. It is well known that

$$[e_\alpha, e_\beta] = \begin{cases} h_\alpha & \text{if } \alpha+\beta=0, \text{ where } 0\neq h_\alpha \in L(T)\otimes\mathbb{C} \\ N_{\alpha\beta} e_{\alpha+\beta} & \text{if } \alpha+\beta \text{ is a root, where } N_{\alpha\beta}\neq 0 \\ 0 & \text{otherwise} \end{cases}$$

and so the choice $\epsilon_\alpha = 1$ for all $\alpha \in \Delta^+$ defines a complex structure on G/T. Any invariant complex structure arises in this way for some fundamental Weyl chamber. Thus, of the $2^{|\Delta^+|}$ almost complex structures, $|W|$ are integrable, where W is the Weyl group. For further details, see [7]. There is a natural invariant metric on G/T given by the negative of the Killing form of G. Relative to this, any other invariant metric may be specified by a choice of positive real numbers $\{r_\alpha\}_{\alpha\in\Delta^+}$ giving the ratios of the two metrics on each $V_\alpha$. A distinguished family of invariant metrics is obtained by realizing G/T explicitly as the generic orbit of the adjoint action of G on L(G); for a regular point $x \in L(T)$ (i.e. a point whose centralizer is T) the orbit of x has a natural symplectic form $\omega_x$, which is preserved by the action of T on G/T. The form $\omega_x$ is compatible with the complex structure determined by the Weyl chamber containing x, and hence one has a Kähler structure. For the associated Kähler metric, $r_\alpha = \alpha(x)$.

With respect to the complex structure determined by some fixed Weyl chamber, one may give holomorphically represented generators of $\pi_2$ G/T in a straightforward manner. We assume that G is simply connected, and shall identify $L(T)^*$ with L(T) by means of the Killing form when convenient. First, observe that $H^2(G/T;\mathbb{Z}) \cong H^1(T;\mathbb{Z}) \cong \hat{T}$ (the character group of T), which may be identified with the integer lattice in $H^2(G/T;\mathbb{R}) \cong H^1(T;\mathbb{R}) \cong L(T)^*$. For a regular point $x \in L(T)^*$, the Kähler form for the structure defined on $Ad(G)x \cong G/T$ corresponds to x itself. Let $\lambda_1,\ldots,\lambda_\ell \in \hat{T}$ be the "basic weights". These are characterized by the equations $\langle\lambda_i,\alpha_j\rangle = \frac{1}{2}\langle\alpha_j,\alpha_j\rangle\, \delta_{ij}$, and are such that the (dual) Weyl chamber consists of all points of the form $x_1\lambda_1+\ldots+x_\ell\lambda_\ell$ with $x_i \geqslant 0$ (see proposition 5.62 of [1]). Dually, one has the group $\pi_2$ G/T $\cong \pi_1$ T, which may be identified with the integer lattice I in T. The root $\alpha_i$ determines a subalgebra of L(G) isomorphic to $L(SU_2)$, namely that which is generated by $V_{\alpha_i}$ and $[V_{\alpha_i}, V_{\alpha_i}]$, and this gives a homomorphism $SU_2 \longrightarrow G$ (which is in fact an embedding–see lemma 3.1 of [6]). A maximal torus $S \subseteq SU_2$ is determined by the line $[V_{\alpha_i}, V_{\alpha_i}]$ and the induced map $f_i\colon \mathbb{C}P^1 \cong SU_2/S \longrightarrow G/T$ is holomorphic. The class $[f_i] \in \pi_2\, G/T$ corresponds to the class $[\lambda_i] \in H^2(G/T;\mathbb{Z})$ under the pairing

$$\begin{array}{ccccc} H^2(G/T;\mathbb{Z}) & \times & \pi_2\, G/T & \longrightarrow & \mathbb{Z} \\ \wr\Vert & & \wr\Vert & & \\ I^* & & I & & \end{array}$$

since $\langle X, [e_{\alpha_i}, e_{-\alpha_i}]\rangle = \langle [X, e_{\alpha_i}], e_{-\alpha_i}\rangle = 2\pi(-1)^{1/2}\alpha(X)\langle e_{\alpha_i}, e_{-\alpha_i}\rangle$ for any $X \in L(T) \otimes \mathbb{C}$, i.e. $\alpha_i \in L(T)^* \otimes \mathbb{C}$ corresponds to (a multiple of) $[e_{\alpha_i}, e_{-\alpha_i}] \in L(T) \otimes \mathbb{C}$ via the Killing form. It follows now by a result of [19] that any homotopy class of the form $\sum_{i=1}^{\ell} n_i [f_i]$, where the $n_i$ are non-negative integers, may be holomorphically represented. By changing the Weyl chamber, we deduce that <u>any</u> element of $\pi_2\, G/T$ may be represented by a map which is holomorphic with respect to <u>some</u> complex structure.

The composition of two holomorphic maps is holomorphic; while this is in general false for harmonic maps, one has the following result (see §4 of [12] for further discussion).

**Proposition 3.** Let $f_1\colon M \longrightarrow N$, $f_2\colon N \longrightarrow P$ be (smooth) maps of (compact) Riemannian manifolds. Assume (a) that $f_2$ is a Riemannian submersion and (b) that $f_1$ is harmonic and horizontal with respect to $f_2$. Then $f_2 \circ f_1$ is harmonic.

**Proof.** The hypotheses mean that $df_2$ is an isometric isomorphism on each horizontal subspace $H_x = (\mathrm{Ker}(df_2)_x)^{\perp} \subseteq T_x N$, and $df_1(T_y M) \subseteq H_{f_1(y)}$ for all $x \in N$, $y \in M$. Applying the formula given at the beginning of this section one has

$$(\nabla_X d(f_2 \circ f_1))Y = \nabla_X df_2(df_1(Y)) - df_2 \circ df_1(\nabla_X Y)$$

$$= (\nabla_{df_1(X)} df_2)(df_1(Y)) + df_2(\nabla_X df_1(Y))$$

$$- df_2 \circ df_1(\nabla_X Y)$$

$$= (\nabla_{df_1(X)} df_2)(df_1(Y)) + df_2((\nabla_X df_1)Y)$$

for $X,Y \in TM$. On taking the trace, the second term disappears as $f_1$ is harmonic. Since $f_2$ is a Riemannian submersion, $\nabla df_2$ vanishes on horizontal vectors. (This geometrical fact is "dual" to the assertion discussed at the beginning of the next section, that for a Riemannian immersion f, $\nabla df$ takes values in the normal bundle. See [28].) Hence the second term is also zero.

**Remark**. The map $f_2$ is not necessarily harmonic here; it is harmonic if and only if the "fibres" are minimal submanifolds (see [12], §4).

## Harmonic Maps and Representations

Let $f: M \longrightarrow N$ be a Riemannian immersion of compact manifolds whose metrics are g,h (thus, $f^*h = g$). Then $\nabla(df)$ is the classical second fundamental form. Note that the projection of $\nabla_X^{f^*TN}(df(Y))$ onto $df(TM) \subseteq f^*TN$ is just $df(\nabla_X Y)$, hence $(\nabla_X df)Y = \nabla_X^{f^*TN}(df(Y)) - df(\nabla_X Y)$ actually lies in the normal bundle $(df(TM))^\perp \subseteq f^*TN$. For $m \in M$, $\mathrm{tr}\nabla(df)_m$ is known as the mean curvature normal at m, and f is said to be a minimal immersion if its mean curvature normals are all zero. Thus, minimal immersions are examples of harmonic maps. It is known that, amongst Riemannian immersions, minimal immersions are characterized as those which are extrema for the volume functional

$$V: f \longmapsto \int_M |\det f^*h|^{1/2}$$

(see [12], §6). Using the third expression for the energy functional E given at the beginning of the last section, one sees that $V(f) \leqslant E(f)$ with equality if and only if $f^*h$ is a scalar multiple of $g$. Let $\theta: G \longrightarrow U_{n+1}$ be a unitary representation of a compact Lie group G on $\mathbb{C}^{n+1}$, and let $\mathbb{P}(\theta): G \longrightarrow PU_{n+1}$ denote the associated projective representation on $\mathbb{C}P^n$. The orbits of $\theta$ and $\mathbb{P}(\theta)$ are homogeneous quotients of G, and these provide natural candidates for minimal embeddings (with respect to suitable metrics). One criterion has been given by W.-Y. Hsiang [23]: if an orbit is of "isolated type", then it is minimal. We shall give an algebraic procedure for constructing minimal embeddings from $\theta$, which may then be modified to produce large families of harmonic maps.

We may choose a maximal torus T of G such that $\theta(T)$ is contained in the standard maximal torus $U_1 x \ldots x U_1 \subseteq U_{n+1}$. Then $\theta\big|_T$ $= (\lambda_0,\ldots,\lambda_n)$ where $\lambda_i \in \hat{T}$, and the associated linear forms $\Lambda_i \in L(T)^*$ are the weights of $\theta$. From now on we assume that $\theta$ is irreducible. Hence there is a unique maximal weight $\Lambda$, and all other weights are obtained by subtracting suitable positive roots from $\Lambda$ (see [24]). Let $v_0,\ldots,v_n$ be weight vectors for $\Lambda_0,\ldots,\Lambda_n$, respectively. There is an induced map

$$\tilde{\theta}: G/T \longrightarrow U_{n+1}/U_1 x \ldots x U_1 \cong F_{1,2,\ldots,n}(\mathbb{C}^{n+1})$$

where $F_{1,2,\ldots,n}(\mathbb{C}^{n+1})$ denotes the manifold of "full" flags $\{0\} = E_0 \subseteq E_1 \subseteq \ldots \subseteq E_n \subseteq E_{n+1} = \mathbb{C}^{n+1}$ in $\mathbb{C}^{n+1}$ ($E_i$ being a complex linear subspace of $\mathbb{C}^{n+1}$ of dimension i). This identification endows $U_{n+1}/U_1 x \ldots x U_1 \cong SU_{n+1}/S(U_1 x \ldots x U_1)$ with a complex structure (namely the invariant complex structure corresponding to the "standard" Weyl chamber for $S(U_1 x \ldots x U_1)$, which is determined by the standard ordered basis of $\mathbb{C}^{n+1}$). Although $\tilde{\theta}$ is not necessarily an embedding, it is certainly harmonic, as it is possible to choose invariant Kähler structures making $\tilde{\theta}$ holomorphic. Indeed, for any complex structure on G/T given by some fundamental Weyl chamber, one may take the complex structure of $F_{1,2,\ldots,n}(\mathbb{C}^{n+1})$ determined by the ordered basis of weight vectors $v_{i_0},\ldots,v_{i_n}$, provided these are arranged so that, if $\Lambda_{i_j} + \alpha = \Lambda_{i_k}$ for some positive root $\alpha$, then $i_k > i_j$. In particular, $\Lambda = \Lambda_{i_n}$. There are projection maps $\pi_i\colon F \longrightarrow \mathbb{C}P^n$ defined by $\pi_i(E_0 \subseteq E_1 \subseteq \ldots \subseteq E_{n+1}) = E_i^{\perp} \cap E_{i+1}$, and these may be combined to give maps from F into various "partial flag manifolds" in the obvious way. For example, each ordered subset $\{i_1,i_2,\ldots,i_k\}$ of $\{0,1,\ldots,n\}$ gives a map denoted $\pi_{i_1} \oplus \ldots \oplus \pi_{i_k}$ from F into the Grassmannian $Gr_k(\mathbb{C}^{n+1})$ of k-planes in $\mathbb{C}^{n+1}$. With respect to compatible invariant metrics, these maps are all Riemannian submersions.

The following result allows one to use proposition 3 to prove that certain maps are harmonic. By $F_{i,j}(\mathbb{C}^{n+1})$ we mean the manifold of partial flags $0 \subseteq E_i \subseteq E_j \subseteq \mathbb{C}^{n+1}$, and we shall also use $E_i, E_j$ to denote the tautologous bundles over $F_{i,j}(\mathbb{C}^{n+1})$ of fibre dimensions i,j respectively.

<u>Proposition</u> <u>4</u>. (a) The map

$$\emptyset\colon G/T \longrightarrow F_{i,j}(\mathbb{C}^{n+1}), \qquad x \longmapsto ((E_i)_{\tilde{\theta}(x)}, (E_j)_{\tilde{\theta}(x)})$$

is horizontal with respect to

$$\pi\colon F_{i,j}(\mathbb{C}^{n+1}) \longrightarrow Gr_{j-i}(\mathbb{C}^{n+1}), \qquad (E_i,E_j) \longmapsto E_i^{\perp} \cap E_j$$

(using any invariant metric on $F_{i,j}(\mathbb{C}^{n+1})$) providing the following condition is satisfied:

(*) $X \cdot \Gamma(\tilde{\theta}^*E_i) \subseteq \Gamma(\tilde{\theta}^*E_j)$ for all $X \in T_{1,0}(G/T)$,
where the vector field X acts on sections of $G/T \times \mathbb{C}^{n+1}$ (of which $\tilde{\theta}^*E_i$ is a subbundle) in the usual way.

(b) Let $\emptyset$ satisfy (*) in (a). Let $P \in M_{n+1}(\mathbb{C})$ be an $(n+1) \times (n+1)$ matrix, such that $P \cdot \emptyset$ is a well defined map of $G/T$ into some flag manifold $F_{i',j'}(\mathbb{C}^{n+1})$. Then $P \cdot \emptyset$ also satisfies (the analogue of) (*).

**Proof.** (a) This depends on the fact that the horizontal and vertical subbundles of $TF_{i,j}(\mathbb{C}^{n+1})$ are homogeneous and are easily identified. A similar result appears in [18]. (b) This is immediate. The matrix P should be considered as an (improper) "projectivity".

Finally, we point out how (*) reduces to an algebraic condition. Let $e_\alpha$ be a nonzero element of $E_\alpha$, so that $T_0(G/T) \cong \sum_{\alpha\in\Delta^+} \mathbb{C}e_\alpha \oplus \sum_{\alpha\in\Delta^+} \mathbb{C}e_{-\alpha}$. The full flag manifold $F_{1,2,\ldots,n}(\mathbb{C}^{n+1})$ has tautologous bundles $E_0 \subseteq E_1 \subseteq \ldots \subseteq E_{n+1}$, and we shall write $D_i = E_i^\perp \cap E_{i+1}$ (i=0,1,...,n) for the associated line bundles.

**Proposition 5.** The action of the tangent vector $X = e_\alpha$ on the subspace $\Gamma(\tilde{\theta}^* D_i)$ of the space of functions $G/T \longrightarrow \mathbb{C}^{n+1}$ satisfies:

$$(X \cdot S)_{(E_i)^\perp} \begin{cases} \in \Gamma(\tilde{\theta}^* D_j) & \text{if there exists a weight } \Lambda_j \\ & \text{of } \theta \text{ with } j>i \text{ and } \Lambda_i+\alpha=\Lambda_j \\ = 0 & \text{otherwise} \end{cases}$$

where $(X \cdot S)_{(E_i)^\perp}$ is the projection of $X \cdot S$ on $(E_i)^\perp$.

**Proof.** The pertinent fact is that, if $v_i$ is a weight vector for $\Lambda_i$, then $\theta(e_\alpha)(v_i)$ is either zero or a weight vector for $\Lambda_i + \alpha$. (Explicitly, for $x \in L(T)$, $\theta(x)(\theta(e_\alpha)v_i) = (\theta[x,e_\alpha] + \theta(e_\alpha)\theta(x))(v_i) = \theta(\alpha(x)e_\alpha)v_i + \theta(e_\alpha)(\Lambda_i(x)v_i) = (\alpha(x) + \Lambda_i(x))\theta(e_\alpha)v_i$, where $\theta$ refers here to the corresponding Lie algebra representation.)

Propositions 4 and 5 permit the construction of many harmonic maps of G/T into Grassmannians and flag manifolds. A more extensive discussion of this algebraic construction will be given elsewhere. We content ourselves here by describing one important application, namely where $G = SU_2$. For each positive integer n, there is (up to equivalence) precisely one irreducible representation $s_n$ of $SU_2$ of dimension n + 1 and it is well known that this may be described as the natural action of $SU_2$ on the space of homogeneous polynomials in two variables $t_0$, $t_1$ of total degree n ($s_1 = \sigma$: $SU_2 \longrightarrow SU_2$ is the identity map, and $s_n = S^n\sigma$, the $n^{th}$ symmetric power of $\sigma$). With respect to the maximal torus T of $SU_2$ consisting of matrices of the form

$$\begin{bmatrix} z & 0 \\ 0 & z^{-1} \end{bmatrix}$$

the weights of $s_n$ are $n, n-2, n-4, \ldots, -n \in \mathbb{R} \cong L(T)^*$, and $t_0^n, t_0^{n-1}t_1, t_0^{n-2}t_1^2, \ldots, t_1^n$ will serve as corresponding weight vectors. It is easy to interpret proposition 5 here, for $T_{1,0}(SU_2/T)_0 \cong T_{1,0}\mathbb{C}P_0^1 \cong \mathbb{C}e$ for some $e \neq 0$, and $s_n(e)(t_0^i t_1^{n-i})$ is a multiple of $t_0^{i-1}t_1^{n-i+1}$.

Thus, each composition

$$\mathbb{C}P^1 \xrightarrow{\tilde{s}_n} F_{1,2,\ldots,n}(\mathbb{C}^{n+1}) \xrightarrow{\pi_i} \mathbb{C}P^n$$

is harmonic. Geometrically, the complex curve $\pi_i \circ \tilde{s}_n$ is the embedding of the projectivized orbit of the i-th weight vector; only the orbit of the maximal weight vector is <u>holomorphically</u> embedded, and this is a rational normal curve of degree n. (Any rational normal curve of degree n is given in this way by a representation equivalent to $s_n$.) Using (b) of proposition 4, further harmonic maps are obtained. These turn out to be branched minimal immersions. It is an interesting fact that <u>all</u> harmonic maps of $\mathbb{C}P^1$ into projective spaces arise through this construction. We shall not pursue this here, other than to say that in [17,34] it is shown that a holomorphic map f of $\mathbb{C}P^1$ into projective space gives rise to a sequence $f = f_0, f_1, f_2, \ldots$ of harmonic maps, $f_i(z)$ being the complement of the (i-1)-th "osculating plane" in the i-th "osculating plane" (at $z \in \mathbb{C}P^1$), or, alternatively, the line spanned by the (i+1)-th vector in a "Frenet frame" (at z). It is shown that any harmonic map arises from some holomorphic map via this procedure. (See also [10,11,20].) Using the fact that f is of the form $P \cdot \emptyset$ for some rational normal curve $\emptyset$: $\mathbb{C}P^1 \longrightarrow \mathbb{C}P^n$ and a projectivity represented by some $P \in M_{n+1}\mathbb{C}$, one may check that the procedure agrees with the one described above. Thus, all harmonic maps $\mathbb{C}P^1 \longrightarrow \mathbb{C}P^n$ arise by direct geometrical constructions from the representations of the group $SU_2$. Two facts should be emphasized before closing this section:

(1) The holomorphic maps $\mathbb{C}P^1 \longrightarrow \mathbb{C}P^n$ determine all the harmonic maps $\mathbb{C}P^1 \longrightarrow \mathbb{C}P^n$.

(2) There are no harmonic maps $\mathbb{C}P^1 \longrightarrow \mathbb{C}P^1$ other than holomorphic or antiholomorphic maps.

The latter was proved directly (i.e. not as a consequence of (1)) in [15].

## Instantons

Let $P \longrightarrow M$ be a principal G-bundle over a compact Riemannian manifold M, where G is a compact Lie group with fixed bi-invariant metric. Let $\mathfrak{A}$ be the affine space consisting of all G-connections in $P \longrightarrow M$. The Yang-Mills functional

$$L: \mathfrak{A} \longrightarrow \mathbb{R},\ \nabla \longmapsto \frac{1}{2} \int_M |F_\nabla|^2$$

assigns to $\nabla$ the norm of its curvature $F_\nabla$, where $F_\nabla$ is considered as a 2-form on M taking values in the adjoint bundle $Ad(P) = P \times_G L(G)$ of P. The Yang-Mills equations are the Euler-Lagrange equations

$$D^*_{Ad}\,(F_\nabla) = 0$$

for L, together with the Bianchi identity

$$D_{Ad}(F_\nabla) = 0$$

where $D_{Ad}$ denotes the operator on $\Omega^*Ad(P)$ (Ad(P)-valued forms) associated to $\nabla$, and $D^*_{Ad}$ is its adjoint with respect to the metric. The equations say that $\nabla$ has "harmonic curvature". If M is an oriented 4-manifold, the Hodge operator $*$ on 2-forms satisfies $*^2 = 1$, and $\nabla$ is said to be <u>self dual</u> (respectively, <u>anti-self dual</u>) if its curvature satisfies $*F_\nabla = F_\nabla$ (respectively, $*F_\nabla = -F_\nabla$). Such connections are called <u>instantons</u>, and these form a subspace of $\mathfrak{A}$ which we denote by A. Since $D^*_{Ad} = \pm * D_{Ad} *$, an instanton is automatically a solution of the Yang-Mills equations (the second equation, which holds for any connection, implies the first in this case). With respect to a fixed orientation of M, instantons are either always self dual or always anti-self dual; changing orientation reverses the type. We refer to [4] or §3 and §4 of [3] for further explanation. It has been known for some time that there are basic similarities between the Yang-Mills functional in this situation, and the energy functional E on $Map(\mathbb{C}P^1, \mathbb{C}P^n)$, and a comparison of some aspects was given in [9]. For example, with respect to the eigenspace decomposition for $*$ we may write $F_\nabla = F^+_\nabla + F^-_\nabla$, and define subsidiary functionals $L^+, L^-$ by the formulae:

$$L^+(F_\nabla) = \frac{1}{2}\int_M |F^+_\nabla|^2, \quad L^-(F_\nabla) = \frac{1}{2}\int_M |F^-_\nabla|^2$$

Thus, $L(F_\nabla) = L^+(F_\nabla) + L^-(F_\nabla)$, and there is the following analogue of proposition 1.

**<u>Proposition</u>** <u>6</u>. $L^+(F_\nabla) - L^-(F_\nabla)$ is a constant multiple of the first Pontrjagin number of $P \longrightarrow M$; in particular, it is independent of $\nabla$.

**<u>Proof</u>**. This follows from the Chern-Weil description of characteristic classes.

As in proposition 2, one deduces that instantons represent the absolute minima of L.

A major new feature of L (compared with E) is that, while its domain $\mathfrak{A} \cong \Omega^1$ AdP is contractible, it is invariant under the gauge group $\mathfrak{G}$ consisting of (smooth) automorphisms of the principal bundle $P \longrightarrow M$ which cover the identity on M, and so it is natural to consider the induced functional on $\mathfrak{A}/\mathfrak{G}$. The group $\mathfrak{G}$ is infinite dimensional, and may be identified with $\Gamma(AD(P))$, the space of sections of the bundle of groups $AD(P) = P \times_G G$ where G acts on the second factor by conjugation. Its Lie algebra may be identified with $\Gamma(Ad(P))$. Although $\mathfrak{G}$ does not act freely, its normal subgroup $\mathfrak{G}_0$ consisting of based automorphisms (i.e. automorphisms which are the identity over a given point) does, and one has $\mathfrak{G}/\mathfrak{G}_0 \cong G$. We shall indicate how the moduli space $A/\mathfrak{G}_0$ is the correct analogue of the space of based "$\mathbb{C}P^n$-instantons" (i.e. basepoint preserving holomorphic maps $\mathbb{C}P^1 \longrightarrow \mathbb{C}P^n$).

The following was proved in [31].

**Theorem 7.** The inclusion $Hol_d(\mathbb{C}P^1,\mathbb{C}P^n) \longrightarrow Map_d(\mathbb{C}P^1,\mathbb{C}P^n)$ of the space of "$\mathbb{C}P^n$-instantons" into the space of all "$\mathbb{C}P^n$-fields" (of "charge" $d > 0$) induces an isomorphism in homotopy groups $\pi_i$ and homology groups $H_i$ for $i < d(2n-1)$, and a surjection for $i = d(2n-1)$.

See [21,25] for generalizations where $\mathbb{C}P^n$ is replaced by other homogeneous Kähler manifolds. In [5], a somewhat analogous result was obtained for the Yang-Mills situation where $P \longrightarrow M$ is an $SU_2$-bundle over $M = S^4$ of first Pontrjagin number $d > 0$. In this case, $\mathfrak{G}$ has the homotopy type of $Map_*(S^4,SU_2)$ (to see this, consider the case $d = 0$), and $\mathfrak{G}_0$ has the homotopy type of the subspace $\Omega^4_* SU_2$ of $Map_*(S^4,SU_2)$ consisting of based (i.e. basepoint preserving) maps. Denote the space of connections here by $\mathfrak{A}_d$, and let $A_d$ be the subspace of instantons. The induced inclusion of $A_d/\mathfrak{G}_0$ into $\mathfrak{A}_d/\mathfrak{G}_0$ may be interpreted as follows. As $\mathfrak{A}_d$ is contractible and $\mathfrak{G}_0$ acts freely, $\mathfrak{A}_d/\mathfrak{G}_0$ has the homotopy type of the classifying space $B\mathfrak{G}_0$, which has the homotopy type of $\Omega^3_* SU_2$. (Note that this is independent of d.)

**Theorem 8.** The inclusion $A_d/\mathcal{D}_0 \longrightarrow \Omega^3_* SU_2$ (defined up to homotopy) of the moduli space of based "$SU_2$-instantons" into the space of all based "$SU_2$-fields" (of "charge" $d > 0$) induces a surjection of homology groups $H_i$, providing d is large compared with i.

The following result of Atiyah and Donaldson [2] (referred to earlier) suggests a closer relationship between theorems 7 and 8.

**Theorem 9.** For any classical group G, the space $A_d/\mathcal{D}_0$ may be identified with the space of based holomorphic maps of degree d from $\mathbb{C}P^1$ into the based loop space $\Omega G$.

In other words, theorem 8 is a version of theorem 7 where the inclusion $\mathrm{Hol}_d(\mathbb{C}P^1,\mathbb{C}P^n) \longrightarrow \mathrm{Map}_d(\mathbb{C}P^1,\mathbb{C}P^n)$ is replaced by $\mathrm{Hol}_d(\mathbb{C}P^1,\Omega G) \longrightarrow \mathrm{Map}_d(\mathbb{C}P^1,\Omega G)$. The loop group $\Omega G$ admits the structure of a homogeneous Kähler manifold, and may be identified with the quotient space $\mathrm{Map}(S^1,G^c)/P$, where $G^c$ is the complex Lie group corresponding to G and P is the subspace consisting of those maps $S^1 \longrightarrow G^c$ which extend to holomorphic maps of the disc. A map $f: \mathbb{C}P^1 \longrightarrow \Omega G$ lifts locally to a map $f_U: U \longrightarrow \Omega G^c$ (over some open subspace $U \subseteq \mathbb{C}P^1$), and f is said to be holomorphic if the corresponding map $U \times S^1 \longrightarrow G^c$ extends to a holomorphic map of $U \times D_U$ for some annulus $D_U$ containing $S^1$, for each U. For further information on the geometry of $\Omega G$, see [29].

Theorems 7 and 8 may be interpreted as generalizations of a classical result of Morse theory for a real valued function on a compact manifold whose critical points occur in nondegenerate critical manifolds [8], namely that the manifold of absolute minima carries the homotopy groups of the whole manifold, up to a dimension which is essentially the lowest index of any non-minimal critical point. The proofs of these theorems do not involve the methods of Morse theory, however, and it does not appear to be true that the range of dimensions appearing is determined by non-minimal critical points. For example, in theorem 7 with $n = 1$, we have already noted that no other critical points (of the energy functional) besides holomorphic maps exist, yet $Hol_d(\mathbb{C}P^1,\mathbb{C}P^1)$ is a finite dimensional submanifold of $Map_d(\mathbb{C}P^1,\mathbb{C}P^1)$, and certainly not a deformation retract. At the time of writing, it is still not known whether other critical points of the Yang-Mills functional exist besides instantons, in situations such as that of theorem 8. As pointed out in [2], it would therefore be of interest to know whether there exist harmonic non-holomorphic maps $\mathbb{C}P^1 \longrightarrow \Omega G$ (of degree $d > 0$). In any case, it remains to understand why Morse theory appears to fail in theorem 7, and it turns out that Yang-Mills theory suggests a possible explanation, at least in the case $n = 1$. Here, holomorphic maps $\mathbb{C}P^1 \longrightarrow \mathbb{C}P^1$ may be interpreted as "monopoles", i.e. absolute minima of the Yang-Mills-Higgs functional defined on the space of connections on the $SU_2$-bundle $\mathbb{R}^3 \times SU_2$ (which satisfy certain boundary conditions), a space whose homotopy type is closely related to that of $Map_*(\mathbb{C}P^1,\mathbb{C}P^1)$. This functional appears to be better behaved (analytically) than the energy functional, and has been shown (see [33]) to have non-minimal critical points. For further information on this we refer to [2,22,27].

## References


[1] J. F. Adams, Lectures on Lie Groups, Benjamin (New York, Amsterdam) 1969.

[2] M. F. Atiyah, Instantons in two and four dimensions, to appear.

[3] M. F. Atiyah, R. Bott, The Yang-Mills equations over Riemann surfaces, Phil. Trans. R. Soc. Lond. A 308 (1982), 523-615.

[4] M. F. Atiyah, N. J. Hitchin, I. M. Singer, Self-duality in four dimensional Riemannian geometry, Proc. R. Soc. Lond. A 362 (1978), 425-461.

[5] M. F. Atiyah, J. D. S. Jones, Topological aspects of Yang-Mills theory, Commun. Math. Phys. 61 (1978), 97-118.

[6] M. F. Atiyah, L. Smith, Compact Lie groups and the stable homotopy of spheres, Topology 13 (1974), 135–142.

[7] A. Borel, F. Hirzebruch, Characteristic classes and homogeneous spaces I, Amer. J. Math. 80 (1958), 458–538.

[8] R. Bott, Nondegenerate critical manifolds, Annals of Math. 60 (1954), 248–261.

[9] J.-P. Bourguignon, Harmonic curvature for gravitational and Yang-Mills fields, (in proceedings of conference on harmonic maps, New Orleans, 1980) Springer Lecture Notes in Mathematics 949, Springer-Verlag (Berlin, Heidelberg, New York) 1982, 35–47.

[10] D. Burns, Harmonic maps from $\mathbb{C}P^1$ to $\mathbb{C}P^n$, (in proceedings of conference on harmonic maps, New Orleans, 1980) Springer Lecture Notes in Mathematics 949, Springer-Verlag (Berlin, Heidelberg, New York) 1982, 48–56.

[11] A. M. Din, W. J. Zakrzewski, General classical solutions in the $\mathbb{C}P^{n-1}$ model, Nuclear Phys. B 174 (1980), 397–406.

[12] J. Eells, L. Lemaire, A report on harmonic maps, Bull. Lond. Math. Soc. 10 (1978), 1–68.

[13] J. Eells, L. Lemaire, Selected Topics in Harmonic Maps, CBMS Regional Conference Series No. 50, American Mathematical Society 1983.

[14] J. Eells, J. H. Sampson, Harmonic maps of Riemannian manifolds, Amer. J. Math. 86 (1964), 109–160.

[15] J. Eells, J. C. Wood, Restrictions on harmonic maps of surfaces, Topology 15 (1976), 263–266.

[16] J. Eells, J. C. Wood, Maps of minimum energy, J. Lond. Math.

Soc. 2 (1981), 303-310.

[17] J. Eells, J. C. Wood, Harmonic maps from surfaces to complex projective spaces, Advances in Math. 49 (1983), 217-263.

[18] S. Erdem, J. C. Wood, On the construction of harmonic maps into a Grassmannian, J. Lond. Math. Soc. (2) 28 (1983), 161-174.

[19] A. Futaki, Nonexistence of minimizing harmonic maps from 2-spheres, Proc. Japan. Acad. 56 (1980), 291-293.

[20] V. Glaser, R. Stora, Regular solutions of the $\mathbb{C}P^n$ models and further generalizations, preprint.

[21] M. A. Guest, Topology of the space of absolute minima of the energy functional, Amer. J. Math. 106 (1984), 21-42.

[22] N. J. Hitchin, Monopoles and geodesics, Commun. Math. Phys. 83 (1982), 579-602.

[23] W.-Y. Hsiang, On the compact homogeneous minimal submanifolds, Proc. Nat. Acad. Sci. 56 (1966), 5-6.

[24] N. Jacobson, Lie Algebras, Interscience Tracts in Pure and Applied Mathematics, Wiley (New York, London) 1962.

[25] F. Kirwan, On spaces of maps from Riemann surfaces to Grassmannians and applications to the cohomology of moduli of vector bundles, to appear.

[26] A. Lichnerowicz, Applications harmoniques et variétés Kählériennes, Symposia Mathematics III, Bologna, 1970.

[27] M. K. Murray, Monopoles and spectral curves for arbitrary Lie groups, Commun. Math. Phys. 90 (1983), 263-271.

[28] B. O'Neill, The fundamental equations of a submersion, Michigan Math. J. 13 (1966), 459-470.

[29] A. N. Pressley, G. B. Segal, Loop Groups, Oxford University Press, to appear.

[30] S. Salamon, Harmonic and holomorphic maps, to appear.

[31] G. B. Segal, The topology of spaces of rational functions, Acta. Math. 143 (1979), 39-72.

[32] Y.-T. Siu, The complex-analyticity of harmonic maps and the strong rigidity of compact Kähler manifolds, Annals of Math. 112 (1980), 73-111.

[33] C. H. Taubes, The existence of a non-minimal solution to the $SU_2$ Yang-Mill-Higgs equations on $\mathbb{R}^3$: Part I, Commun. Math. Phys. 86 (1982), 257-298; Part II, 299-320.

[34] J. G. Wolfson, Minimal surfaces in complex manifolds, Ph.D. thesis, Berkeley, 1982.

# A COXETER GROUP APPROACH TO SCHUBERT VARIETIES

By

Ziad Haddad

One purpose of this note is to sketch the generalization of the cohomological interpretation of the Kazhdan-Lusztig polynomials ([KL]) to the case of arbitrary crystallographic groups W. This generalization of the argument in [KL] requires a study of the intersections $BwB \cap B_-yB$ of Bruhat and Birkhoff cosets of the corresponding Kac-Moody group, and unions of such, from a combinatorial as well as a geometric point of view; one of the main tools for this is a systematic use of W. Specific examples are included at the end.

1. The first step is to associate to a given $n \times n$ generalized Cartan matrix A, and a field K of arbitrary characteristic, a "flag variety over K", which is a union of finite dimensional Schubert varieties. To that end, let $\mathfrak{g}$ be the Kac-Moody algebra over $\mathbb{Q}$ associated to A, W its Weyl group, $\mathfrak{h}$ its Cartan subalgebra, $\Delta$ the set of roots, ... so that $\mathfrak{g}$ is generated by $\mathfrak{h}$ and root vectors $e_i, f_j$, $1 \leq i,j \leq n$, and set $U_{\mathbb{Z}}$ = the $\mathbb{Z}$-subalgebra of the universal enveloping algebra of $\mathfrak{g}$ generated by all $\frac{e_i^s}{s!}$, $\frac{f_j^t}{t!}$ with $1 \leq i,j \leq n$ and $s,t \in \mathbb{N}$. It is not hard to check that $U_{\mathbb{Z}}$ is a $\mathbb{Z}$-form of the universal enveloping algebra of $\mathfrak{g}$, the automorphism $\dot{r}_i = (\exp(\mathrm{ad}e_i))(\exp(-\mathrm{ad}f_i))(\exp(\mathrm{ad}e_i))$ is well-defined and preserves $U_{\mathbb{Z}}$, for every $1 \leq i \leq n$. One can then check that

1.1 a) $\dot{r}_i \cdot e_j = \frac{1}{(-a_{ij})!} \mathrm{ad}e_i^{(-a_{ij})} e_j$ and $\dot{r}_i \cdot f_j = \frac{(-1)^{a_{ij}}}{(-a_{ij})!} \mathrm{ad}f_i^{(-a_{ij})} f_j$ (where $a_{ij}$ is the ij-th entry of A).

b) One can define a map $W \longrightarrow \mathrm{Aut}(U_{\mathbb{Z}})$, $w \longmapsto \dot{w}$, by setting $\dot{w} \cdot v = \dot{r}_{j_1} \ldots \dot{r}_{j_k} \cdot v$ whenever $r_{j_1} \ldots r_{j_k} = w$ is a reduced expression. In particular, for any positive real root $\lambda \in \Delta_+^{re}$, there exist root vectors $e_\lambda, f_\lambda$ with $U_{\mathbb{Z}} \cap \mathfrak{g}_{-\lambda} = \mathbb{Z} f_\lambda$. (For b), one computes that if $R_{ij} = \dot{r}_i \dot{r}_j \dot{r}_i \ldots$ ($m_{ij} - 1$ factors, where $m_{ij}$ is the order of $r_i r_j$ in W), and if $x \in \{e,f\}$, then $R_{ij}^{-1} \cdot x_j = \begin{cases} x_j & \text{if } m_{ij} \text{ is even} \\ x_i & \text{if } m_{ij} = 3 \end{cases}$).

Given $\Lambda \in \mathfrak{g}^*$ dominant integral (i.e. $\langle \Lambda, \alpha_i^\vee \rangle \in \mathbb{N}$ for all i), let $L(\Lambda)$ be the irreducible highest weight $\mathfrak{g}$-module with highest weight $\Lambda$, $v_\Lambda^+$ a highest weight vector. Again, by integrability, the elements $\dot{r}_i(1) = (\exp e_i)(\exp -f_i)(\exp e_i) \in \mathrm{Aut}(L(\Lambda))$ are well-defined, and

$$1.2 \qquad a') \ \dot{r}_i(1) \cdot r_\Lambda^+ = \frac{(-1)^{\langle \Lambda, \alpha_i^\vee \rangle}}{\langle \Lambda, \alpha_i^\vee \rangle !} f_i^{\langle \Lambda, \alpha_i^\vee \rangle} \cdot v_\Lambda^+.$$

b') The map $W \longrightarrow \mathrm{Aug}_{\mathbb{Q}}(L(\Lambda))$ assigning to $w \in W$ the element $\dot{w} = \dot{r}_{j_1} \ldots \dot{r}_{j_k}$ (when $w = r_{j_1} \ldots r_{j_k}$ is a reduced expression for w) is well-defined, so that for each w there exists a vector $v_{w\Lambda}$ of weight $w\Lambda$ satisfying $(U_{\mathbb{Z}} \cdot v_\Lambda^+) \cap L(\Lambda)_{w\Lambda} = \mathbb{Z} v_{w\Lambda}$.

c) $(U_{\mathbb{Z}} \cdot v_\Lambda^+)$ is homogeneous with respect to the $\mathfrak{h}$-weight space gradation of $L(\Lambda)$.

Given any field K, it is now easy to construct a group G(K) which acts on the $U_{\mathbb{Z}} \otimes K$-modules $U_{\mathbb{Z}} \otimes K$ and $(U_{\mathbb{Z}} \cdot v_\Lambda^+) \otimes K$ for $\Lambda$ dominant integral (exactly as in the finite dimensional case [St]; see [PK] for the case char (K) = 0). G(K) is generated by the 1-parameter groups $\{\exp(tx_\lambda), \ t \in K\}$ with $\lambda \in \Delta_+^{re}$, $x \in \{e,f\}$, and

1.3 Given $\alpha, \beta \in \Delta^{re}$ such that $\alpha + \beta \neq 0$ and the set $S = ((\mathbb{N}_\alpha + \mathbb{N}_\beta) \cap \Delta) - \{\alpha, \beta\}$ is finite, there exists integers $c_i$ so that, for all $s,t \in K$,

$$z_\alpha(s)z_\beta(t)z_\alpha(-s)z_\beta(-t) = \prod_{(n_i\alpha+m_i\beta)\in S} z_{n_i\alpha+m_i\beta}(c_i s^{n_i} t^{m_i})$$

where $z_\lambda(u)$ stands for $\exp(ue_\lambda)$ if $\lambda \in \Delta_+^{re}$, and for $\exp(uf_\lambda)$ otherwise (if S is finite, it must consist entirely of real roots), the product taken in any fixed ordering of S.

If $\dot{r}_i(t)$ is defined as $\exp(te_i)\exp(-\frac{1}{t} f_i)\exp(te_i)$, for $t \in K^x$, then $\dot{r}_i(1)\dot{r}_i(-t)$ acts on $v_\mu \otimes 1$ by multiplication by $t^{\langle\mu,\alpha_i^\vee\rangle}$ whenever $v_\mu$ lies in $L(\Lambda)_\mu \cap (U_\mathbb{Z} \cdot v_\Lambda^+)$, so letting

H be the subgroup of G generated by all elements $\dot{r}_i(1)\dot{r}_i(-t)$

N be the subgroup of G generated by $\dot{r}_i(t)$, $1 \leq i \leq n$, $t \in K^x$

U the subgroup generated by all $\exp(te_\lambda)$, $\lambda \in \Delta_+^{re}$, $t \in K$

B the subgroup generated by U and H,

then $(G, B, N, \{r_1,\ldots,r_n\})$ is a Tits system (in particular $G = \bigcup_{w\in W} B\dot{w}B$), and using the subgroup $B_-$ generated by H and $U_-$, where $U_-$ is the subgroup of G generated by all $(\exp tf_\lambda)$, $\lambda \in \Delta_+^{re}$, $t \in K$, then G also admits a Birkhoff decomposition: $G = \bigcup_{w\in W} B_-\dot{w}B$ ([PK] in char (K) = 0). The following refinement of the main lemma used in [PK] to prove both decompositions extends to K of arbitrary characteristic (and is needed for what follows):

1.4 If $g = \dot{w}b\dot{r}_i$, and $b \in B - (B \cap \dot{r}_i B\dot{r}_i^{-1})$ then

$w \cdot \alpha_i > 0$ implies $g \in (B\dot{w}\dot{r}_iB) \cap (B_-\dot{w}B)$ while $w \cdot \alpha_i < 0$ implies that $g \in (B\dot{w}B) \cap (B_-\dot{w}\dot{r}_iB)$.

2. For $w \in W$, write $C(w)$ for the image of the coset $B\dot{w}B$ in $G/B$, $C_-(w)$ for that of $B_-\dot{w}B$, and $wC_-(1)$ for the $\dot{w}$-translate of the "big cell" $C_-(1)$. In all that follows, fix an element $w \in W$, $w = r_{j_1} \ldots r_{j_k}$ a reduced expression.

Then the Deodhar decomposition ([D]) identifies the sets $C(w) \cap C_-(y)$ in the case W is a finite Weyl group. To generalize this decomposition to the case at hand, observe ([PK]) that the group $U_w = U \cap \dot{w}U_-\dot{w}^{-1}$ acts simply transitively on $C(w)$. Using the Birkhoff decomposition, one may thus associate to every $u \in U_w$ a sequence $s_1(u), \ldots, s_k(u)$ of elements of W, by requiring that $u\dot{r}_{j_1} \ldots \dot{r}_{j_i} \in B_-\dot{s}_i(u)B$. Using 1.4, it follows that

2.1 a) $s_i(u) \in \{s_{i-1}(u), s_{i-1}(u)r_{j_i}\}$

b) In either case $s_i(u) \leqslant s_{i-1}(u)r_{j_i}$ in the Bruhat order

on W. Given a sequence $\sigma = (\sigma_1, \ldots, \sigma_k) \in W^k$ satisfying 2.1, set $D_\sigma = \{u \in U_w \mid s_i(u) = \sigma_i\}$. Writing $\varphi(u)$ for the image $u\dot{w}B$ in $C(w) \subset G/B$ of $u \in U_w$, one sees that $C(w) \cap C_-(y)$ is equal to the disjoint union $\cup\varphi(D_\sigma)$ taken over all $\sigma$ with $\sigma_k = y$. Now, using 1.3, $U_w$ is in a natural way a nilpotent algebraic group whose underlying variety is isomorphic to affine k-space, and the argument in [D] generalizes readily using 2.1 to give

2.2 For every $\sigma$, $D_\sigma$ is a locally closed subset of $U_w$, and is isomorphic to $(K)^{m(\sigma)} \times (K^\times)^{n(\sigma)}$, where $m(\sigma) = \#\{i \mid \sigma_{i-1} > \sigma_i\}$, $n(\sigma) = \#\{i \mid \sigma_{i-1} = \sigma_i\}$.

Fix $\Lambda$ integral strictly dominant (i.e. $\langle\Lambda, \alpha_i^\vee\rangle > 0$ for all i), $v_\Lambda^+ \in L(\Lambda)_\Lambda$, and write L for $(U_{\mathbb{Z}} \cdot v_\Lambda^+) \otimes K$, $L_\lambda$ for $(U_{\mathbb{Z}} \cdot v_\Lambda^+ \cap L(\Lambda)_\lambda) \otimes K$, and $L(m)$ for $\sum_{\text{height}(\Lambda-\lambda) \leqslant m} L_\lambda$, whenever $m \in \mathbb{N}$.

Then $G/B$ injects in $\mathbb{P}(L)$ via the map $g \longmapsto [g \cdot v_\Lambda^+ \otimes 1]$, and for each $z \in W$, $C(z)$ maps into $\mathbb{P}(L(m))$ as long as $m \geq \nu(z) = \text{height}(\Lambda - z\Lambda)$. To extend the act that the Zariski closure of $C(w)$ in $\mathbb{P}(L(\nu(w)))$ is $\bigcup_{z \leq w} C(z)$ to algebraically closed fields K of arbitrary characteristic (see [PK] for char (K) = 0), one constructs a complete Bott-Samelson-Demazure variety $Y_w$ and a morphism $Y_w \longrightarrow \mathbb{P}(L(\nu(w)))$, the image of which is identified as $\bigcup_{z \leq w} C(z)$.

As a set, $Y_w = (P_k \times \ldots \times P_1)/\sim$, where $P_{k-i+1} = \dot{r}_{j_k} \dot{r}_{j_{k-1}} \ldots \dot{r}_{j_{i+1}} \cdot$ (the subgroup of G generated by B and $\dot{r}_{j_i}) \cdot \dot{r}_{j_{i+1}}^{-1} \dot{r}_{j_{i+2}}^{-1} \ldots \dot{r}_{j_k}^{-1}$, and the equivalence $\sim$ is given by letting $B_{k-1} = \dot{r}_{j_k} \dot{r}_{j_{k-1}} \ldots \dot{r}_{j_{i+1}} \cdot B \cdot \dot{r}_{j_{i+1}}^{-1} \dot{r}_{j_{i+2}}^{-1} \ldots \dot{r}_{j_k}^{-1}$ (so that $B_i \subset P_i \cap P_{i+1}$) and $(p_k, \ldots, p_1) \sim (p_k b_{k-1}, b_{k-1}^{-1} p_{k-1} b_{k-2}, b_{k-2}^{-1} p_{k-3} b_{k-3}, \ldots, b_1^{-1} p_1 b_0)$ for any $p_i \in P_i$, $b_i \in B_i$. The map $Y_w \longrightarrow \mathbb{P}(L(\nu(w)))$ is given by $(p_k, \ldots, p_1) \longmapsto [\dot{w} p_k \cdot p_{k-1} \ldots \cdot p_1 \cdot v_\Lambda^+ \otimes 1]$.

Using 1.3 and 1.4, it is not hard to reconstruct $Y_w$ as a successive fibration by projective lines over $P_k/B_{k-1} \simeq \mathbb{P}^1$ (in fact, $Y_w$ can be embedded in a natural way in the $2^k - 1 = M$-fold product $(\mathbb{P}^1)^M$), and, using the action of the torus $H \subset G$ on $Y_w$ and $\mathbb{P}(L)$, the image of $Y_w$ is identified as being $X_w = \bigcup_{z \leq w} C(z)$.

Given $y \leq w$, write $V_y$ for the image of the space $\{v \in L \mid$ the $L_\lambda$-coordinate of v is 0 if $((\lambda - w\Lambda) \not\geq 0)$ or $((y\Lambda - \lambda) \not\geq 0)$, and the $L_{y\Lambda}$-coordinate of v is nonzero$\}$ in $\mathbb{P}(L)$, so $V_y$ is a (finite dimensional) linear space. Define an action of the multiplicative group $K^x$ on L by $t \cdot v = t^{\text{height}(\Lambda - \lambda)} \cdot v$ if $v \in L_\lambda$. Then

2.3 a) $(X_w \cap y\, C_-(1)) \simeq C(y) \times (X_w \cap C_-(y))$
$\simeq C(y) \times (X_w \cap V_y)$

b) In the natural identification of $V_y$ with affine space, the point $v_{y\Lambda}$ maps to the origin, the $K^x$ action induced by the one defined above leaves $X_w \cap V_y$ stable, and, as this action decomposes into a sum of positive characters, $X_w \cap V_y$ is a (weighted) cone.

2.3 is the lemma about $X_w$ needed in [KL] to prove the "purity" of the intersection cohomology complex on the Schubert varieties. Suppose char (K) = p. Defining the polynomials $P_{y,w}$ in the ariable $q^{1/2}$ for $y,w \in W$ by $P_{y,w} = \sum_j \dim H^j_{yB}(IC_w)q^{j/2}$, where $IC_w$ is the intersection cohomology complex on $X_w$, the argument in [KL] now carries over to show that $P_{y,w} \in \mathbb{Z}[q]$, and, applying the Lefschetz formula to the Frobenius (i.e. $\mathrm{tr}_{\mathbb{H}^*_c(X,IC_w)} F^n = \sum_{x\in X^{F^n}} \mathrm{tr}_{H^*_x(IC_w)} F^n$, where $X = X_w \cap y\, C\_(1)$, $\mathbb{H}_c$ is (hyper)cohomology with compact support, $X^{F^n} = \{x \in X \mid F^n x = x\}$, and $\mathrm{tr}_{A^*} = \sum_i (-1)^i \mathrm{tr}_{A^i}$) and using the purity of $IC_w$ as well as the Poincaré duality it admits, one obtains as in [KL] the formula

$$2.4 \qquad p^{n(l(w)-l(y))}P_{y,w}(p^{-n}) = \sum_{\substack{z \\ y\leqslant z\leqslant w}} \#(C(z) \cap C\_(y))^{F^n} P_{z,w}(p^n)$$

(where l(z) denotes the length of z), valid for all $n \in \mathbb{N}$. Now $\#(C(z) \cap C\_(y))^{F^n} = \sum_{\sigma \text{ with } \sigma_k = y} (p^n)^{m(\sigma)}(p^n-1)^{n(\sigma)}$ as in 2.2, and this sum is identified in [D] (in the general context of an arbitrary Coxeter group) as the polynomial $R_{y,z}(p^n)$ of [KL]. So 2.4 becomes, in the notation of [KL]

$$q^{l(w)-l(y)}P_{y,w}(q^{-1}) = \sum_{y\leqslant z\leqslant w} R_{y,z}(q)P_{z,w}(q)$$

which characterizes $P_{y,w}$ as the Kazhdan-Lusztig polynomials.

3. It is interesting to note that $Y_w \longrightarrow X_w$ is an isomorphism when the reduced expressions of w consist of distinct simple reflections (i.e. $r_{j_s} \neq r_{j_t}$ if $s \neq t$). In particular $X_w$ is then smooth. For example, if $l(w) = 2$, say $w = r_i r_j$, then $Y_{r_i r_j}$ is the subset

$Y_2 = Y_1 X^{-a_{ij}}$ of $\mathbb{P}^1 \times \mathbb{P}^1 \times \mathbb{P}^1$ where $X, X^{-1}, Y_1, Y_1^{-1}, Y_2, Y_2^{-1}$ are the local coordinates on $\mathbb{P}^1$; so $X_{r_i r_j}$ is indeed the ruled surface whose rigid section $Y_1 = Y_2 = \infty$ has self intersection $a_{ij}$ (see [T]).

When $w' < w$ and length $(w') - 1 = k-1$, there exists subsets $N_{w'} \subset M_{w'}$ of $Y_{w'}$ with $N_{w'} \simeq A^{k-1} \times \{0\} \subset A^k \simeq M_{w'}$, such that $M_{w'} \longrightarrow X_w$ is an isomorphism onto an open subset of $X_{w'}$, $N_{w'}$ maps onto $C(w')$, and $M_{w'} - N_{w'}$ into $C(w)$. This implies that $X_{w'}$ is non-singular in codimension 1. That $X_w$ is actually normal for all finite dimensional groups has recently been proved ([Se], [J]).

Here are a few examples illustrating the singularities of $X_w$: Fix $\Lambda$ such that $\langle \Lambda, \alpha_i^\vee \rangle = 1$ for all i, identify G/B with its image in $\mathbb{P}(L)$, and write $X_w^0$ for $X_w \cap C\_(1)$ (this is an open affine neighborhood of the point C(1), the worst potential singularity):

a) $w = {}_i r_j r_i$: then $X_w^0 = C(\mathbb{P}^1 \subset \mathbb{P}^{-a_{ij}}) \times A^1$, where $C(\mathbb{P}^1 \subset \mathbb{P}^n)$ is the cone in $A^{n+1}$ of the highest-weight orbit of the (n+1) dimensional irreducible representation of $SL_2$ (i.e. the cone over the projective line embedded in $\mathbb{P}^n$ by the complete linear system).

So in this case $X_w^0$ is the maximal spectrum of $K[X, Y_0, \ldots, Y_{-a_{ij}}]/\langle Y_m Y_n - Y_{m'} Y_{n'}$ for any $m+n = m'+n' \rangle$. In this identification,

the subset $X^0_{r_i r_j}$ of $X_w^0$ corresponds to $Y_1 = \ldots = Y_{-a_{ij}} = 0$,

the subset $X^0_{r_j r_i}$ of $X_w^0$ corresponds to $Y_m = X^m Y_0$ (the twisted line)

the subset $X^0_{r_i}$ of $X_w^0$ corresponds to $Y_0 = \ldots = Y_{-a_{ij}} = 0$

the subset $X^0_{r_j}$ of $X_w^0$ corresponds to $Y_1 = \ldots = Y_{-a_{ij}} = X = 0$.

b) $w = r_i r_j r_i r_j$: Let $R_{i,j}$ be the ring $K[X_1, \ldots, X_{-a_{ij}}, Y_1, \ldots, Y_{-a_{ji}}, Z_0, Z_1]/I$, where I is the ideal generated by: $X_m X_{m'} - X_{m''} X_{m'''}$ whenever $m+m' = m''+m'''$, $Y_n Y_{n'} - Y_{n''} Y_{n'''}$ whenever $n+n' = n''+n'''$, $X_m Y_n - X_1^n X_{m-n+1}$ if $m \geqslant n$, $X_m Y_n - Y_1^m Y_{n-m+1}$ if $m \leqslant n$, and $X_1 - Y_1$.

If either $a_{ij} = -1$ (and $a_{ji}$ is arbitrary), or $a_{ij} = -2$ and

$a_{ji} \geqslant -2$, then $X_w^0$ is the maximal spectrum of $R_{i,j}$. It is conjectured that this is the case without restriction on the Cartan integers. In any case,

the subset $X^0_{r_j r_i r_j}$ corresponds to $X_m = Z_1^{m-1} X_1$

the subset $X^0_{r_i r_j r_i}$ corresponds to $Y_n = Z_0^{n-1} Y_1$

the subset $X^0_{r_i r_j}$ corresponds to $X_m = Y_n = 0$, all m,n

the subset $X^0_{r_j r_i}$ corresponds to $\begin{cases} X_m = Z_0 Z_1^m \\ Y_n = Z_1 Z_0^n \end{cases}$

the subset $X^0_{r_i}$ corresponds to $X_m = Y_n = Z_0 = 0$

the subset $X^0_{r_j}$ corresponds to $X_m = Y_n = Z_1 = 0$

c) $w = (r_{j_1} \ldots r_{j_s}) r_i (r_{j_{s+1}} \ldots r_{j_{s+t}}) r_i (r_{j_{s+t+1}} \ldots r_{j_{s+t+u}})$, with $r_{j_2} \neq r_{j_\beta}$ if $\alpha \neq \beta$, and $r_{j_\alpha} \neq r_i$. Then $X_w^0 \cong A^{s+u} \times X^0_{r_i r_{j_{s+1}} \cdots r_{j_{s+t}} r_i}$, and $X^0_{r_i r_{j_{s+1}} \cdots r_{j_{s+t}} r_i}$ is the maximal spectrum of $K[X;\ Y_0^{(1)}, \ldots, Y^{(1)}_{-a_{ij_{s+1}}};\ Y_0^{(2)}, \ldots, Y^{(2)}_{-a_{ij_{s+2}}};\ \ldots;\ Y_0^{(t)}, \ldots, Y^{(t)}_{-a_{ij_{s+t}}}]$ modulo the ideal generated by $Y_m^{(1)} Y_n^{(1')} - Y_{m'}^{(1)} Y_{n'}^{(1')}$ whenever $m+n = m'+n'$. For a better description, assume for simplicity that $-a_{ij_{s+1}} = -a_{ij_{s+2}} = \ldots = -a_{ij_{s+t}} = q-1$, let $E_{t,q}$ be the space of linear maps $A^t \longrightarrow A^q$, $E'_{t,q}$ the highest weight orbit of $GL_q \times GL_t$ on $E_{t,q}$, $C'_{t,q} = \{l \in E_{t,q}$ such that $\mathrm{Image}(l) \subset C(\mathbb{P}^1 \subset \mathbb{P}^{q-1})\}$ (in the notation of example a)). Then $K[Y_0^{(1)}, \ldots, Y_{q-1}^{(1)};\ Y_0^{(2)}, \ldots, Y_{q-1}^{(2)};\ \ldots;\ Y_0^{(t)}, \ldots, Y_{q-1}^{(t)}] / \langle Y_m^{(1)} Y_n^{(1')} - Y_{m'}^{(1)} Y_{n'}^{(1')} \rangle$ is the ring of regular functions on $E'_{t,q} \cap C'_{t,q}$.

d) $w = r_i r_j r_k r_j r_i$: then $X^0_w = X^0_{r_i r_k r_i} \times_{A^2} X^0_{r_i r_j r_i} \times_{A^2} X^0_{r_j r_k r_j}$, where the map $A^1 \times C(\mathbb{P}^1 \subset \mathbb{P}^n) \longrightarrow A^2$ is given by $K[X,Y_0,\ldots,Y_n] \longleftarrow K[X,Y_0]$ in the notation of example a).

It is not hard to check that all the rings described above are integrally closed.

## References


[D] V. Deodhar, On some geometric aspects of Bruhat ordering I; Indiana U., 1982.

[J] A. Joseph, On the Demazure character formula; Preprint.

[KL] D. Kazhdan and G. Lusztig, Schubert varieties and Poincaré duality; Proc. Symp. Pure Math. 36, AMS, 1980.

[PK] D. Peterson and V. Kac, Infinite flag varieties and conjugacy theorems; Proc. Nat. Acad. Sc., 1983.

[Se] C. S. Seshadri, Normality of Schubert varieties; Preprint.

[St] R. Steinberg, Lectures on Chevalley groups; Yale U., 1967.

[T] J. Tits, Théorie des groupes; Res. Cours, Coll. Fr., 1981.

# CONSTRUCTING GROUPS ASSOCIATED TO INFINITE-DIMENSIONAL LIE ALGEBRAS

By

Victor G. Kac

Department of Mathematics, MIT, Cambridge MA 02139

In these notes a representation theoretical approach to the construction of groups associated to (possibly infinite-dimensional) "integrable" Lie algebras is discussed. In the first part a general framework is outlined; here most of the discussion consists of definitions, examples and open problems. Deep results are available only in the case of groups associated to Kac-Moody algebras, which are discussed in the second part; it is based on joint work with Dale Peterson [18], [19], [20], [21], [22], [26]. Extension of these results to other classes of groups, like the group of biregular automorphisms of an affine space, would provide a solution to some very difficult open problems of algebraic geometry.

Throughout the paper the base field is the field $\mathbb{C}$ of complex numbers.

Partially supported by NSF grant MCS-8203739.

# CHAPTER 1. Integrable Lie Algebras and Associated Groups

§1.1 Let V be a (possibly infinite-dimensional) vector space (over $\mathbb{C}$), and let A be an endomorphism of V. A is called <u>locally finite</u> if every $v \in V$ lies in a finite-dimensional A-invariant subspace of V (or, equivalently, $\{A^n(v) \mid n = 0, 1, \dots\}$ are linearly dependent for every $v \in V$). A is called <u>locally nilpotent</u> if for every $v \in V$ there exists $n > 0$ such that $A^n(v) = 0$. A is called <u>semisimple</u> if V admits a basis of eigenvectors for A. Obviously, locally nilpotent and semisimple elements are locally finite.

A locally finite endomorphism A always admits a <u>Jordan decomposition</u>, i.e. A can be represented in the form $A = A_s + A_n$, where $A_s$ is semisimple and $A_n$ is locally nilpotent and $A_sA_n = A_nA_s$; such a decomposition is unique. This follows from the usual Jordan decomposition in the finite-dimensional case.

If A is a locally finite endomorphism of V, we can form the corresponding 1-parameter group of automorphisms of V:

$$\exp tA = \sum_{n\geq 0} \frac{t^n}{n!} A^n, \qquad t \in \mathbb{C}.$$

Let A be a semisimple endomorphism of V so that $A(e_i) = \lambda_i e_i$ for some basis $\{e_i\}$ of V. An endomorphism A' of V, such that $A'(e_i) = \lambda'_i e_i$ for all i, is called a <u>replica</u> of A if a relation $\sum_i n_i\lambda_i = 0$, where $n_i \in \mathbb{Z}$ and all but a finite number of them are 0, implies that $\sum_i n_i\lambda'_i = 0$.

Let A be an arbitrary locally finite endomorphism of V and $A = A_s + A_n$ its Jordan decomposition. All replicas of $A_s$ and the endomorphism $A_n$ are called <u>replicas</u> of A. The linear span of all replicas of A is called the <u>algebraic hull</u> of A.

**Lemma.** Let A be a locally finite endomorphism of V and let A' be a replica of A. Let $U_1 \subset U_2$ be two subspaces of V such that $A(U_1) \subset U_2$. Then $A'(U_1) \subset U_2$.

Indeed, let $v \in U_1$ and let U' be a finite-dimensional

A-invariant subspace containing v. Put $U_i' = U_i \cap U'$ $(i = 1, 2)$. Then $A(U_1') \subset U_2'$. By [29, p. 6-04], A' restricted to U' is a polynomial with zero constant term in A restricted to U'. Hence $A'(v) \in U_2' \subset U_2$.

**Examples.**

(a) Let R be a commutative associative algebra (over $\mathbb{C}$) with no zero divisors. Let $(r_{ij})_{i,j=1}^{n}$ be a matrix over R and let $\det(\lambda\delta_{ij} - r_{ij}) = \lambda^n + a_1\lambda^{n-1} + \dots + a_n$, $a_i \in R$, be its characteristic polynomial. The matrix $(r_{ij})$ acts on the free R-module $R^n$ of n-columns over R by left multiplication; regarding $R^n$ as an infinite-dimensional vector space V over $\mathbb{C}$, we get an endomorphism of V, which we denote by r. Then r is locally finite if and only if $a_i \in \mathbb{C}$, $i = 1, \dots, n$, and is locally nilpotent if and only if $a_i = 0$, $i = 1, \dots, n$. Indeed, if all the $a_i \in \mathbb{C}$, then, for $v \in V$, we have $r^n(v) \in \sum_{j=0}^{n-1} \mathbb{C}r^j(v)$. Conversely, if r is locally finite, it has at most n eigenvalues $\lambda_1, \dots, \lambda_s$ (which are the complex roots of $\det(\delta_{ij}\lambda - r_{ij})$), and for any r-invariant subspace U of V, all the eigenvalues of r on V/U are from the set $\{\lambda_1, \dots, \lambda_s\}$. Taking $U = I^n \subset R^n = V$, where I is a maximal ideal of R (we may assume R to be finitely generated) we deduce that all the $a_i \in \mathbb{C}$.

(b) Let D be a derivation of an algebra R generated by some elements $a_1, a_2, \dots$. Then D is locally finite (resp. locally nilpotent) if and only if $\dim \sum_i \mathbb{C}D^i(a_j) < \infty$ (resp. $D^{n_j}(a_j) = 0$ for some $n_j$) for all j. This follows from the Leibnitz rule.

Furthermore, if D is a locally finite derivation of R, then all replicas of D are derivations of R as well. Indeed, let $R = \bigoplus_{\lambda \in \mathbb{C}} R_\lambda$ be the generalized eigenspace decomposition with respect to D. Then $R_\lambda R_\mu \subset R_{\lambda+\mu}$, which implies that $D_s$, and hence $D_n$, is a derivation of R. It is also clear that all replicas of $D_s$ are derivations of R as well.

(c) Let $A = \sum_{i=1}^{n} P_i \frac{\partial}{\partial x_i}$ be a linear differential operator with polynomial coefficients acting on the vector space $V = \mathbb{C}[x_1, \ldots, x_n]$. In the following two cases, A is evidently locally finite: $\deg P_i \leq 1$ for all i (affine differential operator); $\frac{\partial P_i}{\partial x_j} = 0$ for $j \geq i$ (triangular differential operator). This follows from (b).

§1.2 Let $\mathfrak{g}$ be a (possibly infinite-dimensional) Lie algebra (over $\mathbb{C}$) and let V be a $\mathfrak{g}$-module with action $\pi$. Let $\mathfrak{r}$ denote the center of $\mathfrak{g}$.

An element $x \in \mathfrak{g}$ is called $\pi$-locally finite if $\pi(x)$ is a locally finite endomorphism of the vector space V.

We denote by $F_{\mathfrak{g}}$ the set of all ad-locally finite elements of $\mathfrak{g}$, and by $\mathfrak{g}_{fin}$ the subalgebra of $\mathfrak{g}$ generated by $F_{\mathfrak{g}}$. The Lie algebra $\mathfrak{g}$ is called integrable if $\mathfrak{g} = \mathfrak{g}_{fin}$. Denote by $F_{\mathfrak{g},\pi}$ the set of $\pi$-locally finite elements of $F_{\mathfrak{g}}$.

**Lemma.**

(a) The subalgebra of $\mathfrak{g}$ generated by $F_{\mathfrak{g},\pi}$ is the linear span (over $\mathbb{C}$) of $F_{\mathfrak{g},\pi}$. In particular, $\mathfrak{g}_{fin}$ is spanned by $F_{\mathfrak{g}}$.

(b) Let $\dim \mathfrak{g} < \infty$ and let $\mathfrak{g}$ be generated by $F_{\mathfrak{g},\pi}$. Then V is a locally finite $\mathfrak{g}$-module (i.e. any $v \in V$ is contained in a finite-dimensional $\mathfrak{g}$-submodule).

**Proof.** Let $\mathfrak{g}_\pi$ denote the $\mathbb{C}$-span of $F_{\mathfrak{g},\pi}$. Let $x \in F_{\mathfrak{g},\pi}$. Using that

$$(1) \qquad \pi((\exp \operatorname{ad} x)y) = (\exp \pi(x))\pi(y)(\exp -\pi(x))$$

we deduce that $\mathfrak{g}_\pi$ is invariant with respect to $\exp t(\operatorname{ad} x)$, $t \in \mathbb{C}$. Since

$$\lim_{t\to 0} ((\exp t\,\mathrm{ad}x)(y) - y)/t = [x,y],$$

it follows that $[x,g_\pi] \subset g_\pi$, proving (a). (b) follows from (a) by the PBW theorem.

<u>**Conjecture**</u>. Let $g = \bigoplus_{j\in\mathbb{Z}} g_j$ be an integrable $\mathbb{Z}$-graded Lie algebra (we assume that $\dim g_j < \infty$), which has no nontrivial graded ideals. Then $g$ is isomorphic either to a "centreless" Kac-Moody algebra $g'(A)/\mathfrak{c}$ (see §2.2 for the definition) or to a Lie algebra of the Cartan series $S_n$ or $H_n$ (see e.g. [12] for their definition).

<u>**Problem**</u>. Characterize the <u>"general" integrable Lie algebra</u> $gl(V)_{fin}$.

The $g$-module $(V,\pi)$ is called <u>integrable</u> if $F_{g,\pi} = F_g$. Of course the $g$-module $(g,\mathrm{ad})$ is integrable. In general it is difficult to check that a module is integrable since there is not much information about the set $F_g$. The matter would simplify considerably if one can prove the following conjecture (which is a strengthening of Lemma 1.2(a)):

<u>**Conjecture**</u>. If $F_{g,\pi}$ generates the Lie algebra $g_{fin}$, then $V$ is an integrable $g$-module.

<u>**Problem**</u>. Does any Lie algebra admit a faithful integrable module?

Put $V_{fin} = \{v \in V \mid$ for every $x \in F_g$ there exists a finite-dimensional $\pi(x)$-invariant subspace of $V$ containing $v\}$. It follows from the Leibniz rule that $V_{fin}$ is a $g$-submodule, which is obviously integrable. The functors $g \longmapsto g_{fin}$ and $V \longmapsto V_{fin}$ have many nice properties.

## §1.3 <u>Examples</u>.

(a) Let $R$ be a complex commutative associative algebra with unity and let $g$ be a complex finite-dimensional semisimple Lie algebra. Then the Lie algebra $g_R := R \otimes_{\mathbb{C}} g$ is an integrable Lie algebra (over $\mathbb{C}$). To show this, take a root space decomposition $\mathfrak{g} = \mathfrak{h} \oplus \sum_\alpha \mathbb{C}e_\alpha$. Then all elements of the form $r \otimes e_\alpha$ are ad-locally finite (even nilpotent), and they generate the Lie algebra $\mathfrak{g}_R$.

Experience shows that the universal central extension $\tilde{\mathfrak{g}}_R \xrightarrow{d\tau} \mathfrak{g}_R$ of $\mathfrak{g}_R$ has a much more interesting representation theory than $\mathfrak{g}_R$ itself. This central extension is constructed as follows [23]:

$$0 \longrightarrow \Omega_R^1/dR \longrightarrow \tilde{\mathfrak{g}}_R := \mathfrak{g}_R \oplus (\Omega_R^1/dR) \xrightarrow{d\tau} \mathfrak{g}_R \longrightarrow 0,$$

where $\Omega_R^1$ is the space of all formal differentials (i.e. expressions of the form $fdg$, where $f,g \in R$, with relation $d(fg) = fdg + gdf$), and the bracket on $\tilde{\mathfrak{g}}_R$ is defined by

$$[r_1 \otimes g_1, r_2 \otimes g_2] = r_1r_2 \otimes [g_1,g_2] + (g_1 | g_2) r_2 dr_1 \pmod{dR}$$

where $(\cdot | \cdot)$ is the Killing form on $\mathfrak{g}$. Of course, $\tilde{\mathfrak{g}}_R$ is an integrable Lie algebra as well.

If $V$ is a finite-dimensional $\mathfrak{g}$-module, then one easily shows that $V_R := R \otimes_{\mathbb{C}} V$ is an integrable $\mathfrak{g}_R$-module. The following special case of the problem stated above is open, however: For which $R$ there exists a faithful integrable $\tilde{\mathfrak{g}}_R$-module?

<u>**Remark**</u>. Put $\tilde{\mathfrak{g}}_R = (R \otimes_{\mathbb{C}} \mathfrak{g}) \oplus \Omega_R^1 \oplus \mathrm{Der}\, R$ and define a bracket by: $[r_1 \otimes g_1, r_2 \otimes g_2] = r_1r_2 \otimes [g_1,g_2] + (g_1g_2) r_2 dr_1$; $[D, r \otimes g] = D(r) \otimes g$ for $D \in \mathrm{Der}\, R$; $[\tilde{\mathfrak{g}}_R, \tilde{\Omega}_R^1] = 0$. Then $\tilde{\mathfrak{g}}_R$ is a Lie algebra mod $dR \subset \Omega_R^1$. Define on $\tilde{\mathfrak{g}}_R$ an $R$-valued symmetric bilinear form $(\cdot | \cdot)_R$ by: $(r_1 \otimes g_1 | r_2 \otimes g_2) = r_1r_2(g_1 | g_2)$; $(fdg | D) = fD(g)$ for $D \in \mathrm{Der}\, R$; $(\Omega_R^1 + \mathrm{Der}\, R \mid R \otimes_{\mathbb{C}} \mathfrak{g}) = 0$; $(\Omega_R^1 | \Omega_R^1) = 0$; $(\mathrm{Der}\, R | \mathrm{Der}\, R) = 0$. It is non-degenerate, and invariant, i.e. $([a,b] | c)_R = (a | [b,c])_R$. Let $F$ be a linear function on $R$; put $(\mathrm{Der}\, R)_F = \{D \in \mathrm{Der}\, R \mid F(D(\varphi)) = 0 \text{ for all } \varphi \in R\}$. Then $\tilde{\mathfrak{g}}_{R,F} := (R \otimes_{\mathbb{C}} \mathfrak{g}) \oplus (\Omega_R^1/dR) \oplus (\mathrm{Der}\, R)_F$ is a Lie algebra, and $(a | b) := F((a | b)_R)$ is an invariant bilinear form on it. For example, let $M$ be a compact manifold with a volume form $\Omega$, let $R$ be the algebra of complex valued $C^\infty$-functions on $R$ and let $F(\varphi) = \int_M \varphi\Omega$. Then DerR is the

Lie algebra of vector fields on M, $(\mathrm{Der}\ R)_F$ is the subalgebra of vector fields with zero divergence, and the bilinear form $(\cdot|\cdot)$ on $\tilde{g}_{R,F}$ is non-degenerate.

(b) Let $W_n$ be the Lie algebra of all linear differential operators with polynomial coefficients in n indeterminates $x_1, \dots, x_n$. It carries a $\mathbb{Z}$-gradation $W_n = \bigoplus_{j \geq -1} (W_n)_j$, where $(W_n)_j = \{\Sigma P_i \frac{\partial}{\partial x_i} \in W_n \mid P_i$ are homogeneous of degree $j + 1\}$; then $W_n^0 = \bigoplus_{j \geq 0} (W_n)_j$ is a maximal subalgebra of $W_n$. Put $CS_n = \{D \in W_n \mid \mathrm{div}\ D \in \mathbb{C}\}$. This is a subalgebra of $W_n$ (recall that $\mathrm{div}\ \Sigma P_i \frac{\partial}{\partial x_i} = \Sigma \frac{\partial P_i}{\partial x_i}$ and that $\mathrm{div}\ [D_1,D_2] = D_1 \cdot \mathrm{div}\ D_2 - D_2 \cdot \mathrm{div}\ D_1$). Put $(CS_n)_j = CS_n \cap W_n$; we have a $\mathbb{Z}$-gradation $CS_n = \bigoplus_{j \geq -1} (CS_n)_j$, where

$$(CS_n)_{-1} = (W_n)_{-1} = \Sigma\mathbb{C} \frac{\partial}{\partial x_i}, \quad (CS_n)_0 = (W_n)_0 \simeq gl_n(\mathbb{C}).$$

Furthermore, the $(CS_n)_0$-module $(CS_n)_j$ is irreducible if $j \neq 0$ with a highest weight vector $x_1^{j+1} \frac{\partial}{\partial x_n}$, if $n > 1$ (see e.g. [12]). Since the elements of $(CS_n)_0$ and the elements $x_1^j \frac{\partial}{\partial x_n}$ are locally finite if $n > 1$, we obtain that $CS_n \subset (W_n)_{fin}$ and that $CS_n$ is an integrable Lie algebra.

It is clear that $(W_1)_{fin} = CS_1 = \mathbb{C} \frac{\partial}{\partial x} + \mathbb{C}x \frac{\partial}{\partial x}$. Let me show that $(W_n)_{fin} = CS_n$ for any n.

Denote by $\pi$ the action of $W_n$ on the vector space $\mathbb{C}[x_1, \dots, x_n]$. Then $D \in W_n$ is ad-locally finite if and only if it is $\pi$-locally finite. Indeed, if D is ad-locally finite, then, applying the Leibnitz rule to $D^N(P \frac{\partial}{\partial x_1})$, we see that D is $\pi$-locally finite. Conversely, if D is $\pi$-locally finite, then exp tD is an automorphism of the polynomial algebra $\mathbb{C}[x_1, \dots, x_n]$ such that $(\exp tD)x_i = P_i(x_1, \dots, x_n, t)$, where the degrees of the $P_i$ in the $x_j$ are bounded uniformly for all t. Since the change of indeterminates $\varphi_t$: $x_i \longmapsto P_i$ is invertible, denoting the inverse by $x_i \longmapsto \bar{P}_i$, we get $(\exp tD) \frac{\partial}{\partial x_i} = \sum_j \frac{\partial \bar{P}_j}{\partial x_i} \frac{\partial}{\partial x_j}$ and the degrees of the $\bar{P}_i$ in the indeterminates $x_j$ are bounded uniformly for all t. It follows that $\sum_j \mathbb{C}(\mathrm{ad} D)^j \frac{\partial}{\partial x_i}$ is finite-dimensional. Hence, D is ad-locally finite. In other words, $\mathbb{C}[x_1, \ldots, x_n]$ is an integrable $W_n$- (and $CS_n$-) module.

Furthermore, if D is ad-locally finite, then it is $\pi$-locally finite, and we have the change of indeterminates $\varphi_t$. But its Jacobian $J(\varphi_t) := \det \left(\frac{\partial P_i}{\partial x_j}\right)$ is an invertible polynomial, hence $J(\varphi_t) \in \mathbb{C}^{\times}$. Therefore, div $D \in \mathbb{C}$. Thus, $(W_n)_{fin} = CS_n$.

**Problem**. Is it true that any ad-semisimple element of $CS_n$ is conjugate (by a change of indeterminates) to an element of the form $\sum_i \lambda_i x_i \frac{\partial}{\partial x_i}$, where $\lambda_i \in \mathbb{C}$?

This problem is equivalent to the well-known problem, whether a regular action of $\mathbb{C}^{\times}$ on $\mathbb{C}^n$ is biregularly equivalent to a linear action.

As we shall see in Chapter 2, the conjugacy problem is intimately related to the problem of existence of non-trivial closed orbits in the projectivized space. Unfortunately, there is no such orbits for the action of Aut $\mathbb{C}^n$ on $\mathbb{C}[x_1, \ldots, x_n]$.

**Problem**. Compute the closure of the orbit of $x_1$ in $\mathbb{C}[x_1, \ldots, x_n]$ under the action of Aut $\mathbb{C}^n$ (a set is closed if its intersection with any finite-dimensional subspace U is closed in U).

Finally, let $\varphi: x_i \longrightarrow P_i$ be a polynomial change of indeterminates with $J(\varphi) \in \mathbb{C}^{\times}$; we can assume that $P_i(0) = 0$. Then we have the induced (non-zero) homomorphism $\varphi: W_n \longrightarrow W_n$ which maps $W_n^0$ into itself. Conversely, any non-zero homomorphism $\varphi$: $W_n \longrightarrow W_n$ that maps $W_n^0$ into itself induces an isomorphism $\hat{\varphi}$: $\hat{W}_n \longrightarrow \hat{W}_n$ of the formal completion and hence is given by a formal change of indeterminates $x_i \longrightarrow P_i$ with $P_i(0) = 0$ [28]. Since $\hat{\varphi}\left[\frac{\partial}{\partial x_i}\right] \in W_n$, we obtain that the inverse change of indeterminates is polynomial. Since $\hat{\varphi}\left[x_i^k \frac{\partial}{\partial x_i}\right] \in W_n$, the $P_i$ are polynomials and $J(\varphi) \in \mathbb{C}^{\times}$.

Thus, the Jacobian conjecture is equivalent to the question whether a non-zero homomorphism $W_n \longrightarrow W_n$ which maps $W_n^0$ into itself is an isomorphism (one can replace $W_n$ by $CS_n$).

§1.4 Let V be a faithful integrable $\mathfrak{g}$-module, so that $\mathfrak{g} \subset \mathfrak{gl}(V)$. If all replicas of any element of $F_{\mathfrak{g}}$ lie in $\mathfrak{g}$, the linear Lie algebra $\mathfrak{g}$ is called <u>algebraic</u>. An integrable $\mathfrak{g}$-module $(U,\varphi)$ over an algebraic Lie algebra $\mathfrak{g}$ is called <u>rational</u> if for any $x \in F_{\mathfrak{g}}$ one has $\varphi(\bar{x}) = \overline{\varphi(x)}$, where $\bar{x}$ denotes the algebraic hull of x, and $\varphi(x_s) = \varphi(x)_s$, where $x = x_s + x_n$ is the Jordan decomposition of $\pi(x)$. If $\mathfrak{g} \subset \mathfrak{gl}(V)$ is not algebraic, we let $\bar{\mathfrak{g}}$ be the subalgebra of $\mathfrak{gl}(V)$ generated by algebraic hulls of all $x \in F_{\mathfrak{g}}$. Then $\bar{\mathfrak{g}} \subset \mathfrak{gl}(V)$ is an algebraic Lie algebra called the <u>algebraic hull</u> of $\mathfrak{g}$.

Let $\mathfrak{g} \subset \mathfrak{gl}(V)$ be an algebraic Lie algebra. Then its adjoint representation is rational. Indeed, let $x \in F_{\mathfrak{g}}$ and let $\pi(x) = A_s + A_n$ be the Jordan decomposition. Let $V = \bigoplus_{\lambda \in \Lambda} V_\lambda$ be the eigenspace decomposition for $A_s$; it is $A_n$-invariant. Since $\pi(\mathfrak{g}) \subset \operatorname{End} V = \prod_{\lambda,\mu \in \Lambda} \operatorname{Hom}(V_\lambda, V_\mu)$ and $x \in F_{\mathfrak{g}}$, we deduce that $\operatorname{ad} x = \operatorname{ad} A_s + \operatorname{ad} A_n$ is the Jordan decomposition of ad x, and that all the eigenvalues of ad x and ad $A_s$ are $\lambda - \mu$, where $\lambda, \mu \in \Lambda$.

Note that the definition of an algebraic Lie algebra, the Jordan decomposition, etc., are independent of the choice of the rational $\mathfrak{g}$-module. Thus, if the center of $\mathfrak{g}$ is trivial, we can start with its adjoint representation and talk about the Jordan decomposition of $x \in F_{\mathfrak{g}}$.

It follows from Lemma 1.1, that if A is a locally finite endomorphism of V and A' is a replica of A, and if $U_1 \subset U_2$ are two subspaces of $\mathfrak{gl}(V)$, such that $(\text{ad } A)U_1 \subset U_2$, then $(\text{ad } A')U_1 \subset U_2$. As in [29, p. 6-06], one deduces the following easy facts:

(a) Every ideal of $\mathfrak{g}$ remains an ideal in $\bar{\mathfrak{g}}$.

(b) Center of $\mathfrak{g}$ lies in the center of $\bar{\mathfrak{g}}$.

(c) $[\mathfrak{g},\mathfrak{g}] = [\bar{\mathfrak{g}},\bar{\mathfrak{g}}]$, $\mathfrak{g}$ is an ideal in $\bar{\mathfrak{g}}$ and $\bar{\mathfrak{g}}/\mathfrak{g}$ is abelian.

(d) If $\mathfrak{a}$ is an ideal of $\mathfrak{g}$, then $[\bar{\mathfrak{g}},\bar{\mathfrak{a}}] \subset \mathfrak{g}$.

**Problem**. Is it true that $[\mathfrak{g},\mathfrak{g}]$ is an algebraic Lie algebra? This is true if $\dim \mathfrak{g} < \infty$. The proof of this and other deeper facts of the theory of finite-dimensional algebraic groups uses the Noetherian property of finite-dimensional algebraic varieties (see e.g. [2]).

The Lie algebra $\mathfrak{g}_R$ acting on $V_R$ (see Example 1.3(a)), where V is a faithful (finite-dimensional) $\mathfrak{g}$-module, is an algebraic Lie algebra, and all $\mathfrak{g}_R$-modules $U_R$, where U is a finite-dimensional $\mathfrak{g}$-module, are rational. To see this, consider $\mathfrak{g}_{\text{Fr } R}$, where Fr R is the field of fractions of R, and use the uniqueness of the Jordan decomposition.

The Lie algebra Der R of all derivatives of an algebra R is an algebraic linear Lie algebra. This follows from Example 1.1(b). In particular, $W_n$ is an algebraic Lie algebra. Since $(W_n)_{fin} = CS_n$, it follows that $CS_n$ is an algebraic Lie algebra as well.

§1.5 Let $\mathfrak{g}$ be an integrable Lie algebra. We associate to $\mathfrak{g}$ a group G as follows. Let $G^*$ be a free group on the set $F_{\mathfrak{g}}$. Given an integrable $\mathfrak{g}$-module $(V,d\pi)$, we define a $G^*$-module $(V,\tilde{\pi})$ by

$$\tilde{\pi}(x) = \exp d\pi(x) := \sum_{n\geqslant 0} (d\pi(x))^n/n!, \qquad x \in F_{\mathfrak{g}}.$$

We put $G = G^*/\cap \text{Ker } \tilde{\pi}$, where the intersection is taken over all integrable $\mathfrak{g}$-modules $d\pi$. Thus, the $G^*$-module $(V,\tilde{\pi})$ is naturally a G-module $(V,\pi)$, the integrable $\mathfrak{g}$-module $(V,d\pi)$ being its "differential". We call G the <u>group associated to the Lie algebra</u> $\mathfrak{g}$ and $(V,\pi)$ the G-module <u>associated</u> to the integrable $\mathfrak{g}$-module.

Given an element $x \in F_{\mathfrak{g}}$, we denote its image in G under the canonical homomorphism $G^* \longrightarrow G$ by exp x. Thus, we have by definition:

$$\pi(\exp x) = \exp d\pi(x), \qquad x \in F_{\mathfrak{g}}$$

for an integrable $\mathfrak{g}$-module $(V,d\pi)$. Note also that $\{\exp tx \mid t \in \mathbb{C}\}$ is a 1-parameter subgroup of G.

Put $F_G = \{\exp x \mid x \in F_{\mathfrak{g}}\} \subset G$. A G-module $(V,\pi)$ is called <u>differentiable</u> if all elements of $F_G$ act locally finitely on V and exp tx restricted to any invariant finite-dimensional subspace is analytic in t ($x \in F_{\mathfrak{g}}$). This definition is justified by the following:

<u>**Conjecture**</u>. Let $(V,\pi)$ be a differentiable G-module. Then there exists a unique action $d\pi$ of $\mathfrak{g}$ on V such that $\pi(\exp x) = \exp d\pi(x)$ for all $x \in F_{\mathfrak{g}}$. $(V,d\pi)$ is an integrable $\mathfrak{g}$-module.

Uniqueness follows from Lemma 1.2(a). To show the existence put

$$d\pi(x) := \frac{d}{dt}\, \pi\,(\exp tx)\Big|_{t=0} \text{ for } x \in F_{\mathfrak{g}}.$$

The difficulty is to show that $d\pi$ is linear. This granted, one would have by (1):

$$\pi(\exp tx)\; d\pi(y)\; \pi(\exp -tx) = d\pi(\exp(\operatorname{ad} tx)y), \text{ for } x \in F_{\mathfrak{g}},$$

and therefore,

$$(1 + td\pi(x) + o(t))\; d\pi(y)\; (1 - td\pi(x) + o(t)) =$$

$$= d\pi(y) + td\pi([x,y]) + o(t),$$

which would yield $[d\pi(x),d\pi(y)] = d\pi[x,y])$.

Of course, the G-module $(V,\pi)$ associated to an integrable g-module $(V,d\pi)$ is differentiable. Thus, we would have an invertible functor between the categories of integrable g-modules and differentiable G-modules.

A homomorphism $d\varphi$: $g_1 \longrightarrow g$ of integrable Lie algebras is called <u>integrable</u> if $d\varphi(F_{g_1}) \subset F_g$; then $d\varphi(g_1)$ is called an <u>integrable subalgebra</u> of g. Given an integrable homomorphism $d\varphi$ of Lie algebras, we have a canonically defined homomorphism of the associated groups $\varphi$: $G_1 \longrightarrow G$, so that $d(\pi|_{G_1}) = (d\pi)|_{g_1}$. The subgroup $\varphi(G_1)$ of G is called the subgroup corresponding to the integrable subalgebra $\varphi(g_1)$ of g. It is generated by the exp x with $x \in \varphi(g_1) \cap F_g$.

Of course, any isomorphism of integrable Lie algebras is integrable and lifts to an isomorphism of the associated groups.

§1.6 Let $\mathfrak{r}_0$ denote the intersection of kernels of all integrable g-modules; then $\mathfrak{r}_0 \subset \mathfrak{r}$. Replacing g by $g/\mathfrak{r}_0$ we can (and will) assume that $\mathfrak{r}_0 = 0$. Let C denote the center of G. Associated to the integrable g-module (g,ad), we have the <u>adjoint</u> G-module (g,Ad). We denote by Ad G the image of the action of G on g and call it the adjoint group associated to g. Then we have the following exact sequence

$$(2) \qquad 1 \longrightarrow C \longrightarrow G \longrightarrow \mathrm{Ad}\, G \longrightarrow 1.$$

This is because, given a faithful integrable g-module $(V,d\pi)$, we can compute the adjoint G-module, thanks to formula (1), by

$$d\pi((\mathrm{Ad}\ g)x) = \pi(g)d\pi(x)\pi(g)^{-1}, \qquad g \in G.$$

Hence Ad $g = 1$ iff $\pi(g)$ commutes with $d\pi(g)$, iff $\pi(g)$ commutes with $d\pi(F_g)$, iff $\pi(g)$ commutes with $\pi(F_G)$, iff $\pi(g)$ commutes with $\pi(G)$. Choosing $(V,d\pi)$ such that $(V,\pi)$ is a faithful G-module, we get that Ker Ad = C. Note that we have shown at the same time that if the g-module $(V,d\pi)$ is faithful, then Ker $\pi \subset C$.

**Problem.** It is true that for a faithful differentiable G-module $(V,\pi)$ the corresponding (integrable) $\mathfrak{g}$-module $(V,d\pi)$ is also faithful?

§1.7 Let $(V,\pi)$ be a $\mathfrak{g}$-module and let $\sigma$ be an automorphism of the Lie algebra $\mathfrak{g}$. Then we have a new $\mathfrak{g}$-module $(V,\pi_\sigma)$ defined by

$$\pi_\sigma(g)v = \pi(\sigma \cdot g)v \qquad \text{for} \qquad g \in \mathfrak{g},\ v \in V.$$

Define the big adjoint group $\widetilde{\text{Ad}}\ G$ by: $\widetilde{\text{Ad}}\ G = \{\sigma \in \text{Aut}\ \mathfrak{g} \mid \pi_\sigma$ is isomorphic to $\pi$ for any integrable $\mathfrak{g}$-module $\pi\}$.

We have an obvious inclusion $\text{Ad}\ G \subset \widetilde{\text{Ad}}\ G$. It is also clear that Ad G is a normal subgroup of $\widetilde{\text{Ad}}\ G$. We define the group $K_1(\mathfrak{g})$ by the exact sequence

$$(3) \qquad 1 \longrightarrow \text{Ad}\ G \longrightarrow \widetilde{\text{Ad}}\ G \longrightarrow K_1(\mathfrak{g}) \longrightarrow 1.$$

We put $K_2(\mathfrak{g}) = C$ and define $K_0(\mathfrak{g})$ as the Grothendieck group of the category of all integrable $\mathfrak{g}$-modules. Note that $K_i(\mathfrak{g}_R)$ are closely related to the usual K-functors $K_i(R)$, $i = 1, 2$.

**Problem.** Compute the groups $K_i(\mathfrak{g}_R)$ and $K_i(\tilde{\mathfrak{g}}_R)$, $i = 0, 1, 2$.

**Conjecture.** $K_i(CS_n)$ for $i = 1, 2$ are trivial, i.e. the group associated to the Lie algebra $CS_n$ is Aut $\mathbb{C}^n$, the group of biregular automorphisms of $\mathbb{C}^n$.

This conjecture is closely related to the well-known question whether the group Aut $\mathbb{C}^k$ is generated by affine and triangular automorphisms (cf. [30]).

**Remarks.**

(a) Note that, given an integrable $\mathfrak{g}$-module $(V,\pi)$, and $\sigma \in \widetilde{\text{Ad}}\ \mathfrak{g}$, we have

$$\pi(\sigma\cdot g) = A_\sigma \pi(g) A_{\sigma^{-1}} \text{ for some } A_\sigma \in \text{Aut } V.$$

Denote by $G_V$ the group generated by all such $A_\sigma$. Then we, clearly, have the following exact sequence:

$$1 \longrightarrow \text{Aut}_{\mathfrak{g}} V \longrightarrow G_V \longrightarrow \widetilde{\text{Ad}}\, G \longrightarrow 1.$$

Assuming that $(V,\pi)$ is a <u>Schur</u> <u>module</u>, i.e. that $\text{Aut}_{\mathfrak{g}} V = \mathbb{C}^\times$, we get a central extension of $\widetilde{\text{Ad}}\, G$.

(b) Given an arbitrary Lie algebra $\mathfrak{g}$ and a category of $\mathfrak{g}$-modules, one can define the associated (adjoint) group as above.

§1.8 If $\mathfrak{g} \subset \mathfrak{gl}(U)$ is a linear algebraic integrable Lie algebra, one can make the definition of the associated group G more algebraic as follows. Let $F^n_{\mathfrak{g}}$ denote the set of all locally nilpotent elements and $F^s_{\mathfrak{g}}$ the set of all semisimple elements with integral eigenvalues; put $F^{alg}_{\mathfrak{g}} = F^n_{\mathfrak{g}} \cup F^s_{\mathfrak{g}}$. Let $G^*$ be the free product of a collection of copies of the additive group $\mathbb{C}$ indexed by $F^n_{\mathfrak{g}}$ and a collection of copies of the multiplicative group $\mathbb{C}^\times$ indexed by $F^s_{\mathfrak{g}}$. Given a rational integrable $\mathfrak{g}$-module $(V,d\pi)$, we define a $G^*$-module $(V,\tilde{\pi})$ by $\tilde{\pi}(t) = \exp d\pi(tx)$ if $t \in \mathbb{C}_x$, $\tilde{\pi}(t)v_i = t^{k_i}v_i$ if $t \in \mathbb{C}^\times_y$, where $v_i$ is an eigenvector of $d\pi(y)$ with eigenvalue $k_i$. Put $G = G^*/\cap \text{Ker } \tilde{\pi}$, where intersection is taken over all rational integrable $\mathfrak{g}$-modules $d\pi$. This definition of G coincides with the one in §1.5.

Note that for every $x \in F^n_{\mathfrak{g}}$ (resp. $x \in F^s_{\mathfrak{g}}$) we have a homomorphism $\psi_x\colon \mathbb{C} \longrightarrow G$ (resp. $\mathbb{C}^\times \longrightarrow G$). Given an ordered finite set $\bar{x} = \{x_1, \ldots, x_n\}$ of $F^{alg}_{\mathfrak{g}}$, we have a map $\psi_{\bar{x}}$ from the product of several copies of $\mathbb{C}$ and $\mathbb{C}^\times$ into G defined by

$$\psi_{\bar{x}}(t_1, \ldots, t_n) = \psi_{x_1}(t_1) \ldots \psi_{x_n}(t_n),$$

where $t_i \in \mathbb{C}$ if $x_i \in F^n_{\mathfrak{g}}$ and $t_i \in \mathbb{C}^\times$ if $x_i \in F^s_{\mathfrak{g}}$. In the general case, given an ordered set $\bar{x} = \{x_1, \ldots, x_n\}$ of elements of $F_{\mathfrak{g}}$, one defines $\psi'_{\bar{x}} : \mathbb{C}^n \longrightarrow G$ by $\psi'_{\bar{x}}(t_1, \ldots, t_n) = (\exp t_1x_1) \ldots$

$(\exp t_n x_n)$.

Now we are in a position to discuss how one introduces various structures on G.

A function f: G $\longrightarrow$ $\mathbb{C}$ is called <u>regular</u> if all the functions f $\circ$ $\psi_{\bar{x}}$ are polynomial. Denote by $\mathbb{C}[G]$ the algebra of all regular functions on G. (Similarly one defines an analytic function on G in the general setup.)

Let $(V, d\pi)$ be an integrable $\mathfrak{g}$-module. Given v $\in$ V and a linear function $v^* \in V^*$, we get a regular function $f_{v^*,v}$ on G, called a <u>matrix coefficient</u>:

$$f_{v^*,v}(g) = \langle\pi(g)v, v^*\rangle.$$

Since the direct sum and tensor product of two integrable $\mathfrak{g}$-modules is again an integrable $\mathfrak{g}$-module, the set of matrix coefficients forms a subalgebra $\mathbb{C}[G]_{m.c.}$ of the algebra $\mathbb{C}[G]$. Note that $\mathbb{C}[G]$ is a G $\times$ G-module under the action $\pi_{reg}$ defined by $(\pi_{reg}(g_1,g_2)f)(g) = f(g_1^{-1}gg_2)$, the subalgebra $\mathbb{C}[G]_{m.c.}$ being a submodule.

Note also that matrix coefficients separate the orbits of G. For if $g \neq 1$, there exists a differentiable G-module $(V,\pi)$ such that $\pi(g)v \neq v$ for some v $\in$ V; choosing $v^* \in V^*$ such that $\langle\pi(g)v,v^*\rangle = 0$ and $\langle v,v^*\rangle = 1$, we get that $f_{v^*,v}(g) = 0$ and $f_{v^*,v}(1) = 1$.

Let $\mathbb{C}[G]_{\bar{x}} = \{f \in \mathbb{C}[G] \mid f \text{ vanishes on the image of } \psi_{\bar{x}}\}$. Taking the $\mathbb{C}[G]_{\bar{x}}$ for a basis of neighborhoods of 0 makes $\mathbb{C}[G]$ into a Hausdorff complete topological ring. We have the canonical inclusion G $\longrightarrow$ Specm $\mathbb{C}[G]$ (= set of all closed ideals of codimension one).

<u>**Problem**</u>. Compute Specm $\mathbb{C}[G]$ and Specm $\mathbb{C}[G]_{m.c.}$.

Let $\mathcal{M}$ be a subcategory of the category of integrable $\mathfrak{g}$-modules, closed under taking finite direct sums and tensor products. We denote by $\mathbb{C}[G]^{\mathcal{M}}_{s.r.}$ the subalgebra of $\mathbb{C}[G]_{m.c.}$ consisting of functions $f_{v^*,v}$ with v $\in$ V, $v^* \in (V^*)_{fin}$, where V is a module from M, and call elements of $\mathbb{C}[G]^{\mathcal{M}}_{s.r.}$ <u>strongly</u> <u>regular</u> <u>functions</u> (with respect to the category $\mathcal{M}$).

Returning to the general setup, we introduce a topology on G as follows. Fix a subset $X \subset F_{\mathfrak{g}}$ such that the set $\{\exp tx \mid x \in X, t \in \mathbb{C}\}$ generates G. We call a subset U of G open if $(\psi'_{\bar{x}})^{-1}(U) \subset \mathbb{C}^n$ is open in the metric topology of $\mathbb{C}^n$ for all $\bar{x}$ such that the $x_i$ of $\bar{x}$ are from X. With this topology G is a Hausdorff topological space (since the matrix coefficients are continuous); it is obviously connected. The inversion map $g \longmapsto g^{-1}$ is obviously continuous. The multiplication map is not continuous in general, however (a counterexample will be given below). One can show (using Milnor's lemma) that if X is countable, then G is a topological group. (It should not be difficult to show that if $\mathfrak{g}$ is countably-dimensional, then G is a topological group for $X = F_{\mathfrak{g}}$.)

§1.9 Let M be a set and let $\mathbb{C}^M$ denote the direct sum of a collection of copies of $\mathbb{C}$ indexed by M. By metric (resp. Zariski) topology on $\mathbb{C}^M$ we mean the finest topology that induces metric (resp. Zariski) topology on finite-dimensional subspaces (i.e. $U \subset \mathbb{C}^M$ is open iff $U \cap V$ is open in V for any finite-dimensional subspace V of $\mathbb{C}^M$).

The additive group of $\mathbb{C}^M$ is the group associated to $\mathbb{C}^M$ viewed as a commutative Lie algebra. If the set M is countable, then the metric topology on $\mathbb{C}^M$ is equivalent to the box topology and hence $\mathbb{C}^M$ is a topological group. If M is uncountable, then $\mathbb{C}^M$ is not a topological group (this has been pointed out to me by D. Wigner).

Let $V = \mathbb{C}^M$ and let $x_i$, $i \in M$, denote the linear coordinate functions on V. The algebra $\mathbb{C}[V]$ of <u>regular functions</u> on V consists of $\mathbb{C}$-valued functions whose restriction to any finite-dimensional subspace is a polynomial function. The subalgebra $\mathbb{C}[V]_{s.r.}$ of $\mathbb{C}[V]$ of <u>strongly regular functions</u> consists of polynomials in a finite number of the $x_i$. These definitions agree with the ones in §1.8 for the additive group of V.

The set X of zeros of an ideal of $\mathbb{C}[V]$ in V is called an <u>affine variety</u>; the intersection of X with a finite-dimensional subspace is called a <u>finite</u> subvariety of X. A map $\varphi: X \longrightarrow Y$ of affine varieties is a <u>morphism</u> if for any finite subvariety F of X there exists a finite subvariety F' of Y such that $\varphi(F) \subset F'$ and the map $\varphi: F \longrightarrow F'$ is a morphism of finite-dimensional algebraic varieties. A group in this category is called an affine algebraic group of Shafarevich type [18], [30].

It is easy to see that given an algebra R with a fixed basis $\{v_i\}$, the group Aut R is naturally an affine algebraic group of Shafarevich type. For we have

$$(4) \qquad v_i v_j = \sum_k c_{ij}^k v_k, \qquad c_{ij}^k \in \mathbb{C},$$

and $g \in$ Aut R if and only if $g(v_s) = \sum_t x_{st} v_t$, $g^{-1}(v_s) = \sum_t y_{st} v_t$, with the $x_{st}$, $y_{st}$ satisfying the following system of equations: $g$ and $g^{-1}$ preserve (4) and $g \cdot g^{-1} = 1$.

**Problem**. For which integrable Lie algebras the associated group is an affine algebraic group of Shafarevich type? Is it true that the Lie algebra of a group of Shafarevich type (defined in [30]) is an integrable Lie algebra?

**Problem**. Let R be an arbitrary algebra. Then the Lie algebra Der R contains the following three subalgebras: the Lie algebra of the group Aut R (viewed is an affine algebraic group), the Lie algebra of endomorphism which are locally finite on R and the Lie algebra $(\text{Der } R)_{\text{fin}}$. How these subalgebras are related to each other? Interesting examples are: (a) R is a Lie algebra, (b) R is a coordinate ring of a (finite-dimensional) affine algebraic variety, (c) R is the universal enveloping algebra of a finite-dimensional Lie algebra.

# CHAPTER 2. Groups Associated to Kac-Moody Algebras

§2.1 Let $A = (a_{ij})_{i,j=1}^{n}$ be a generalized Cartan matrix, i.e. $a_{ii} = 2$, $a_{ij}$ are non-positive integers for $i \neq j$, and $a_{ij} = 0$ implies $a_{ji} = 0$. For a pair of indices i,j such that $i \neq j$ put $m_{ij} = 2, 3, 4$ or $6$ if $a_{ij}a_{ji} = 0, 1, 2$ or $3$ respectively and put $m_{ij} = 0$ otherwise; put $m_{ii} = 1$.

We associate to A a discrete group $\bar{W}(A)$ on n generators $\bar{r}_1, \ldots, \bar{r}_n$ and the following defining relations (r1) and (r2) ($i,j = 1, \ldots, n$):

(r1) $$\bar{r}_j \bar{r}_i^2 \bar{r}_j^{-1} = \bar{r}_i^2 \bar{r}_j^{-2a_{ij}}.$$

(r2) $$\bar{r}_i \bar{r}_j \bar{r}_i \ldots = \bar{r}_j \bar{r}_i \bar{r}_j \ldots \text{ ($m_{ij}$ factors on each side).}$$

Conjugating both sides of (r1) by $\bar{r}_j$ we get $\bar{r}_j^2 \bar{r}_i^2 \bar{r}_j^{-2} = \bar{r}_i^2$, i.e. the subgroup $T_{(2)} = \langle \bar{r}_i^2 \mid i = 1, \ldots, n \rangle$ of $\bar{W}(A)$ is a normal commutative subgroup. Also, it follows from (r1) for $i = j$ that $\bar{r}_i^4 = 1$.

Let W(A) be the corresponding Coxeter group, i.e. the group on generators $r_1, \ldots, r_n$ and the following defining relations ($i,j = 1, \ldots, n$):

$$(r_i r_j)^{m_{ij}} = 1.$$

Then we have a homomorphism $\bar{W}(A) \longrightarrow W(A)$ defined by $\bar{r}_i \longmapsto r_i$ and the exact sequence

$$1 \longrightarrow T_{(2)} \longrightarrow \bar{W}(A) \longrightarrow W(A) \longrightarrow 1.$$

Let $w = r_{i_1} \ldots r_{i_m}$ be a reduced expression of $w \in W$ (i.e. a shortest expression in the $r_i$); one defines $\ell(w) := m$. Deleting some of the $r_i$ from this expression one gets a new element w' and writes $w' \leq w$. The partial ordering $\leq$ on W(A) is called the Bruhat order.

One constructs a section of the map $\bar{W}(A) \longrightarrow W(A)$ putting $\bar{w} = \bar{r}_{i_1} \ldots \bar{r}_{i_m}$; one can show that $\bar{w} \in \bar{W}(A)$ is independent of the choice of the reduced expression of w (see e.g. [20]).

We shall construct connected topological groups $G(A) \supset K(A)$ such that they contain $\bar{W}(A)$ as a discrete subgroup and $W(A)$ is their "Weyl group".

§2.2 We first present the necessary material on Kac-Moody algebras and their representations. One may consult the book [14] for details.

Let $(\mathfrak{h},\Pi,\Pi^\vee)$ be a realization (unique up to isomorphism) of the matrix A, i.e. $\mathfrak{h}$ is a vector space of dimension 2n-rank A, and $\Pi = \{\alpha_1, \ldots, \alpha_n\} \subset \mathfrak{h}^*$, $\Pi^\vee = \{h_1, \ldots, h_n\} \subset \mathfrak{h}$ are linearly independent sets satisfying $\alpha_j(h_i) = a_{ij}$.

The Kac-Moody algebra $\mathfrak{g}(A)$ associated to the generalized Cartan matrix A is the Lie algebra generated by the vector space $\mathfrak{h}$ and symbols $e_i$ and $f_i$ ($i = 1, \ldots, n$), with the following defining relations:

(A1) $[\mathfrak{h},\mathfrak{h}] = 0$; $[h,e_i] = \alpha_i(h)e_i$, $[h,f_j] = -\alpha_j(h)f_j$ $(h\in\mathfrak{h})$

(A2) $[e_i,f_j] = \delta_{ij}h_i$; $(\mathrm{ad}\, e_i)^{1-a_{ij}}e_j = 0$, $(\mathrm{ad}\, f_i)^{1-a_{ij}}f_j = 0$ $(i \neq j)$.

The derived Lie algebra $\mathfrak{g}'(A)$ is also called the Kac-Moody algebra; it coincides with the subalgebra of $\mathfrak{g}(A)$ generated by $e_i$, $f_i$, $h_i$ ($i = 1, \ldots, n$) and its defining relations are (A2) and

(A'1) $[h_i,h_j] = 0$; $[h_i,e_j] = a_{ij}e_j$, $[h_i,f_j] = -a_{ij}f_j$.

We have the canonical embedding $\mathfrak{h} \subset \mathfrak{g}(A)$ and $\mathfrak{h}' \subset \mathfrak{g}'(A)$, where $\mathfrak{h}' = \sum \mathbb{C}h_i = \mathfrak{h} \cap \mathfrak{g}'(A)$. Let $\mathfrak{n}_+$ (resp. $\mathfrak{n}_-$) be the subalgebra of $\mathfrak{g}(A)$ and $\mathfrak{g}'(A)$ generated by the $e_i$ (resp. $f_i$), $i = 1, \ldots, n$. Then we have the triangular decompositions $\mathfrak{g}(A) = \mathfrak{n}_- \oplus \mathfrak{h} \oplus \mathfrak{n}_+$ and $\mathfrak{g}'(A) = \mathfrak{n}_- \oplus \mathfrak{h}' \oplus \mathfrak{n}_+$.

The center of $\mathfrak{g}(A)$ and $\mathfrak{g}'(A)$ is $\mathfrak{c} = \{h \in \mathfrak{h}' \mid \alpha_i(h) = 0$ for all $i = 1, \ldots, n\}$. (In the non-affine case this follows from the fact that any root $\alpha \in \mathfrak{h}^*$ of $\mathfrak{g}(A)$ restricted to $\mathfrak{h}'$ remains non-zero [14, Chapter 5]; in the affine case this is a consequence of the Gabber-Kac theorem [14, §9.11].) Note that $\mathfrak{c} = 0$ iff $\mathfrak{h} = \mathfrak{h}'$

(which happens iff det $A \neq 0$).

Both $\mathfrak{g}(A)$ and $\mathfrak{g}'(A)$ are integrable Lie algebras since the $e_i$ and $f_i$ are ad-locally nilpotent and elements from $\mathfrak{h}$ are ad-semisimple.

Furthermore, the subalgebras $\mathfrak{g}_i := \mathbb{C}f_i + \mathbb{C}h_i + \mathbb{C}e_i$ and any subspace of $\mathfrak{h}$ are, clearly, integrable subalgebras of $\mathfrak{g}(A)$. This is also true for the subalgebras $\mathfrak{h}'' + \mathfrak{n}_+$ and $\mathfrak{h}'' + \mathfrak{n}_-$, where $\mathfrak{h}''$ is a subspace of $\mathfrak{h}$, since such a subalgebra, say $\mathfrak{p}$, has the property that for any $x \in \mathfrak{p}$ and $y \in \mathfrak{g}$, $(\mathrm{ad}\ x)^N y \in \mathfrak{p}$ for sufficiently large N.

Given $\Lambda \in \mathfrak{h}'^*$, we extend it in some way to a linear function $\tilde{\Lambda} \in \mathfrak{h}^*$ and define the <u>highest weight module</u> $L(\Lambda)$ over $\mathfrak{g}(A)$ with action $d\pi_\Lambda$ by the properties

(L1) $L(\Lambda)$ is irreducible;

(L2) there exists a non-zero vector $v_\Lambda \in L(\Lambda)$ such that

$$d\pi_\Lambda(e_i)v_\Lambda = 0,\ i = 1, \ldots, n;\ d\pi_\Lambda(h)v_\Lambda = \tilde{\Lambda}(h)v_\Lambda,\ h \in \mathfrak{h}.$$

The module $L(\Lambda)$ remains irreducible when restricted to $\mathfrak{g}'(A)$ and is independent of the extension $\tilde{\Lambda}$ of $\Lambda$.

It is easy to see that if $L(\Lambda)$ is an integrable module (in the sense of §1.2), then the $\Lambda(h_i)$ are non-negative integers; we denote the set of such $\Lambda$ by $P_+$ ($\subset \mathfrak{h}'^*$). We put $P_{++} = \{\Lambda \in P_+ \mid \Lambda(h_i) > 0,\ i = 1, \ldots, n\}$. Define fundamental weights $\Lambda_1, \ldots, \Lambda_n \in P_+$ by $\Lambda_i(h_j) = \delta_{ij}$.

A much deeper result is that conversely, if $\Lambda \in P_+$, then $L(\Lambda)$ is an integrable module [26, Corollary 9]. It follows that $\mathfrak{r}_0 = 0$. This will be discussed in §2.3.

Incidentally, provided that A is a symmetrizable matrix and $\Lambda \in P_+$, the $\mathfrak{g}'(A)$-module $L(\Lambda)$ is characterized by (L1) and (L2). For the annihilator of $v_\Lambda \in L(\Lambda)$ is a left ideal in the enveloping algebra of $\mathfrak{g}'(A)$ generated by $e_i$, $f_i^{\Lambda(h_i)+1}$ and $h_i - \Lambda(h_i)$, $i = 1, \ldots, n$ [14, (10.4.6)]; on the other hand, if $(V,\pi)$ is a $\mathfrak{g}'(A)$-module satisfying (L1) and (L2), then, using the gradation of V by eigenspaces of $h_i$, one checks that $\pi(f_i)^{\Lambda(h_i)+1}v_\Lambda = 0$, $i = 1, \ldots, n$, to get a surjective $g'(A)$-module homomorphism $L(\Lambda) \longrightarrow V$.

It is not difficult to show (by making use of the structure of $\mathrm{Der}\; g'(A)$) that the linear Lie algebra $g'(A)$, acting on $\bigoplus_{\Lambda\in P_+} L(\Lambda)$, is algebraic.

Similarly, one defines the <u>lowest weight module</u> $(L^*(\Lambda), d\pi^*_\Lambda)$ over $g(A)$ as the irreducible module for which there exists a non-zero vector $v^*_\Lambda$ such that

$$d\pi^*_\Lambda(f_i)v^*_\Lambda = 0,\; i = 1, \ldots, n;\; d\pi^*_\Lambda(h)v^*_\Lambda = -\tilde{\Lambda}(h)v^*_\Lambda,\; h \in \mathfrak{h}.$$

This module is integrable if and only if $\Lambda \in P_+$. Actually, one has:

$$L^*(\Lambda) \simeq (L(\Lambda)^*)_{\mathrm{fin}}.$$

§2.3 In the remainder of the notes we shall study the group $G(A)$ associated to the (integrable) Lie algebra $g'(A)$. (This is a more "canonical" object than the group associated to $g(A)$). We have the associated $G(A)$-modules $(L(\Lambda), \pi_\Lambda)$, $\Lambda \in P_+$, and the adjoint $G(A)$-modules $(g(A), \mathrm{Ad})$ and $(g'(A), \mathrm{Ad})$. The correspondence between the integrable $g'(A)$-modules and differentiable $G(A)$-modules (conjectured in §1.5 for an arbitrary integrable Lie algebra) has been established in [18].

Denote by $G_i$, $H_i$, $H$, $U_+$, $U_-$, $B_+$ and $B_-$ the subgroups of $G(A)$ corresponding to the integrable subalgebras $g_i$, $\mathbb{C}h_i$, $\mathfrak{h}'$, $\mathfrak{n}_+$, $\mathfrak{n}_-$, $\mathfrak{h}' + \mathfrak{n}_+$ and $\mathfrak{h}' + \mathfrak{n}_-$ respectively of $g(A)$. We proceed to give a more explicit description of these groups.

We have an integrable homomorphism $d\varphi_i$: $\mathfrak{sl}_2(\mathbb{C}) \longrightarrow g(A)$ defined by

$$d\varphi_i \begin{bmatrix} a & b \\ c & -a \end{bmatrix} = ah_i + be_i + cf_i.$$

Let $\varphi_i$: $SL_2(\mathbb{C}) \longrightarrow G(A)$ be the corresponding homomorphism of groups. Put $H_i(t) = \varphi_i \begin{bmatrix} t & 0 \\ 0 & t^{-1} \end{bmatrix}$. The homomorphisms $\varphi_i$ are injective and one has: $G_i = \varphi_i(SL_2(\mathbb{C}))$; $H_i = \{H_i(t) \mid t \in \mathbb{C}^\times\}$;

$\exp te_i = \varphi_i\begin{bmatrix} 1 & t \\ 0 & 1 \end{bmatrix}$, $\exp tf_i = \varphi_i\begin{bmatrix} 1 & 0 \\ t & 1 \end{bmatrix}$, $t \in \mathbb{C}$. Furthermore, H is an abelian group equal to the direct product of the subgroups $H_i$. We also have $B_\pm = H \ltimes U_\pm$.

The map $\bar{r}_i \longmapsto \varphi_i\begin{bmatrix} 0 & 1 \\ -1 & 0 \end{bmatrix}$ $(= (\exp e_i)(\exp -f_i)(\exp e_i))$ extends to an injective homomorphism $\psi$: $\bar{W}(A) \longrightarrow G(A)$. We denote by $\bar{W}$ its image and denote the image of $\bar{r}_i$ again by $\bar{r}_i \in G(A)$.

The image of $T_{(2)}$ is a subgroup $\bar{W} \cap H = \{h \in H \mid h^2 = 1\}$. It follows that $T_{(2)} \simeq (\mathbb{Z}/2\mathbb{Z})^n$. The group $\bar{W}$ normalizes H. Denote by N the subgroup of G generated by H and $\bar{W}$. The group N acts on $\mathfrak{h}$ and $\mathfrak{h}'$ via the adjoint action, H acting trivially. The map $r_i \longrightarrow \bar{r}_i H$ extends to an isomorphism $W(A) \xrightarrow{\sim} W := N/H$; the image of $r_i$ is again denoted by $r_i \in W$. The group W is called the Weyl group of G(A) and the $r_i$ its fundamental reflections. Put $S = \{r_1, \ldots, r_n\}$. The adjoint action of W on $\mathfrak{h}'$ is

$$r_j \cdot h_i = h_i - a_{ij}h_j \qquad (i,j = 1, \ldots, n).$$

All the above facts of this subsection are easily checked by calculating in the adjoint and the integrable highest weight modules. More involved is the proof of the following fundamental result:

**Lemma [26, Corollary 8].** An element of a Kac-Moody algebra $\mathfrak{g}(A)$ is ad-locally finite (resp. locally nilpotent, resp. semisimple) if and only if it can be conjugated to an (ad-locally finite) subalgebra $\mathfrak{h} + (\mathfrak{n}_+ \cap (\mathrm{Ad}w)\mathfrak{n}_+)$ (resp. $\mathfrak{n}_+ \cap (\mathrm{Ad}w)\mathfrak{n}_+$, resp. $\mathfrak{h}$) for some $w \in \bar{W}$.

The proof of this lemma is based on Borel's fixed point theorem [2] and the Theorem 2.3 stated below.

It follows immediately from the lemma that a $\mathfrak{g}'(A)$-module is integrable if and only if all the $e_i$ and $f_i$ are locally finite (in particular, the $L(\Lambda)$ and $L^*(\Lambda)$ with $\Lambda \in P_+$ are integrable). Therefore, the present definition of G(A) coincides with that of [18]-[21], [26]. Another application of this lemma is the conjugacy of Cartan subalgebras of $g'(A)$ and the description of Aut $g'(A)$:

**Corollary** [26].

(a) Every ad-diagonalizable subalgebra of the Kac-Moody algebra $g(A)$ (resp. $g'(A)$) is Ad $G(A)$-conjugate to a subalgebra of $\mathfrak{h}$ (resp. $\mathfrak{h}'$).

(b) Any automorphism of the Kac-Moody algebra $g'(A)$ can be written in the form $\lambda\sigma$ or $\omega\lambda\sigma$ where $\sigma \in$ Ad $G$; $\lambda(e_i) = \lambda_{i_k} e_{i_k}$, $\lambda(f_i) = \lambda_{i_k}^{-1} f_{i_k}$, $i = 1, \ldots, n$, for some $\lambda_i \in \mathbb{C}^{\times}$ and a permutation $i \longmapsto i_k$ preserving the matrix $A$; $\omega(e_i) = -f_i$, $\omega(f_i) = -e_i$, $i = 1, \ldots, n$.

Put $V_\Lambda = \{c\pi_\Lambda(g)v_\Lambda \mid g \in G(A, c \in \mathbb{C}\}$. The following is the key result.

**Theorem** [26]. $V_\Lambda$ is a closed affine subvariety of $L(\Lambda)$ (more precisely, $V_\Lambda$ is the set of zeros of an ideal of $S(L^*(\Lambda))$.

In the case of a symmetrizable generalized Cartan matrix A, one can write down explicit equations for $V_\Lambda$. For that choose a non-degenerate invariant bilinear form $(\cdot \mid \cdot)$ on $g(A)$ ([14, Chapter 2]), choose a basis $\{x_i\}$ of $g(A)$ consistent with the triangular decomposition (i.e. a union of bases of $n_-$, $n_+$ and $\mathfrak{h}$) and let $\{y_i\}$ be the dual basis of $g(A)$, i.e. $(x_i \mid y_j) = \delta_{ij}$. Then $v \in V_\Lambda$ if and only if it satisfies in $L(\Lambda) \otimes L(\Lambda)$ [18]:

(1) $$(\Lambda \mid \Lambda)v \otimes v = \sum_i d\pi_\Lambda(x_i)v \otimes d\pi_\Lambda(y_i)v$$

The equations (1) are called <u>generalized Plücker relations</u> (they are identical with the usual Plücker relations in the classical case of the $SL_n(\mathbb{C})$-modules $\Lambda^k\mathbb{C}^n$). One can show that generalized Plücker relations generate the ideal of $V_\Lambda$ in the symmetric algebra of $L^*(\Lambda)$ [18].

The following proposition summarizes the key results on the

structure of the group G(A):

Proposition [20], [26].

(a) The group G(A) is generated by the 1-parameter subgroups exp $te_i$ and exp $tf_i$, $i = 1, \dots, n$.

(b) $(G(A), B_+, N, S)$ is a Tits system (see [4] for the background on Tits systems).

(c) $C = \{H_1(t_1) \dots H_n(t_n) \mid t_1^{a_{i1}} \dots t_n^{a_{in}} = 1 \text{ for } i = 1, \dots, n\}$.

(d) $U_+$ is generated by the 1-parameter subgroups exp $t(w \cdot e_i)$, where $\bar{w} \in \bar{W}$ is such that $(\mathrm{Ad}\,w)e_i \in \mathfrak{n}_+$, $i = 1, \dots, n$.

(e) N is the normalizer of H in G(A).

One is referred to [26] for the details of the proof of the Lemma, Theorem and Corollary.

A standard consequence of Proposition 2.3(b) is

(2) $G(A) = \coprod_{w \in W} B_+ \bar{w} B_+$ (Bruhat decomposition).

Here and further on $\bar{w}$ denotes a preimage of w in $\bar{W}$ (for example one may take $\bar{w}$ constructed in §2.1). Somewhat less standard is

(3) $G(A) = \coprod_{w \in W} B_- \bar{w} B_+$ (Birkhoff decomposition).

To prove (3) we check that [26]

$$B_- \bar{w} B_+ \bar{r}_i \subset B_- \bar{w} B_+ \cup B_- \bar{w} \bar{r}_i B_+.$$

Since also $B_- \bar{w}\,_+$ exp $te_i = B_- \bar{w} B_+$, and since the $\bar{r}_i$ and exp $te_i$ generate G(A) (by Proposition 2.3(a)), we get that the right-hand side of (3) is stable under right multiplication by G(A) and hence coincides with G(A). The disjointness in (2) and (3) is easily proved by making use of the G(A)-modules $L(\Lambda)$. For example, if $B_-\bar{w}B_+ = B_-\bar{w}_+B_+$, applying to $v_\Lambda \in L(\Lambda)$, we get $\pi_\Lambda(B_-)\pi_\Lambda(\bar{w})v_\Lambda = \pi_\Lambda(B_-)\pi_\Lambda(\bar{w}_1)v_\Lambda$ and therefore $\mathbb{C}\pi_\Lambda(\bar{w})v_\Lambda = \mathbb{C}\pi_\Lambda(\bar{w}_1)v_\Lambda$. Taking $\Lambda \in P_{++}$, we get $w = w_1$.

We conclude this section by a discussion on presentation problems. It is clear that N is a group on generators $\bar{r}_i$ and $H_i(t)$ where $i = 1, \ldots, n$, $t \in \mathbb{C}^\times$, with the following defining relations:

$$H_i(t)H_j(t') = H_j(t')H_i(t);$$

$$\bar{r}_iH_j(t)\bar{r}_i^{-1} = H_j(t)H_i(t^{-a_{ji}});$$

$$\bar{r}_i^2 = H_i(-1);$$

$$\bar{r}_i\bar{r}_j\bar{r}_i \ldots = \bar{r}_j\bar{r}_i\bar{r}_j \ldots \text{ ($m_{ij}$ factors on each side).}$$

Thus, N, as well as $\bar{W}$ and W are amalgamated products of subgroups of "rank" 1 and 2. We will see in §2.5 that this is also the case for the "unitary form" K(A). It is also known to be the case for the finite-dimensional G(A) [5] but seems unlikely for general G(A).

**Problem.** Find a presentation of G(A) and $U_+$. (These two questions are closely related to each other; a solution is known in the rank 2 case only [20].)

**Problem.** For which indecomposable A the group G(A)/C is simple. (It is simple if A is of finite type and it is not if A is of affine type. As shown in [25], the formal completion of G(A)/C is always simple.)

**§2.4** Recall that a function f: $G(A) \longrightarrow \mathbb{C}$ is regular (weakly regular in the terminology of [18]) iff the functions $f\circ\psi_{\bar{x}}: \mathbb{C}^n \longrightarrow \mathbb{C}$ are polynomial functions for all $\bar{w} = (x_1, \ldots, x_n)$ with the $x_i$ taken from the set $\{e_1, \ldots, e_n, f_1, \ldots, f_n\}$ (see §1.8). Very little is known about the structure of the algebra $\mathbb{C}[G(A)]$ of regular functions and the $G(A) \times G(A)$-module $(\mathbb{C}[G(A)], \pi_{reg})$.

**Conjecture.** The inclusion $G(A) \longrightarrow \operatorname{Specm} \mathbb{C}[G(A)]$ is surjective.

The subalgebra $\mathbb{C}[G(A)]_{s.r.}$ of strongly regular functions is understood better. Recall that a regular function f is called strongly regular [18] if for any $g \in G(A)$ there exists a subgroup $U_+'$ of $U_+$ (resp. $U_-'$ of $U_-$) which is an intersection of $U_+$ (resp. $U_-$) with a finite number of its conjugates, such that $f(u_-gu_+) = f(g)$ whenever $u_+ \in U_+'$, $u_- \in U_-'$. One has the following analogue of the Peter-Weyl theorem.

**Theorem** [18]. The linear map $\Phi: \bigoplus_{\Lambda \in P_+} L^*(\Lambda) \otimes L(\Lambda) \longrightarrow \mathbb{C}[G(A)]$ defined by $\Phi(v^* \otimes v) = f_{v^*,v}$, is a well-defined injective $G(A) \times G(A)$-module homomorphism onto $\mathbb{C}[G(A)]_{s.r.}$.

The proof of this theorem is fairly simple: we check that the $G(A) \times G(A)$-module $\mathbb{C}[G(A)]_{s.r.}$ is differentiable and hence can be viewed as a $\mathfrak{g}(A) \times \mathfrak{g}(A)$-module, to which we apply a version of the complete reducibility theorem from [17].

Note that $\mathbb{C}[G(A)]_{s.r.} = \mathbb{C}[G(A)]^{\mathcal{M}}_{s.r.}$ (in the terminology of 1.8), where $\mathcal{M}$ is the subcategory of integrable $\mathfrak{g}(A)$-modules from the category $\mathcal{O}$.

For a subgroup P of G(A), let $\mathbb{C}[G(A)]^P$ denote the algebra of all $f \in \mathbb{C}[G(A)]$ such that $f(gp) = f(g)$ for all $p \in P$. G(A) acts on it by $(g \cdot f)(x) = f(g^{-1}x)$. Let $\theta_\Lambda$ be the character of $B_+$ defined by $\theta_\Lambda((\exp h)u) = e^{\Lambda(h)}$ for $h \in \mathfrak{h}$, $u \in U_+$, and let, for $\Lambda \in P_+$, $S_\Lambda = \{f \in \mathbb{C}[G(A)]_{s.r.} \mid f(gb) = \theta_\Lambda(b)f(g)$ for $g \in G$, $b \in B_+\}$. Then we have an immediate corollary of the Theorem 2.

**Corollary.**

(a) (Borel-Weil-type theorem) The map $L^*(\Lambda) \longrightarrow S_\Lambda$ defined by $v \longmapsto f_{v,v^*}$ is a G-module isomorphism.

(b) $\mathbb{C}[G(A)]^{U_+}_{s.r.} = \bigoplus_{\Lambda \in P_+} S_\Lambda$ and this algebra is isomorphic to $\bigoplus_{\Lambda \in P_+} L^*(\Lambda)$ with the Cartan product $\mu$: $L^*(\Lambda) \otimes L^*(M) \longrightarrow L^*(\Lambda + M)$ characterized by the properties that $\mu$ is a $\mathfrak{g}(A)$-module homomorphism and $\mu(v^*_\Lambda \otimes v^*_M) = v^*_{\Lambda+M}$.

The main result of [18] about the algebra structure of $\mathbb{C}[G(A)]_{s.r.}$ is that this algebra is a unique factorization domain. An immediate corollary of this is that the algebra $\mathbb{C}[G(A)]^{U_+}_{s.r.}$ is a unique factorization domain and that the coordinate ring of strongly regular functions on $V_\Lambda$ is integrally closed.

It is also shown in [18] that G(A) can be given a structure of an affine algebraic group of Shafarevich type by constructing an embedding of G(A) in a vector space as a closed affine subvariety. It has many nice properties, for example, G(A) acts morphically on $L(\Lambda)$, $\Lambda \in P_+$ and on $\mathfrak{g}(A)$. It is still an open problem, however, whether the coordinate ring for this embedding coincides with $\mathbb{C}[G(A)]$.

Finally, I want to mention one striking difference between the finite- and infinite-dimensional cases. Let A be an indecomposable generalized Cartan matrix. If A is of finite type, then $\mathbb{C}[G(A)] = \mathbb{C}[G(A)]_{s.r.}$ is the coordinate ring of the finite-dimensional affine algebraic variety G(A); in particular, G(A) = Specm $\mathbb{C}[G(A)]_{s.r.}$.

Now let A be of infinite type. Then, of course, $\mathbb{C}[G(A)]$ is much larger than $\mathbb{C}[G(A)]_{s.r.}$. Moreover, the set R: = Specm $\mathbb{C}[G(A)]_{s.r.} \setminus G(A)$ is always non-empty. Namely, the $G(A) \times G(A)$-invariant subspace $\mathfrak{m} = \theta(\bigoplus_{\Lambda \in P_+ \setminus \{0\}} L^*(\Lambda) \otimes L(\Lambda))$ is an ideal (in a sharp contrast to the finite-dimensional case), which is an element of R [18]. Recently, D. Peterson has computed the set R. In particular, it turned out that $R = \{\mathfrak{m}\}$ in the affine case. A discussion of the "partial compactification" Specm $\mathbb{C}[G(A)]_{s.r.}$ of G(A) in connection to the theory of singularities of algebraic surfaces may be found in this volume [31].

§2.5 In this section we study the algebraic structure of the <u>unitary form</u> K(A) of the group G(A). If A is a generalized Cartan matrix of finite type, then K(A) is the compact real form of the complex semisimple Lie group G(A). Thus, the groups K(A) are infinite-dimensional analogs of compact Lie groups.

The Kac-Moody algebra g'(A) admits an antilinear involution $\omega_0$ determined by $\omega_0(e_i) = -f_i$, $\omega_0(f_i) = -e_i$, $i = 1, \ldots, n$. Since $\omega_0$ preserves the set of locally finite elements, it can be lifted uniquely to an involution of G(A), which we also denote by $\omega_0$. Let K(A) be the fixed point set of this involution in G(A).

Provided that A is symmetrizable and indecomposable, the Kac-Moody algebra g'(A) carries (a unique up to a constant factor) invariant bilinear form $(\cdot \mid \cdot)$ such that $(e_i \mid f_i) > 0$. Put $(x \mid y)_0 = -(x \mid w_0(y))$. The triangular decomposition is othogonal with respect to the Hermitian form $(\cdot \mid \cdot)_0$ and the main result of [19] is that it is positive definite on $n_+$ and $n_-$. Using this, one easily deduces [19] that any G(A)-module $L(\Lambda)$, $\Lambda \in P_+$, carries a unique positive definite Hermitian K(A)-invariant form $H(\cdot \mid \cdot)$ such that $H(v_\Lambda \mid v_\Lambda) = 1$. This is a justification for the term "unitary form".

For an arbitrary generalized Cartan matrix A, it is a simple fact that $L(\Lambda)$, $\Lambda \in P_+$, carries a unique Hermitian form $H(\cdot \mid \cdot)$ such that $H(v_\Lambda \mid v_\Lambda) = 1$. It remains an open problem whether it is positive definite in the non-symmetrizable case.

The involution $\omega_0$ preserves the subgroups $G_i$, $H_i$ and H; we denote by $K_i$, $T_i$ and T respectively the corresponding fixed point subgroups. Then $K_i = \varphi_i(SU_2)$, $T_i = \varphi_i(\{\mathrm{diag}(\lambda,\lambda^{-1}) \mid |\lambda| = 1\})$ is a maximal torus of $K_i$ and $T = \prod_i T_i$ is a maximal commutative subgroup of K(A). Put $H_i^+ = \varphi_i(\{\mathrm{diag}(\lambda,\lambda^{-1}) \mid \lambda \in \mathbb{R}, \lambda > 0\})$, $H_+ = \prod_i H_i^+$; then $H = T \times H^+$.

Let D (resp. $\mathring{D}$) $= \{u \in \mathbb{C} \mid |u| \leq 1$ (resp. $|u| < 1)\}$ be the unit disc (resp. its interior) and let $S^1 = D \setminus \mathring{D}$ be the unit circle. Given $u \in D$, put

$$z(u) = \begin{bmatrix} u & (1-|u|^2)^{1/2} \\ -(1-|u|^2)^{1/2} & \bar{u} \end{bmatrix} \in SU_2,$$

and put $z_i(u) = \varphi_i(z(u))$. We have $\bar{r}_i = z_i(0) \in K_i$, hence $\bar{W} \subset K(A) \subset G(A)$. Put

$$Y_i = \{z_i(u) \mid u \in \mathring{D}\} \subset K_i.$$

The same argument as in [32, Lemma 43(b)] gives

(4) $$B\bar{r}_i B = Y_i B \text{ (uniquely).}$$

(Here and further on "uniquely" means that any element from the set on the left-hand side is uniquely represented as a product of elements from the factors on the right-hand side.)

Let $w = r_{i_1} \dots r_{i_s}$ be a reduced expression of $w \in W$ and let $\bar{w}$ be its preimage in $\bar{W}$ defined in §2.1. Using (4), the same argument as in [32, Lemma 15], gives

(5) $$B_+ \bar{w} B_+ = Y_{i_1} \dots Y_{i_m} B_+ \text{ (uniquely).}$$

Put $K_w = K(A) \cap B_+ \bar{w} B_+$. Put $Y_w = Y_{i_1} \dots Y_{i_m}$; this is independent of the choice of the reduced expression for w, as follows from the following formula [20]:

$$Y_w = \{k \in K_w \mid H(\pi_{\Lambda_i}(k)v_{\Lambda_i} \mid \pi_{\Lambda_i}(\bar{w})v_{\Lambda_i}) > 0,\ i = 1, \dots, n\}.$$

We have by (5):

(6) $$K_w = Y_w T \text{ (uniquely).}$$

Put $\bar{K}_w = K_{i_1} \dots K_{i_m} T$; this is also independent of the reduced expression of w, as follows from

(7) $$\bar{K}_w = \coprod_{w' \leqslant w} K_{w'}.$$

Finally, by the Bruhat decomposition, we have

(8) $$K(A) = \coprod_{w \in W} K_w.$$

We obtain, in particular that $K(A)$ is generated by the $K_i$, $i = 1, \ldots, n$, and the Iwasawa decomposition [26]:

(9) $$G(A) = K(A)H_+U_+ \text{ (uniquely).}$$

We proceed to establish a presentation of the group $K(A)$, which may be viewed as a "real analytic continuation" of the presentation of the group $\bar{W}(A)$.

We have the following relations coming from $SU_2$:

(R1) (i) $z_i(u_1)z_i(u_2) = z_i(u_1u_2)$ if $u_1, u_2 \in S^1$,

(ii) $z_i(u)z_i(-\bar{u}) = z_i(-1)$ if $u \in \mathring{D}$,

(iii) $z_i(u_1)z_i(u_2) = z_i(u_1')z_i(u_2')$ if $u_1, u_2 \in \mathring{D}$ and $u_1 \neq -\bar{u}_2$, for some unique $u_1' \in \mathring{D}$ and $u_2' \in S^1$.

Furthermore, $T_i$ normalizes $K_j$ and the conjugation is given by

(R2) $z_i(u_1)z_j(u_2)z_i(u_1)^{-1} = z_j(u_1^{a_{ij}}u_2)z_j(u_1^{-a_{ij}})$ if $u_1 \in S^1$, $u_2 \in D$.

Finally, if $m_{ij} \neq 0$, then $r_ir_jr_i \ldots = r_jr_ir_j \ldots$ ($m_{ij}$ factors on each side). Hence $Y_iY_jY_i \ldots = Y_jY_iY_j \ldots$ (uniquely). In other words, we have

(R3) $z_i(u_1)z_j(u_2)z_i(u_3) \ldots = z_j(u_1')z_i(u_2')z_j(u_3') \ldots$ ($m_{ij}$ factors on each side), if $u_1, u_2, \ldots \in \mathring{D}$, for some unique $u_1', u_2', \ldots \in \mathring{D}$.

**Theorem** [20]. The group $K(A)$ is a group on generators $z_i(u)$ for $i = 1, \ldots, n$; $u \in D$, with defining relations (R1), (R2) and (R3).

Let $\tilde{K}(A)$ be the group on generators $z_i(u)$ ($i = 1, \ldots, n$; $u \in D$) with defining relations (R1), (R2), (R3), let $\alpha$: $\tilde{K}(A) \longrightarrow K(A)$ be the canonical homomorphism, let $w = r_{i_1} \dots r_{i_m} \in W$ be a reduced expression and let $\tilde{K}_w = \alpha^{-1}(K_w)$. It is not hard to show that any element of $\tilde{K}_w$ can be brought to the form $z_{i_1}(u_1) \dots z_{i_m}(u_m) z_1(v_1) \dots z_n(v_n)$, where $u_i \in \overset{\circ}{D}$, $v_i \in S^1$. Then (6) completes the proof of the Theorem. (The details may be found in [20].)

Note that the groups $\tilde{K}(A)$ have been introduced (in a somewhat different form) in [13] and it was proved there, by a topological argument, that $\mathrm{Ker}\,\alpha$ is a finite central subgroup if A is of finite type.

§2.6 Since G(A) is generated by a finite number of 1-parameter subgroups exp tx, where $x \in X = \{e_i, f_i \mid i = 1, \dots, n\}$, it is a (connected Hausdorff) topological group in the topology defined in §1.8. In this section we discuss some of the results of [21] on the topology of the groups G(A) and K(A) and of the associated flag varieties. The reader is referred to [21] for details.

All the subgroups which have appeared in the discussion are closed. The bijection $K(A) \times H_+ \times U_+ \xrightarrow{\sim} G(A)$ provided by the Iwasawa decomposition is a homeomorphism. Furthermore, $H_+$ and $U_+$ are contractible. Thus (as in the finite-dimensional case) G(A) is homotopically equivalent to K(A).

The topology on K(A) can be described explicitly as follows. Given $w \in W$, take its reduced expression $w = r_{i_1} \dots r_{i_m}$ and define a map $(SU_2)^m \times T \longrightarrow K(A)$ by $(k_1, \dots, k_m, t) \longmapsto \varphi_{i_1}(k_1) \dots \varphi_{i_m}(k_m)t$. The image of this map is $\bar{K}_w$, and we take the quotient topology on it. This topology is independent of the choice of the reduced expression and makes $\bar{K}_w$ a connected Hausdorff compact topological space. Then a subset F of K(A) is closed iff $F \cap \bar{K}_w$ is closed in $\bar{K}_w$ for all $w \in W$. It follows that $\bar{K}_w$ is the closure of $K_w$ and that $\bar{K}_{w'} \leqslant \bar{K}_w$ iff $w' \leqslant w$. Thus, as a topological space, K(A) is the inductive limit with respect to the Bruhat order of the compact spaces $\bar{K}_w$.

The most natural way to study the topology of K(A) is to

consider the fibration

$$\pi\colon K(A) \longrightarrow K(A)/T.$$

The topological space $\mathcal{F}(A) := K(A)/T$ is called the flag variety of the group $K(A)$ and of $G(A)$. Put $C_w = \pi(Y_w)$. Then by (6) and (8) we get a cellular decomposition

$$\mathcal{F}(A) = \coprod_{w \in W} C_w$$

To show that this is a CW-complex one has only to construct attaching maps (for some reason this point is routinely omited in the literature on finite-dimensional groups, see e.g. [2]). For that, given $w \in W$, choose a reduced expression $w = r_{i_1} \ldots r_{i_s}$ and define a map $\alpha_w\colon D^s \longrightarrow \mathcal{F}(A)$ by $\alpha_w(u_1, \ldots, u_s) = z_{i_1}(u_1) \ldots z_{i_s}(u_s) \bmod T$. This gives a homeomorphism of $\overset{\circ}{D}{}^s$ onto $Y_w$ by (5). Since $\bar{K}_w$ is the closure of $K_w$, by (7) we have:

$$\bar{C}_w = \coprod_{w' \leqslant w} C_{w'} \tag{10}$$

where $\bar{C}_w$ is the closure of $C_w$. It is clear that $\alpha_w(D^{k-1} \times S^1 \times D^{s-k}) \subset \bar{C}_{w'}$, where $w'$ is obtained from $w$ by dropping $r_{i_k}$. Thus, by (10) the image of the boundary under the map $\alpha_w$ lies in the union of cells of lower dimension (this argument is taken from [21]).

Since $\dim C_w = 2\ell(w)$, there are no cells of odd dimension. Thus $H_*(\mathcal{F}(A),\mathbb{Z})$ and $H^*(\mathcal{F}(A),\mathbb{Z})$ are free $\mathbb{Z}$-modules on generators of degree $2\ell(w)$, $w \in W$. Putting $W(q) = \sum_{w \in W} q^{\ell(w)}$, we obtain that the Poincaré series for homology and cohomology of $\mathcal{F}(A)$ over any field is $W(q^2)$. (A simple inductive procedure for computing $W(q)$ may be found in [4].)

Actually, as in the finite-dimensional case, $\mathcal{F}(A)$ can be given a natural structure of a complex projective manifold. For that note that, by the Iwasawa decomposition, we have a homeomorphism $G(A)/B$

$\xrightarrow{\sim} \mathcal{F}(A)$. But $G(A)/B$ can be identified with the orbit $G \cdot v_\Lambda$ in the projective space $\mathbb{P}L(\Lambda)$ for $\Lambda \in P_{++}$. This is a closed subvariety of $\mathbb{P}L(\Lambda)$ by Theorem 2.3. An equivalent definition, independent of the choice of $\Lambda \in P_{++}$, is $G(A)/B = \mathrm{Proj} \bigoplus_{\Lambda \in P_+} L^*(\Lambda)$ (cf. Corollary 2.4).

As a result, the $\bar{C}_w$ become finite-dimensional projective varieties, called <u>Schubert varieties</u>, and $\mathcal{F}(A)$ is their inductive limit with respect to Bruhat order [18]. The study of singularities of these varieties has many interesting applications. Some of them are discussed in this volume [11].

P. Deligne kindly provided a proof of the following result:

Let X be a projective algebraic variety over $\mathbb{C}$ which is decomposed into a finite disjoint union of subvarieties $X^i_j$ with $\dim_{\mathbb{C}} X^i_j = i$, such that $\bar{X}^i_j \setminus X^i_j \subset \coprod_j \coprod_{s<i} X^s_j$ and there exist morphisms $\coprod^j \mathbb{C}^i \longrightarrow \coprod_j X^i_j$ which are homeomorphisms. Then the topological space X is rationally formal (in the sense of [6]).

Applying this to our situation, we deduce that $\mathcal{F}(A)$ is a rationally formal topological space.

Let $Q^\vee = \sum_i \mathbb{Z}h_i$ and let $P = \{\lambda \in \mathfrak{h}'^* \mid \lambda(h_i) \in \mathbb{Z},\ i = 1, \dots, n\}$ be the dual lattice. Let $S(P) = \bigoplus_{j \geq 0} S^j(P)$ be the symmetric algebra over the lattice P, and $S(P)^+ = \bigoplus_{j>0} S^j(P)$ the augmentation ideal. Given a field $\mathbb{F}$, we denote $S(P)_{\mathbb{F}} := \mathbb{F} \otimes_{\mathbb{Z}} S(P)$, etc. In order to study the multiplicative structure of $H^*(\mathcal{F}(A), \mathbb{F})$, we define the <u>characteristic homomorphism</u> $\psi: S(P) \longrightarrow H^*(\mathcal{F}(A), \mathbb{Z})$ as follows. Given $\lambda \in P$, we have the corresponding character of T and the associated line bundle $\mathcal{L}_\lambda$ on $\mathcal{F}(A)$. Put $\psi(\lambda) \in H^2(\mathcal{F}(A), \mathbb{Z})$ equal to the Chern class of $\mathcal{L}_\lambda$ and extend by multiplicativity to the whole S(P). Denote by $\psi_{\mathbb{F}}$ the extension of $\psi$ by linearity to $S(P)_{\mathbb{F}}$.

In order to describe the properties of $\psi_{\mathbb{F}}$ define operators $\Delta_i$ for $i = 1, \dots, n$ on S(P) by

$$\Delta_i(f) = (f - r_i(f))/\alpha_i,$$

and extend by linearity to $S(P)_{\mathbb{F}}$. Put $I_{\mathbb{F}} = \{f \in S(P)_{\mathbb{F}}^{+} \mid \Delta_{i_1} \dots \Delta_{i_m}(f) \in S(P)_{\mathbb{F}}^{+}$ for every sequence $i_1, \dots, i_m\}$. This is a graded ideal of $S(P)_{\mathbb{F}}^{+}$.

<u>**Proposition**</u> [21], [22]. Let $\mathbb{F}$ be a field. Then

(a) Ker $\psi_{\mathbb{F}} = I_{\mathbb{F}}$ (this holds for an arbitrary ring $\mathbb{F}$).

(b) $H^*(\mathcal{F}(A),\mathbb{F})$ is a free module over Im $\psi_{\mathbb{F}}$.

(c) Any minimal system of homogeneous generators of the ideal $I_{\mathbb{F}}$ is a regular sequence.

Let $CH(G(A),\mathbb{F})$ denote the quotient (graded) algebra of $H^*(\mathcal{F}(A),\mathbb{F})$ by the ideal generated by $\psi(P_{\mathbb{F}})$; this is called the <u>Chow algebra</u> of $G(A)$ over $\mathbb{F}$. Notice that, by Theorem 2.6(b) below, $CH(G(A),\mathbb{F}) = \pi^*(H^*(\mathcal{F}(A),\mathbb{F}))$. The terminology is justified by the fact that for A of finite type, the Chow ring of the complex semisimple group $G(A)$ is isomorphic to $CH(G(A),\mathbb{Z})$ (A. Grothendieck).

Denote the degrees of the elements of a minimal system of homogeneous generators of the ideal $I_{\mathbb{F}}$ by $d_1, \dots, d_s (s \leq n)$. These degrees are well-defined; we will call them the <u>degrees of basic generators of</u> $I_{\mathbb{F}}$. Note that $s = n$ if char $\mathbb{F} = p \neq 0$ since W acts on $P \otimes_{\mathbb{Z}} \mathbb{F}$ via a finite group.

Actually, Proposition 2.6 holds in a much more general situation [22]. For example, the part (c) holds for any group generated by reflections over a field $\mathbb{F}$ of arbitrary characteristic. For W finite and $\mathbb{F} = \mathbb{C}$ we recover the classical result of Chevalley-Shepard-Todd.

It is not difficult to deduce from Proposition 2.6 the following results.

<u>**Theorem**</u> [21]. Let $\mathbb{F}$ be a field. Then:

(a) $CH(G(A),\mathbb{Q})$ is a polynomial algebra on (in general infinite number of) homogeneous generators. The Poincaré series of $CH(G(A),\mathbb{F})$ is equal to $W(q^2)(1-q^2)^n/\prod_{i=1}^{s}(1-q^{2d_i})$. The (graded) algebra $H^*(K/T,\mathbb{Q})$ is (non-canonically) isomorphic to the tensor product of Im $\psi_{\mathbb{Q}}$ and $CH(G(A),\mathbb{Q})$.

(b) The cohomology spectral sequence $E_r(K(A),\mathbb{F})$ of the fibration $\pi$: $K(A) \longrightarrow \mathfrak{F}(A)$ degenerates at $r = 3$, i.e. $E_3(K(A),\mathbb{F}) = E_\infty(K(A),\mathbb{F})$.

(c) $\pi^*$ induces an injective homomorphism of $CH(G(A),\mathbb{F})$ into $H^*(K(A),\mathbb{F})$ and into $E_\infty(K(A),\mathbb{F})$, the image being a Hopf subalgebra of $H^*(K(A),\mathbb{F})$.

(d) The algebra $E_\infty(K(A),\mathbb{F})$ is isomorphic to a tensor product of $C(G(A),\mathbb{F})$ and the cohomology algebra of the Koszul complex $(\Lambda(P) \otimes \text{Im } \psi_{\mathbb{F}},d)$, where $d(\lambda \otimes u) = \psi(\lambda) \cup u$. The latter algebra is an exterior algebra on homogeneous generators of degrees $2d_1-1, \ldots, 2d_s-1$. The Poincaré series of $H^*(K(A),\mathbb{F})$ is equal to the product of the Poincaré series of $CH(G(A),\mathbb{F})$ and the polynomial $\prod_{i=1}^{s}(1+q^{2d_i-1})$.

As an immediate corollary of Theorem 2.6(a) and (d), we deduce the following classical results.

<u>Corollary</u>. Let K be a connected compact Lie group, T its maximal torus, $\mathfrak{h}$ the complexified Lie algebra of T, W the Weyl group, and let $d_1, \ldots, d_n$ be the degrees of the basic homogeneous invariants for the action of W on $S(\mathfrak{h})$. Then:

(a) $W(q) = \prod_{i=1}^{n}((1-q^{d_i})/(1-q))$.

(b) $H^*(K/T,\mathbb{C})$ is generated by $H^2(K/T,\mathbb{C})$ and is isomorphic to the quotient of $S(\mathfrak{h})$ by the ideal generated by $(S(\mathfrak{h})^+)^W$.

(c) $H^*(K,\mathbb{C})$ is a Grassmann algebra on homogeneous generators of degrees $2d_1-1, \ldots, 2d_n-1$.

(d) The Chow ring of a complex reductive group is finite.

In fact, using explicit formulas or the cup product [21] (see also the next section), it is easy to show that the third term of the cohomology (resp. homology) spectral sequence over $\mathbb{Z}$ of the fibration $\pi$ is isomorphic to the homology of the complex $(C^*,d^*)$ (resp. $(C_*,d_*)$), where $C^* = \mathbb{Z}[W] \otimes_{\mathbb{Z}} \Lambda(P)$, $C_* = \mathbb{Z}[W] \otimes_{\mathbb{Z}} \Lambda(Q^V)$, $\deg \delta_w = \deg \delta^w = 2\ell(w)$, $\deg h_i = \deg \Lambda_i = 1$, and

$$d^*(\delta^w \otimes p) = \sum_{w \xrightarrow{\gamma} w'} \delta^{w'} \otimes (\partial_\gamma p),$$

$$d_*(\delta_w \otimes q) = \sum_{w' \xrightarrow{\gamma} w} \delta_{w'} \otimes (\gamma \wedge q).$$

Here $w' \xrightarrow{\gamma} w$ means that $\ell(w') = \ell(w) - 1$ and there exists a positive real coroot $\gamma \in \sum \mathbb{Z} h_i$ such that $w = w' r_\gamma$, where $r_\gamma$ is the reflection with respect to $\gamma$; $\partial_\gamma$ is an antiderivation of $\Lambda(P)$ such that $\partial_\gamma \lambda = \langle \lambda, \gamma \rangle$ for $\lambda \in P$.

<u>Remark</u>. If we take a standard cellular decomposition of T, then (8) together with (6) gives us a cellular decomposition of K(A). Unfortunately, it is not a CW-complex; but if it were, then, as one can easily see, the complex $(C_*,d_*)$ would be the corresponding homology complex.

<u>Conjecture</u>. $E_\infty(K(A),\mathbb{Z}) = E_3(K(A),\mathbb{Z})$.

Let me state also some corollaries of Theorem 2.6 for arbitrary K(A).

<u>Corollary</u>.

(a) K(A) is a connected simply connected topological group;

$H^2(K(A),\mathbb{Z}) = 0$.

(b) Let A be indecomposable and let $\epsilon = 1$ or 0 according as A is symmetrizable or not. Then $H^3(K(A),\mathbb{Q}) = \mathbb{Z}^{\epsilon}$; $\dim_{\mathbb{Q}} H^4(K(A),\mathbb{Q}) = \#$(cycles of the Dynkin diagram of A) + 1 - $\epsilon$. $H^*(K(A),\mathbb{Q})$ is completely determined (as a graded vector space) by the Weyl group W regarded as a Coxeter group and by $\epsilon$.

(c) The minimal model (in the sense of [6]) of the topological space $\mathcal{F}(A)$ is a tensor product of an exterior algebra on generators $\xi_1, \ldots, \xi_s$ of degrees $2d_1-1, \ldots, 2d_s-1$, and of a polynomial algebra on n generators $\Lambda_1, \ldots, \Lambda_n$ of degree 2, $a_j$ generators of degree 2j, j = 2, 3, ..., where $d_1, \ldots, d_s$ are the degrees of basic generators of $I_{\mathbb{Q}}$ and the $a_j$ are determined by

$$W(q)(1-q)^n = \prod_{i=1}^{s} (1-q^{d_i}) \prod_{j\geq 2} (1-q^j)^{-a_j}.$$

The differential d of this minimal model is 0 on all even generators and $d\xi_i = P_i(\Lambda_1, \ldots, \Lambda_n)$, where the $P_i$ are basic generators of $I_{\mathbb{Q}} \subset \mathbb{Q}[\Lambda_1, \ldots, \Lambda_n]$.

(d) The minimal model of K(A) is isomorphic to $H^*(K(A),\mathbb{Q})$ with trivial differential, and is a tensor product of an exterior algebra on generators of degrees $2d_1-1, \ldots, 2d_s-1$, and a polynomial algebra on $a_j$ generators of degrees 2j, j = 2, 3, ... .

(e) The dimension of the k-th rational homotopy group of $\mathcal{F}(A)$ and K(A) is equal to the number of generators of degree k of their minimal models.

Cohomology and the Chow ring in the finite-dimensional case and arbitrary field $\mathbb{F}$ are discussed in detail (from the presented point of view), in [15]. The affine case will be discussed in the next section. Here I will discuss briefly the case when A is an indecomposable generalized Cartan matrix of non-finite and non-affine type and $\mathbb{F} = \mathbb{Q}$. Put = 1 or 0 according as the matrix A is symmetrizable or not. Then $I_{\mathbb{Q}}$ is generated by $\epsilon$ elements of degree 2. Put

$$C(q) = W(q)(1 - q)^n(1 - q^2)^{-\epsilon}.$$

Then we have by Corollary 2.6(c):

$$(11) \qquad C(q) = \prod_{j \geq 2} (1 - q^j)^{-a_j}, \text{ where } a_j \geq 0.$$

It would be interesting to find a purely combinatorial proof of this result. By Theorem 2.6(a), the Chow algebra $CH(G(A),\mathbb{Q})$ is a polynomial algebra on $a_j$ generators of degree $2j$, $j = 2, 3, \ldots$ . By Theorem 2.6(d), $H^*(K(A),\mathbb{Q})$ is a tensor product of $CH(G(A),\mathbb{Q})$ with the exterior algebra on $\epsilon$ generators of degree 3.

A stronger form of (11) is the following:

Conjecture. $C(q) = \frac{1}{1-B(q)}$, where $B(q) = b_2q^2 + b_3q^3 + \ldots$ and $b_i \geq 0$.

For example, if $n = 2$, then $C(q) = 1$. For the matrix $A = \begin{pmatrix} 2 & -1 & 0 \\ -1 & 2 & -2 \\ 0 & -2 & 2 \end{pmatrix}$ one has $C(q) = (1 - q^2)(1 - q^3)/(1 - q^2 - q^3)$, and $B(q) = q^5/(1 - q^2)(1 - q^3)$.

If $n = 2$, then $E_3(K(A),\mathbb{Z}) = E_\infty(K(A),\mathbb{Z})$ for trivial reasons, and it is not difficult to compute the homology of the complex $(C^*,d^*)$ explicitly, obtaining the additive structure of $H^*(K(A),\mathbb{Z})$. I state here the result for $A = \begin{pmatrix} 2 & -a \\ -a & 2 \end{pmatrix}$, where $a \geq 2$. Define a sequence of integers $c_j$ for $j \in \mathbb{Z}$ by the following recurrent formula:

$$c_0 = 0,\ c_1 = 1,\ c_{j+2} = ac_{j+1} - c_j.$$

Then $H^{2j}(K(A),\mathbb{Z}) \simeq H^{2j+3}(K(A),\mathbb{Z}) \simeq \mathbb{Z}/c_j\mathbb{Z}$. Notice that $c_j = j$ if $a = 2$, and $c_j = \Phi_{2j}$, the 2j-th Fibonacci number, if $a = 3$.

§2.7 The basic tool in the study of the cohomology of flag varieties

$\mathcal{F}(A)$ are certain operators introduced in [21] which "extend" the action of the operators $\Delta_i$ from the image of $\psi$ to the whole cohomology algebra. (This seems to be a new ingredient even in the finite-dimensional case, cf. [1], as far as "bad primes" are concerned [15].)

The Weyl group W acts by right multiplication on $\mathcal{F}(A) = K(A)/T$, which induces an action of W on homology and cohomology of $\mathcal{F}(A)$. On the other hand, since the odd cohomology of $K_i/T_i$ and $K(A)/K_iT$ is trivial, the spectral sequence of the fibration $p_i$: $K(A)/T \longrightarrow K(A)/K_iT$ degenerates after the second term. It follows that $H^*(\mathcal{F}(A),\mathbb{Z})$ is generated by Im $p_i^*$, which is $r_i$-fixed and the element $\psi(\Lambda_i)$.

We deduce that for each i = 1, ..., n there exists a unique $\mathbb{Z}$-linear operator $A^i$ on $H^*(\mathcal{F}(A),\mathbb{Z})$, lowering the degree by 2, such that $r_i$ leaves the image of $A^i$ fixed and

$$u - r_i(u) = A^i(u) \cup \psi(\alpha_i) \text{ for } u \in H^*(\mathcal{F}(a),\mathbb{Z}).$$

Similarly, we introduce operators $A_i$ on $H_*(\mathcal{F}(A),\mathbb{Z})$, raising the degree by 2, such that $r_i(A_i(z)) = -A_i(z)$ and

$$z + r_i(z) = A_i(z) \cap \psi(\alpha_i) \text{ for } z \in H_*(\mathcal{F}(A),\mathbb{Z}).$$

The operators $A^i$ and $A_i$ are dual to each other with respect to the intersection form. One has:

(12) $$A^i(u \cup v) = A^i(u) \cup r_i(v) + u \cup A^i(v);$$

(13) $$A_i(u \cap z) = r_i(u) \cap A_i(z) + A^i(u) \cap z.$$

(14) $$A^i(\psi(\lambda)) = \langle\lambda,h_i\rangle.$$

The operators $A_i$ have the following simple geometric interpretation. Recall the map $\alpha_w$: $D^{\ell(w)} \longrightarrow \mathcal{F}(A)$ defined for $w \in W$ in §2.6. The relative homology map $\alpha_{w*}$ gives us an element $\delta_w \in H_{2\ell(w)}(\mathcal{F}(A),\mathbb{Z})$. Then $\{\delta_w\}_{w\in W}$ is a $\mathbb{Z}$-basis of

$H_*(\mathcal{F}(A),\mathbb{Z})$; let $\{\delta^w\}_{w\in W}$ be the dual basis of $H^*(\mathcal{F}(A),\mathbb{Z})$. We have the following formulas for the action of the Weyl group in these bases generalizing that from [1] (see [21]):

$$(15)\qquad r_i(\delta^w) = \begin{cases} \delta^w \text{ if } \ell(wr_i) > \ell(w), \\ \delta^w - \sum_{wr_i \xrightarrow{\gamma} w'} \langle \alpha_i, \gamma\rangle \delta^{w'} \text{ otherwise} \end{cases}$$

$$(16)\qquad r_i(\delta_w) = \begin{cases} -\delta_w \text{ if } \ell(w) > \ell(wr_i), \\ -\delta_w + \sum_{w' \xrightarrow{\gamma} wr_i} \langle \alpha_i, \gamma\rangle \delta_{w'}, \text{ otherwise} \end{cases}$$

The basic fact that is used to prove these and other formulas is the following lemma which describes the action of the operators $A_i$ and $A^i$ on Schubert cycles $\delta_w$ and cocycles $\delta^w$.

__Lemma__ [21].

(a) $A_i(\delta_w) = \delta_{wr_i}$ if $\ell(wr_i) > \ell(w)$ and $= 0$ otherwise.

(b) $A^i(\delta^w) = \delta^{wr_i}$ if $\ell(w) > \ell(wr_i)$ and $= 0$ otherwise.

__Corollary.__

(a) The subalgebra of W-invariants on $H^*(\mathcal{F}(A),\mathbb{Z})$ coincides with $H^0(\mathcal{F}(A),\mathbb{Z})$.

(b) The operators $A^i$ generate a Hecke algebra, i.e. an associative algebra on the $A^i$ with defining relations: $(A^i)^2 = 0$; $A^iA^jA^i\ldots = A^jA^iA^j\ldots$ ($m_{ij}$ factors on each side).

Note that Corollary 2.7(a) (which means that $A^i(u) = 0$ for all i implies $u \in H^0(\mathcal{F}(A),\mathbb{Z})$) together with (12), (14) and (15) completely determines the multiplicative structure of the algebra $H^*(\mathcal{F}(A),\mathbb{Z})$. Formulas are especially simple when one of the factors is of degree 2; then we get the following formulas, which generalize that in [1] (see

[21]):

$$(17) \qquad \psi(\lambda) \cup \delta^{w} = \sum_{w \xrightarrow{\gamma} w'} \langle\lambda,\gamma\rangle \delta^{w'};$$

$$(18) \qquad \psi(\lambda) \cap \delta_{w} = \sum_{w' \xrightarrow{\gamma} w} \langle\lambda,\gamma\rangle \delta_{w'}.$$

Note that Proposition 2.6(a) follows immediately from the fact that $\psi \circ \Delta_i = A^i \circ \psi$, which is clear from the construction of the $A^i$.

Furthermore, using the operators $A^i$, we can compute by induction on the degree of u the action of the total Steenrod power $\mathcal{P}$ on $H^*(\mathcal{F}(A),\mathbb{F}_p)$ by the following formula [21]:

$$(19) \qquad A^i(\mathcal{P}(u)) = \mathcal{P}(A^i(u))(1 + \psi(\alpha_i)^{p-1}).$$

Finally note that the same approach allows us to compute the Lie algebra cohomology $H^*(\mathfrak{g}'(A),\mathbb{C})$ and to show that it is isomorphic to $H^*(K(A),\mathbb{C})$. A differential forms approach to the study of $\mathcal{F}(A)$ is developed by Kumar in [24] and in a paper of this volume.

**§2.8** A Kac-Moody algebra $\mathfrak{g}(A)$ is finite-dimensional if and only if A is of finite type (i.e. all principal minors of A are positive). The class of these algebras coincides with the class of finite-dimensional semisimple Lie algebras. The associated group G(A) is the Lie group of $\mathbb{C}$-points of the connected simply connected algebraic group whose Lie algebra is $\mathfrak{g}(A)$. The group K(A) is the compact form of G(A), H is the Cartan subgroup of G(A), $B_+$ and $B_-$ are "opposite" Borel subgroups, etc. In this case most of the results of Chapter 2, except for some results of §2.6 and 2.7, are well-known.

In this section we discuss in more detail the case when the matrix A is of affine type, i.e. all proper principal minors of A are positive, but det A = 0 (A is then automatically indecomposable and symmetrizable). An example of such a matrix is the extended Cartan matrix of a simple finite-dimensional Lie algebra. This is the "non-twisted" case we will be dealing with. The "twisted" case is then routinely deduced by taking a fixed point set of an automorphism of order 2 or 3 (see [14, Chapter 8] for details).

Let $\mathring{g}$ be a complex simple finite-dimensional Lie algebra with Chevalley generators $\mathring{e}_i$, $\mathring{f}_i$, $\mathring{h}_i$, $i = 1, \ldots, \ell$, and let $M = \sum \mathbb{Z}\mathring{h}_i$, $\mathring{h} = \mathbb{C} \otimes_{\mathbb{Z}} M$. Let $\mathring{A} = (a_{ij})_{i,j=1}^{\ell}$ be the Cartan matrix and $A = (a_{ij})_{i,j=0}^{\ell}$ the extended Cartan matrix of $\mathring{g}$. We may identify the affine Lie algebra $g'(A)$ with the Lie algebra $\tilde{\mathring{g}}_{\mathbb{C}[z,z^{-1}]}$ (see §1.3 for its definition) via the isomorphism determined by:

$$e_i \mapsto 1 \otimes \mathring{e}_i,\ f_i \mapsto 1 \otimes \mathring{f}_i,\ i = 1, \ldots, \ell;$$

$$e_0 \mapsto z \otimes e_{-\theta},\ f_0 \mapsto z^{-1} \otimes e_{\theta},$$

where $\theta$ is the highest root of $\mathring{g}$, and $e_{-\theta}$ and $e_{\theta}$ are root vectors normalized such that for $\mathring{h}_0 := [e_{\theta}, e_{-\theta}]$ one has: $\theta(\mathring{h}_0) = 2$. Since $\Omega^1_{\mathbb{C}[z,z^{-1}]} = \mathbb{C}\frac{dz}{z} + d\mathbb{C}[z,z^{-1}]$, this construction coincides with the customary one (see e.g. [14, Chapter 7]). In particular dim $\mathfrak{c} = 1$ and $\mathfrak{c} = \mathbb{C}c$, where $c = \sum_{i=0}^{\ell} a_i^{\vee} h_i$, $a_i^{\vee}$ are positive relatively prime integers. Thus, we have an exact sequence:

$$(20) \qquad 0 \longrightarrow \mathbb{C}c \longrightarrow g'(A) \xrightarrow{d\tau} \mathring{g}_{\mathbb{C}[z,z^{-1}]} \longrightarrow 0.$$

Taking $F(P(z))$ = constant term of $P(z) \in \mathbb{C}[z,z^{-1}]$, one easily sees that $g(A) = \hat{\mathring{g}}_{\mathbb{C}[z,z^{-1}],F}$ (see §1.3 for the definition).

As in the case of the affine Lie algebra theory, our objective is to describe the structure of the affine group $G(A)$ in terms of the "underlying" finite-dimensional group $G(\mathring{A})$.

Let $G$ be connected simply connected algebraic group over $\mathbb{C}$ whose Lie algebra is $\mathring{g}$. We will denote by $G_R$ the group of points of $G$ over a commutative algebra $R$ in a fixed finite-dimensional faithful $G$-module $V$.

First of all, we identify the group $G(\mathring{A})$ with the group $\mathring{G} :=$

$G_{\mathbb{C}}$. Using the notation of §2.3, we have injective homomorphisms $\mathring{\varphi}_i$: $SL_2(\mathbb{C}) \longrightarrow \mathring{G}$, the subgroups $\mathring{G}_i$, $\mathring{H}_i$, $\exp t\mathring{e}_i$, $\exp t\mathring{f}_i$ and elements $\mathring{\bar{r}}_i$ for $i = 1, \ldots, \ell$. Then $\mathring{H} := \prod_{i=1}^{\ell} \mathring{H}_i$ is the Cartan subgroup of $\mathring{G}$, the subgroup $\mathring{U}_+$ (resp. $\mathring{U}_-$), generated by the $\exp t\mathring{e}_i$ (resp. $\exp t\mathring{f}_i$), $t \in \mathbb{C}$, $i = 1, \ldots, n$, are maximal unipotent subgroups of $\mathring{G}$.

Let $\mathring{\bar{W}}$ (resp. $\mathring{N}$) be the subgroup of $\mathring{G}$ generated by the $\mathring{\bar{r}}_i$, $i = 1, \ldots, n$ (resp. by $\mathring{\bar{W}}$ and $\mathring{H}$). Then $\mathring{N}$ is the normalizer of $\mathring{H}$ in $\mathring{G}$ and $\mathring{N}/\mathring{T} = \mathring{W}$, the Weyl group of $\mathring{G}$. Let $\mathring{C}$ denote the center of $\mathring{G}$ (it is finite).

It is not difficult to see that the group associated to the integrable Lie algebra $\mathring{\mathfrak{g}}_{\mathbb{C}[z,z^{-1}]}$ is $\tilde{G} := G_{\mathbb{C}[z,z^{-1}]}$, and that associated to the exact sequence (20), we have an exact sequence of groups:

$$(21) \qquad 1 \longrightarrow \mathbb{C}^{\times} \xrightarrow{\ \mu\ } G(A) \xrightarrow{\ \tau\ } \tilde{G} \longrightarrow 1.$$

We have a canonical embedding $\mathring{G} \longrightarrow \tilde{G}$; the exact sequence (21) splits uniquely over $\mathring{G}$, hence we have a canonical embedding $\mathring{G} \longrightarrow G(A)$, so that $\varphi_i = \mathring{\varphi}_i$, $G_i = \mathring{G}_i$, $H_i = \mathring{H}_i$ and $\bar{r}_i = \mathring{\bar{r}}_i$ for $i = 1, \ldots, \ell$. Furthermore, associated to the integrable homomorphism $s\ell_2(\mathbb{C}) \longrightarrow \mathring{\mathfrak{g}}_{\mathbb{C}[z,z^{-1}]}$ defined by $\begin{pmatrix} a & b \\ c & -a \end{pmatrix} \longmapsto -a\mathring{h}_0 + bz^{-1}e_{\theta} + cze_{-\theta}$, we have an injective homomorphism $SL_2(\mathbb{C}) \longrightarrow \tilde{G}$, which lifts uniquely to $\varphi_0$: $SL_2(\mathbb{C}) \longrightarrow G(A)$. The homomorphism $\mu$ is defined by $\mu(t) = \prod_{i=0}^{\ell} H_i(t)$, $t \in \mathbb{C}^{\times}$, and we have $C = \mu(\mathbb{C}^{\times}) \times \mathring{C}$.

Define an embedding $M \longrightarrow \tilde{G}$ by $\mathring{h}_i \longmapsto \mathring{H}_i(z)$, $i = 1, \ldots, \ell$. Then we get the subgroup $\mathring{\bar{W}} \ltimes M$ of $\tilde{G}$. Restricting $\tau$ to the subgroup $\bar{W}$ of G(A), we get from (21) the following exact sequence:

$$1 \longrightarrow \{\pm 1\} \longrightarrow \bar{W} \longrightarrow \mathring{\bar{W}} \ltimes M \longrightarrow 1.$$

This sequence of course splits over $\mathring{\bar{W}}$, but over M it gives a non-split exact sequence

$$1 \longrightarrow \{\pm 1\} \longrightarrow L \longrightarrow M \longrightarrow 1.$$

It is not hard to show using the results of [7], that this central extension is determined by the property that for any preimages $\tilde{\alpha}$ and $\tilde{\beta}$ of $\alpha, \beta \in M$, one has

$$\tilde{\alpha}\tilde{\beta}\tilde{\alpha}^{-1}\tilde{\beta}^{-1} = (-1)^{(\alpha \mid \beta)},$$

when the bilinear form $(\cdot \mid \cdot)$ is the W-invariant form on $\mathring{\mathfrak{h}}^*$ normalized by the condition $(\theta \mid \theta) = 2$. Of course, $W(A) \simeq \mathring{W}(A) \ltimes M$.

The invariant bilinear form on $\mathfrak{g}(A)$ (defined in §1.3) is non-degenerate and invariant under Ad G(A) and the adjoint action via $\tilde{G}$ is (see e.g. [19]):

$$(\mathrm{Ad}\ a(z))x(z) = a(z)x(z)a(z)^{-1} + \mathrm{Res\ tr}\ \frac{da(z)}{dz}\ x(z)a(z)^{-1}.$$

Put $\tilde{U}_+ = \{a(z) \in G_{\mathbb{C}[z]} \mid a(0) \in \mathring{U}_+\}$, $\tilde{U}_- = \{a(z^{-1}) \in G_{\mathbb{C}[z^{-1}]} \mid a(\infty) \in \mathring{U}_-\} \subset \tilde{G}$. The exact sequence (21) splits over $\tilde{U}_+$ and $\tilde{U}_-$, but not uniquely. The subgroups $U_+$ and $U_-$ of G(A) are the (unique) sections which fix $v_\Lambda \in L(\Lambda)$ for all $\Lambda \in P_+$. Put $\tilde{U}^k = \{a(z^{\pm 1}) \in G_{\mathbb{C}[z^{\pm 1}]} \mid a(z^{\pm 1}) - I_V \in z^{\pm k} G_{\mathbb{C}[z^{\pm 1}]}\}$, and let $U^k$ be the preimage of $\tilde{U}^k$ in $U_{\pm}$.

The Bruhat and Birkhoff decompositions (2) and (3) give the following decompositions:

$$G_{\mathbb{C}[z,z^{-1}]} = G_{\mathbb{C}[z^{\pm 1}]}\ M\ G_{\mathbb{C}[z]},$$

various versions of which play an important role in geometry and analysis (see e.g. [9], [10]).

Among the integrable highest weight modules the basic module $L(\Lambda_0)$ is especially important. It is realized in [16] in the space of polynomials in infinitely many indeterminates. The main idea behind the work of the Kyoto school on the KdV-type hierarchies is that the generalized Plücker relations can be written in this realization in terms of Hirota bilinear equations, which are PDE of certain special form which include many important PDE of mathematical physics; the variety $V_{\Lambda_0}$ thus becomes the totality of polynomial solutions of these PDE (see [14] for a discussion of these results). A somewhat different approach is discussed in this volume by A. Pressley [27].

Of course, the matrix coefficients of the G(A)-module $V_{\mathbb{C}[z,z^{-1}]}$ are regular functions. None of them, except constants, are strongly regular functions, however, since by Theorem 2.4, a strongly regular function f, such that $f(cg) = f(g)$ for all $c \in C$ and $g \in G(A)$, is constant. Notice that f is a strongly regular function iff for every $g \in G(A)$ there exists $k > 0$ such that $f(u_-gu_+) = f(g)$ for any $u_\pm \in U^k$.

The topology on G(A) is the unique topology such that (20) is an exact sequence of topological groups, $\mathbb{C}^\times$ carries the metric topology and $\tilde{G}$ the topology induced by the box topology on $\mathbb{C}[z,z^{-1}]$.

Now we turn to the discussion of the unitary form K(A) of G(A). Let $\mathring{\omega}_0$ be the involution of the group $\mathring{G}$ which leaves the $\mathring{G}_i$ invariant and induces on it the standard involution of $SL_2(\mathbb{C})$: $a \longmapsto {}^t\bar{a}^{-1}$. The fixed point set of $\mathring{\omega}_0$ is a compact form of $\mathring{G}$ denoted by $\mathring{K}$. The involution $\mathring{\omega}_0$ lifts to an involution $\tilde{\omega}_0$ of $\tilde{G}$ via the antilinear involution of the algebra $\mathbb{C}[z,z^{-1}]$ which maps z to $z^{-1}$. In turn, $\tilde{\omega}_0$ lifts (uniquely) to the involution $\omega_0$ of G(A) by requiring $\omega_0(\mu(t)) = \mu(\bar{t}^{-1})$, $t \in \mathbb{C}^\times$.

Note that $\tilde{G}$ may be viewed as the group of polynomial maps $\mathbb{C}^\times \longrightarrow \mathring{G}$. The fixed point set of $\tilde{\omega}_0$ on $\tilde{G}$ are those maps for which the image of the unit circle is contained in $\mathring{K}$; these are called polynomial loops on $\overset{\circ}{K}$. We denote the group of polynomial loops $S^1 \longrightarrow \overset{\circ}{K}$ by $\tilde{K}$. Exact sequence (20) gives, by restriction, the following exact sequence:

$$1 \longrightarrow S^1 \xrightarrow{\mu} K(A) \xrightarrow{\tau} \tilde{K} \longrightarrow 1.$$

Identifying $\overset{\circ}{K}$ with the subgroup of constant loops of $\tilde{K}$ and denoting by $\Omega(\overset{\circ}{K})$ the subgroup of based loops (i.e. 1 goes to 1), we have $\tilde{K} = \overset{\circ}{K} \ltimes \Omega(\overset{\circ}{K})$.

Consider the map $K(A) \longrightarrow \mathbb{P}V_{\Lambda_0}$ defined by $k \longmapsto \pi_{\Lambda_0}(k)v_{\Lambda_0}$. It is not difficult to see that $\overset{\circ}{K}$ is the stabilizer of $v_{\Lambda_0}$ and hence the above map induces a homeomorphism $\Omega(\overset{\circ}{K}) \xrightarrow{\sim} \mathbb{P}V_{\Lambda_0}$. It is a well-known fact (see [8]) that the space of all continuous based loops on a compact Lie group $\overset{\circ}{K}$ is homotopically equivalent to the space of poynomial loops $\Omega(\overset{\circ}{K})$. Thus, classical results on loop space cohomology [3] fall into the general framework of §2.6. Moreover using that $\pi_i(\Omega(X)) \simeq \pi_{i+1}(X)$, we deduce from Corollary 2.6(d) and (e) that for the affine Weyl group W one has [3]:

$$W(q) = \overset{\circ}{W}(q) \prod_{i=1}^{\ell} (1-q^{2m_i})^{-1},$$

where $m_1 + 1 < m_2 + 1 \leq \dots < m_\ell + 1$ are the degrees of the basic W-invariants, and $H^*(\Omega\overset{\circ}{K},\mathbb{Q})$ is a polynomial algebra on generators of degrees $2m_1, \dots, 2m_\ell$.

Put $\Omega(\overset{\circ}{K})\langle 2\rangle := \tau^{-1}(\Omega(\overset{\circ}{K}))$. This is a standard notation of the 2-connected cover of $\Omega(\overset{\circ}{K})$. This means that the map $\tau$: $\Omega(\overset{\circ}{K})\langle 2\rangle \longrightarrow \Omega(\overset{\circ}{K})$ kills the second homotopy group (which is $\mathbb{Z}$) and induces isomorphism of higher homotopy groups (this property of $\tau$ can be easily checked). Thus, we have

$$K(A) = \overset{\circ}{K} \ltimes \Omega(\overset{\circ}{K})\langle 2\rangle.$$

Since the cohomology of $\overset{\circ}{K}$ is by now well understood [15], it remains (and is of independent interest) to compute the cohomology of $\Omega(\overset{\circ}{K})\langle 2\rangle$. Theorem 2.6 leads to the following result.

**Theorem** [21]. Let $\overset{\circ}{K}$ be a connected simply connected simple compact Lie group, and let $m_1 + 1, \ldots, m_\ell + 1$ be the degrees of the basic invariants of its Weyl group. Then

(a) $H^*(\Omega(\overset{\circ}{K})\langle 2\rangle, \mathbb{Q})$ is a polynomial algebra on generators of degrees $2m_2, \ldots, 2m_\ell$.

(b) The Poincaré polynomial of $H^*(\Omega(\overset{\circ}{K})\langle 2\rangle, \mathbb{F})$, where $\mathbb{F}$ is a field of characteristic $p > 0$, is

$$(1 + q^{2p^a-1})(1 - q^{2p^a})^{-1} \prod_{i=2}^{\ell} (1 - q^{2m_i})^{-1}.$$

Here a is the minimal positive integer such that $\Lambda_0^{p^a} \in I_{\mathbb{F}}$. One has: $a = 1$ if $p > m_\ell$. The number a for $p \leqslant m_\ell$ has been computed recently (at my request) by A. Kono using topological arguments.

## References


[1] Bernstein, I.N., Gelfand, I.M. and Gelfand, S.I., Schubert cells and flag space cohomology, Uspechi Matem. Nauk 28 (1973), 3-26.

[2] Borel, A., Linear algebraic groups, Benjamin, New York, 1969.

[3] Bott, R., An application of the Morse theory to the topology of Lie groups, Bull. Soc. Math. France 84 (1956), 251-281.

[4] Bourbaki, N., Groupes et Algebres de Lie, Chap. 4, 5 and 6, Hermann, Paris, 1968

[5] Curtis, C.W., Central extensions of groups of Lie type, Journal für die Reine und angewandte Math., 220 (1965), 174-185.

[6] Deligne, P., Griffits, P., Morgan, J. and Sullivan, D., Real homotopy theory of Kähler manifolds, Inventiones Math. 29 (1975), 245-274.

[7] Garland, H., Arithmetic theory of loop groups, Publ. Math. IHES 52 (1980), 5-136.

[8] Garland, H. and Raghunathan, M.S., A Bruhat decomposition for the loop space of a compact group: a new approach to results of Bott, Proc. Natl. Acad. Sci. USA 72 (1975), 4716-4717.

[9] Gohberg, I. and Feldman, I.A., Convolution equations and projection methods for their solution, Transl. Math. Monography 41, Amer. Math. Soc., Providence 1974.

[10] Grothendieck, A., Sur la classification des fibres holomorphes sur la sphere de Riemann, Amer. J. Math. 79 (1957), 121-138.

[11] Haddad, A., A Coxeter group approach to Schubert varieties, these proceedings.

[12] Kac, V.G., Simple irreducible graded Lie algebras of finite growth, Math. USSR-Izvestija 2 (1968), 1271-1311.

[13] Kac, V.G., Algebraic definition of compact Lie groups, Trudy MIEM 5 (1969), 36-47 (in Russian).

[14] Kac, V.G., Infinite dimensional Lie algebras, Progess in Math. 44, Birkhäuser, Boston, 1983.

[15] Kac, V.G., Torsion in cohomology of compact Lie groups and Chow rings of algebraic groups, Invent. Math., 80 (1985), 69-79.

[16] Kac, V.G., Kazhdan, D.A., Lepowsky, J. and Wilson, R.L., Realization of the basic representation of the Euclidean Lie algebras, Advances in Math., 42 (1981), 83-112.

[17] Kac, V.G. and Peterson, D.H., Infinite-dimensional Lie algebras, theta functions and modular forms, Adv. in Math. 53 (1984), 125-264.

[18] Kac, V.G. and Peterson, D.H., Regular functions on certain infinite-dimensional groups. In: Arithmetic and Geometry, pp. 141-166. Progress in Math. 36, Birkhäuser, Boston, 1983.

[19] Kac, V.G. and Peterson, D.H., Unitary structure in representations of infinite-dimensional groups and a convexity theorem, Invent. Math. 76 (1984), 1-14.

[20] Kac, V.G. and Peterson, D.H., Defining relations of infinite-dimensional groups, Proceedings of the E. Cartan conference, Lyon, 1984.

[21] Kac, V.G. and Peterson, D.H., Cohomology of infinite-dimensional groups and their flag varieties, to appear.

[22] Kac, V.G., Peterson, D.H., Generalized invariants of groups generated by reflections, Proceedings of the conference "Giornate di Geometria", Rome, 1984.

[23] Kassel, C., Kähler differentials and coverings of complex simple Lie algebras extended over a commutative algebra, J. Pure Applied Algebra (1984).

[24] Kumar, S., Geometry of Schubert cells and cohomology of Kac-Moody Lie algebras, Journal of Diff. Geometry, (1985).

[25] Moody, R., A simplicity theorem for Chevalley groups defined by generalized Cartan matrices, preprint.

[26] Peterson, D.H. and Kac, V.G., Infinite flag varieties and conjugacy theorems, Proc. Natl. Acad. Sci. USA 80 (1983), 1778-1782.

[27] Pressley, A., Loop groups, Grassmanians and KdV equations, these proceedings.

[28] Rudakov, A.N., Automorphism groups of infinite-dimensional simple Lie algebras, Izvestija ANSSSR, (Ser. Mat.) 33 (1969), 748-764.

[29] Séminair "Sophus Lie", 1954/55. Ecole Normale Supérieure, 1955.

[30] Shafarevich, I.R., On some infinite-dimensional groups II, Izvestija AN SSSR (Ser. Mat.) 45 (1981), 216-226.

[31] Slodowy, P., An adjoint quotient for certain groups attached to Kac-Moody algebras, these proceedings.

[32] Steinberg, R., Lectures on Chevalley groups, Yale University Lecture Notes, 1967.

[33] Tits, J., Resumé de cours, College de France, Paris, 1981.

[34] Tits, J., Resumé de cours, College de France, Paris, 1982.

# HARISH-CHANDRA MODULES OVER THE VIRASORO ALGEBRA

By

Irving Kaplansky* and L. J. Santharoubane**

## §1. Introduction

The universal central extension V of the Lie algebra W of vector fields on the circle with finite Fourier series is called by physicists the Virasoro algebra. However, W was known in characteristic p as the Witt algebra, in characteristic 0 as the infinite-dimensional Witt algebra, and Gelfand and Fuks [3] determined the second cohomology group of W with trivial coefficients, thereby describing V. The algebra V and an associated superalgebra play a fundamental role in the study of elementary particles [6].

Let $\mathfrak{h}$ be a subalgebra of a Lie algebra $\mathfrak{g}$. As in [1] we define a Harish-Chandra module M over $(\mathfrak{g},\mathfrak{h})$ to be a $\mathfrak{g}$-module which is completely decomposable as an $\mathfrak{h}$-module into simple finite-dimensional $\mathfrak{h}$-modules.

Kac [4] conjectured that if M is an irreducible Harish-Chandra module over $(V, \mathfrak{h}_0)$ with V = the Virasoro algebra and $\mathfrak{h}_0$ = the Cartan subalgebra of V and if all the simple finite-dimensional $\mathfrak{h}_0$-modules occur in M with finite multiplicity then M is either of the first kind (i.e. M or its restricted dual is an object of the category of Bernstein-Gelfand-Gelfand [4]) or M is of the second kind (i.e. all the simple finite-dimensional $\mathfrak{h}_0$-modules occur with multiplicity at most one).

*Mathematical Sciences Research Institute, Berkeley, California, USA

**Mathematical Sciences Research Institute, Berkeley, California, USA and Department of Mathematics, University of California, Berkeley, California, USA

(Permanent: Department of Mathematics, University of Poitiers, Poitiers, France)

In this paper we investigate modules of the second kind; it is a continuation of [5] and completes the study begun in Theorem 2 of [5].

Our final result is as follows. If A is an indecomposable Harish-Chandra module over $(V, \mathfrak{h}_0)$ such that all the simple finite-dimensional $\mathfrak{h}_0$-modules occur with multiplicity exactly one then A can be identified with one of the following:

(1) The module of tensor fields $A_{a,b}$ of the form $Q(z)dz^b$ where Q(z) is a Laurent polynomial in z divisible by $z^{a-b}$ (with a,b complex parameters). The action of the center of V is trivial and the action of $P(z)\frac{d}{dz} \in V$ (P a Laurent polynomial in z) is given by

$$(P(z)\frac{d}{dz})\ (Q(z)dz^b) = [P(z)Q(z)]_b'dz^b,$$

where $[P(z)Q(z)]_b' = P(z)Q'(z) + bP'(z)Q(z)$ is the b-twisted derivative of P(z)Q(z).

(2) Certain modules $A(\alpha)$, $B(\beta)$ where $\alpha$ and $\beta$ are complex parameters. (When $\alpha$ is 0, $A(\alpha)$ reduces to $A_{0,0}$, and when $\beta$ is 0, $B(\beta)$ reduces to $A_{0,1}$.)

We are greatly indebted to Victor Kac for advice and suggestions. In addition the junior author wishes to warmly acknowledge his indebtedness to his colleagues: B. Bernat, J. Borowzyc, C. Deal, F. Ducloux, J. Duchet, B. Grimonprez, M. Lazard, O. Mathieu, M. Rais, and P. Torasso.

## §2. Preliminaries

The Virasoro algebra V can also be defined as follows, with $x_i$ corresponding to $z^{i+1}\frac{d}{dz}$:

$$V = \sum_{i\in\mathbb{Z}} \mathbb{C}x_i + \mathbb{C}C,$$

$$x_ix_j = (j-i)x_{i+j} + \frac{i^3-i}{12}\,\delta_{i,-j}C,\quad Cx_i = 0.$$

(As in [5] we are shortening the writing by omitting brackets. Also, the coefficient $i - j$ has been replaced by $j - i$ and the one-dimensional center has been inserted.)

Let A be a Harish-Chandra module over $(V, \mathbb{C}x_0)$, i.e.

$$A = \sum_{\alpha \in \mathbb{C}} A_\alpha \text{ with } A_\alpha = \{y \in A;\ x_0 y = \alpha y\}.$$

Throughout this paper we assume that A is indecomposable and that the components are at most one-dimensional.

Since C and $x_0$ commute, C leaves each $A_\alpha$ invariant. But $A_\alpha$ is at most one-dimensional, so the action of C is semi-simple. By the indecomposability hypothesis there exists $c \in \mathbb{C}$ such that C acts as multiplication by c. From the relation $x_0 x_i = i x_i$ it follows that $x_i$ sends $A_\alpha$ into $A_{\alpha+i}$. Therefore, again by indecomposability, we can assume that all the $\alpha$'s are congruent modulo $\mathbb{Z}$. Pick $a \in \mathbb{C}$ so that all $\alpha$'s are congruent to a mod $\mathbb{Z}$; then A is the sum of the subspaces $A_{a+j}$ for $j \in \mathbb{Z}$.

## §3. <u>The case where $x_1$ and $x_{-1}$ do not annihilate</u>

In this case we have the setup of Theorem 2 of [5].

We take this opportunity to rectify the discussion in [5]. In the first place there is an unfortunate typographical error: the last factor in the numerator of (14) should be $z - 3b - 3$ rather than $z - 3b + 3$. More important: the formulas given on page 53 <u>do</u> define a representation of V. But by a suitable change of basis, it turns out that this representation can be reverted to the form

(2) $$x_i v_j = (a + bi + j)\, v_{i+j}.$$

These modules coincide with the modules $A_{a,b}$ mentioned in the introduction, with $v_j$ corresponding to $z^{j+a-b} dz^b$. For a reference concerning the modules $A_{a,b}$ see [2].

(The observation that the representations on page 53 of [5] take the form (2) relative to a suitable basis is due to Arne Meurman and Alvani Rocha-Caridi; we are incorporating it with their kind permission.)

The details are as follows. In the representation in question we have

$$x_2 v_j = \frac{(z+b)(z+b-1)(z-2b-2)}{(z-b-1)(z-b-2)} v_{j+2},$$

$$x_{-2} v_j = \frac{(z-b)(z-b+1)(z+2b+2)}{(z+b+1)(z+b+2)} v_{j-2}$$

where $z = a + j$. (Note that there is a change of sign in j here and in (2); this occurs because in (1) we are using $j - i$ in place of the $i - j$ of [5].)

Change basis by $w_j = \lambda_j v_j$, where the $\lambda_j$'s are nonzero constants satisfying

$$\frac{\lambda_j}{\lambda_{j+1}} = \frac{z-b-1}{z+b};$$

the denominator $z + b = a + j + b$ is not 0 since $u_1 v_j = (a + b + j)v_{j+1} \neq 0$, and similarly the numerator is not 0. This can be done, for instance, by setting $\lambda_0 = 1$ and determining the other $\lambda$'s in succession. Then one easily verifies:

$$x_0 w_j = z w_j,\ u_1 w_j = (z - b - 1) w_{j+1},\ x_{-1} w_j = (z + b + 1) w_{j-1}$$

$$x_2 w_j = (z - 2b - 2) w_{j+2},\ x_{-2} w_j = (z + 2b + 2) w_{j-2}.$$

These formulas show that we have achieved the form (2), with the parameter b replaced by $-b - 1$.

With the proof thus amended, Theorem 2 of [5] stands as correct.

## §4. The case where $x_1$ or $x_{-1}$ annihilates

Recall that our module A is a direct sum of one-dimensional submodules $A_{a+j}$, where j ranges over a subset of $\mathbf{Z}$. We shall now assume that the range of j is all of $\mathbf{Z}$.

So A has a basis $\{v_j\}$, j ranging over all of $\mathbf{Z}$, such that $x_i v_j$

is a scalar multiple of $v_{i+j}$. In particular,

(3) $$x_0 v_j = (a + j)\ v_j.$$

Because of §3, in continuing the investigation we may assume that either $x_1$ or $x_{-1}$ annihilates some $v_j$. In fact, we may assume that (say) $x_{-1}$ annihilates some $v_j$. The reason is that we are free to pass to the so-called inverted module (this is the terminology used in [2]). In detail, if the given module has $x_i v_j = f(i,j) v_{i+j}$, in the inverted module we have $x_i v_j = -f(-i,-j) v_{i+j}$. The roles of $x_1$ and $x_{-1}$ are thereby interchanged. (Of course, in stating the final theorem we shall take account of this normalization.)

By a harmless translation of the indexing of the v's we may assume $x_{-1} v_0 = 0$. One then knows that

(4) $$x_1 \cdot x_{-1} v_j = j(j + 2a - 1)\ v_j,$$

(5) $$x_{-1} \cdot x_1 v_j = (j + 1)\ (j + 2a)\ v_j.$$

This is seen by the argument that led to equation (4) on page 51 of [5], together with the simplifying fact that here we know that $d_0 = 0$, where $x_1 \cdot x_{-1} v_j = d_j v_j$.

From (4) it follows that there is at most one other $v_j$ annihilated by $x_{-1}$. We shall now prove that if $x_{-1}$ does annihilate two v's they must be consecutive. Later in the paper it will be convenient to normalize two such consecutive v's as $v_{-1}$ and $v_0$ (rather than $v_0$ and $v_1$). In preparation for this, we shall therefore suppose, as we may, that in addition to $x_{-1} v_0 = 0$ we have $x_{-1} v_{-r} = 0$ with r a positive integer; we proceed to prove r = 1. We have that $x_r \cdot x_{-1} v_{-r}$ and $x_{-1} \cdot x_r v_{-r}$ are both 0. Hence $x_{r-1} v_{-r} = 0$. If r = 1 we have $x_0 v_{-r} = 0$. If r > 1 we note that $x_0$ is obtainable from $x_{r-1}$ by repeated applications of $x_{-1}$, so that we again have $x_0 v_{-r} = 0$. From (3) we deduce that a - r = 0, a = r. On the other hand, from (4) we see that j + 2a - 1 must vanish for j = -r, i.e. -r + 2a - 1 = 0. In conjunction with a = r, this implies r = 1, as required.

Of course it is likewise true that $x_1$ annihilates at most two v's, and if it annihilates two they must be consecutive. We proceed to rule out the possibility that both $x_{-1}$ and $x_1$ annihilate two v's. Suppose, on the contrary, that this does happen. As above, we may assume $x_{-1}v_{-1} = x_{-1}v_0 = 0$, and we observe (as implicitly noted above) that (4) implies a = 1. Equation (5) shows that the two v's annihilated by $x_1$ must be $v_{-1}$ and $v_{-2}$. Now $x_2 \cdot x_{-1}v_{-1}$ and $x_2x_{-1} \cdot v_{-1}$ both vanish. Hence $x_{-1} \cdot x_2v_{-1} = 0$, whence $x_2v_{-1} = 0$ (since $x_{-1}v_1 \neq 0$). Similarly $x_{-2}v_{-1} = 0$. It follows that $\mathbb{C}v_{-1}$ is invariant under V. Next we argue that $x_2v_{-3} = 0$. Since $x_{-1}v_{-2} \neq 0$, it suffices to check that $x_2 \cdot x_{-1}v_{-2} = 0$, and this follows from

(6) $$3x_1v_j = x_{-1} \cdot x_2v_j - x_2 \cdot x_{-1}v_j$$

with j = -2. Similarly $x_{-2}v_1 = 0$ follows from

(7) $$3x_{-1}v_j = x_{-2} \cdot x_1v_j - x_1 \cdot x_{-2}v_j$$

with j = 0. It follows that $A' = \bigoplus_{j \neq -1} \mathbb{C}v_j$ is invariant under $x_{\pm 1}$ and $x_{\pm 2}$ and hence is a V-submodule of A. We have the direct sum decomposition $A = A' \oplus \mathbb{C}v_0$, contradicting the assumed indecomposability of A.

## §5. <u>The action of $x_2$ and $x_{-2}$</u>

The elements $x_{-2}$, $x_2$ and $x_{-2}x_2 = 4x_0 - (1/2)C$ span a copy of the three-dimensional simple Lie algebra. The $v_j$'s with j even span a representation space for this subalgebra. One then knows (as in (4) and (5)) that the coefficient of $v_j$ in $x_{-2} \cdot x_2v_j$ is a quadratic polynomial in j. However, since $x_2$ or $x_{-2}$ do not necessarily annihilate any $v_j$'s, we are not able to identify this polynomial as easily as was the case in (4) and (5).

The same remarks apply to the action of $x_2$ and $x_{-2}$ on the odd $v_j$'s. A priori, a different quadratic polynomial might arise. In this section we shall show that the two polynomials are the same and we shall compute the polynomial.

Recall that we are assuming $x_{-1}v_0 = 0$, and that $x_{-1}$

annihilates at most one other $v_j$. Furthermore we know that if there is such a second annihilated element it must be adjacent to $v_0$. As in the preceding section we take the second one (if it exists) to be $v_{-1}$. We can normalize the $v_j$'s with $j \geq 0$ (by multiplying them by suitable nonzero scalars) so as to satisfy

$$(8) \qquad x_{-1}v_j = j\, v_{j-1} \qquad (j \geq 0).$$

By applying (8) to (5) we then get

$$(9) \qquad x_1 v_j = (j + 2a)\, v_{j+1} \qquad (j \geq 0).$$

On putting $j = 0$ in (6) and using (8) and (9) we evaluate $x_2 v_0$ as $3av_2$. Then (6) can be used inductively to deduce

$$(10) \qquad x_2 v_j = (j + 3a)\, v_{j+2} \qquad (j \geq 0).$$

We proceed to study the action of $x_{-2}$, the argument following closely page 52 of [5]. Write

$$(11) \qquad x_{-2}v_j = (h(j) + j - a)\, v_{j-2}.$$

We insert (11) into (7) and make use of (8) and (9). A brief computation yields

$$(12) \qquad (j + 2a)\, h(j + 1) - (j + 2a - 2)\, h(j) = 0 \qquad (j \geq 2).$$

Take $j > 2|a| + 2$, thereby insuring that the coefficients $j + 2a$ and $j + 2a - 2$ in (12) do not vanish. Then we conclude that, for large positive $j$, $h$ is either 0 or the reciprocal of a quadratic polynomial. For $j$ even, the latter alternative, in conjunction with (10), is incompatible with the fact that the coefficient of $v_j$ in $x_{-2} \cdot x_2 v_j$ is a polynomial in $j$, that is, $x_{-2}v_j = (j - a)\, v_{j-2}$ for large positive $j$.

The identification of the action of $x_2$ and $x_{-2}$ on $v_j$ for large positive $j$ makes it possible to compute the action of $x_{-2}x_2 = 4x_0 -$

(1/2)C. We find that it sends $v_j$ into $4(a + j)\, v_j$ just as $4x_0$ does. Knowing this for a single j suffices to show that c = 0 (recall from §2 that we know that C acts on A as multiplication by the constant c). Thus CA = 0 and C will play no further role in the discussion.

Note that we have

$$x_{-2} \cdot x_2 v_j = (j + 3a)\,(j + 2 - a)\, v_j \tag{13}$$

for large positive j. First for even j, and then for odd j, we now see that the quadratic polynomial occurring as the coefficient of $v_j$ in $x_{-2} \cdot x_2 v_j$ has been identified as $(j + 3a)\,(j + 2 - a)$. We have proved that (13) holds for all j.

## §6. Impossibility of the second case

If we again take into account the option of switching to the inverted module (see §4), we see that there are two cases to consider:

I. $x_{-1}$ annihilates $v_0$ and no other $v_j$,

II. $x_{-1}$ annihilates precisely $v_{-1}$ and $v_0$, while $x_1$ does not annihilate any $v_j$.

In this section we shall show that the second case leads to a contradiction.

We first put $j = -1$ in (4). The left side is 0 since $x_{-1}v_{-1} = 0$. Hence $j(j + 2a - 1)$ must vanish for $j = -1$, and $a = 1$ follows. On putting $a = 1$ in (13) we deduce

$$x_{-2} \cdot x_2 v_j = (j + 3)\,(j + 1)\, v_j \tag{14}$$

for all j.

We examine (6) for $j = -1$. The left side is nonzero since $x_1$ annihilates no $v_j$. The final term is 0 since $x_{-1}v_{-1} = 0$. Hence $x_2 v_{-1} \neq 0$. Use this in (14) with $j = -1$ to deduce $x_{-2}v_1 = 0$. Next set $j = -2$ in (14) to get $x_{-2}v_0 \neq 0$. When all this information is inserted into (7) with $j = 0$ we get the desired contradiction, since the left side and the first term on the right vanish, while the final term does not vanish.

## §7. Analysis of the first case

In this case $x_{-1}$ annihilates only $v_0$. By appropriately normalizing the $v_j$'s with j negative we can strengthen (8) to

$$x_{-1}v_j = jv_{j-1} \qquad \text{(all j)}. \tag{15}$$

then (15) and (5) yield a strengthening of (9):

$$x_1v_j = (j + 2a)\, v_{j+1} \qquad (j \neq -1). \tag{16}$$

We have no information on $x_1v_{-1}$ and write

$$x_1v_{-1} = Hv_0. \tag{17}$$

It is to be noted that a degree of freedom remains: we can multiply the $v_j$'s with $j \leqslant -1$ by a nonzero scalar and those with $j \geqslant 0$ by a different nonzero scalar. This possibility will be appropriately exploited in the discussion that lies ahead.

We proceed to strengthen (10). We put $j = -2$ in (6) and evaluate $x_2v_{-3}$ as $(3a - 3)\, v_{-1}$. Then induction gives us (10) for $j \leqslant -3$. In sum, we have

$$x_2v_j = (j + 3a)\, v_{j+2} \qquad (j \neq -1 \text{ or } -2). \tag{18}$$

Let us write the missing values as $x_2v_{-2} = Dv_0$, $x_2v_{-1} = Ev_1$. From (6) with $j = -1$ and (16) we get

$$3H = D + E. \tag{19}$$

An immediate consequence of (18) and (13) is

$$x_{-2}v_j = (j - a)\, v_{j-2} \qquad (j \neq 1,\ 0,\ 2 - 3a). \tag{20}$$

Write $x_{-2}v_0 = Fv_{-2}$, $x_{-2}v_1 = Gv_{-1}$. By setting $j = -2$ and $-1$ in (13) we find

$$DF = a(2 - 3a), \tag{21}$$

$$EG = (3a - 1)(1 - a). \tag{22}$$

Next we return to (7) and insert the values $j = -1, 1$ not hitherto used. In using (20) to make the computation we must observe the excluded value $2 - 3a$; this results in the values of $a$ excluded in (23) and (24). From $j = -1$ we derive

$$HF = a(1 - 2a) \qquad (a \neq 1), \tag{23}$$

and from $j = 1$

$$HG = (2a - 1)(1 - a) \qquad (a \neq 0). \tag{24}$$

In continuing the discussion there are four cases.

I. $a \neq 0, 1, 1/2$. Here we shall identify the representation as being of the form (2), with $b = a$.

We note that $H \neq 0$ for otherwise (23) and (24) show that $a = 1/2$. The residual degree of freedom mentioned right after equation (17) can be used to normalize $H$ to be $2a - 1$. Cancellation in (23) and (24) then shows that $F = -a$, $G = 1 - a$. Insert these values in (21) and (22) to get $D = 3a - 2$, $E = 3a - 1$. We have now achieved (16) and (18) for all $j$, and (20) for all $j$ except $2 - 3a$, in the event that $3a$ is an integer. However, with the information now available this gap can be filled by setting $j = 1 - 3a$ in (7). (It is to be noted that $a \neq 1$ is again used here, for if $a = 1$ then $x_1v_{1-3a}$, which is $(1 - a)\, v_{2-3a}$ by (16), would vanish and the computation would fail to catch $x_{-2}v_{2-3a}$.) In sum, $x_{\pm 1}$ and $x_{\pm 2}$ act exactly as they should to ensure that (2) is satisfied, with $a = b$. Thus $A$ is $A_{a,a}$.

II. $a = 1/2$. Equation (21) shows that $F \neq 0$ and then

(23) shows that $H = 0$. Equations (19), (21), and (22) become $D + E = 0$, $DF = EG = 1/4$. The residual degree of freedom is still available; we use it to normalize D to be $-1/2$. Then $E = 1/2$, $F = -1/2$, $G = 1/2$. We have fulfilled (16), (18), and (20) for all j (note that 3a is not an integer, so there is no further problem about (20).) Again the representation has the form (2), with $a = b = 1/2$, and A is $A_{1/2,1/2}$.

III. $a = 1$. Equation (23) is not available. Equations (21), (22), and (24) take the form $DF = -1$, $EG = 0$, $HG = 0$. We assert that $G = 0$. Otherwise E and H both vanish, whence $D = 0$ by (19), contrary to $DF = -1$. The residual degree of freedom can be used to normalize F to be $-1$; this implies $D = 1$. H remains at liberty, with E determined as $3H - 1$. We note that (18) and (20) are now fulfilled with $a = 1$ except for $j = -1$. The element $x_2v_{-1}$ is at hand (as $Ev_1$). As for $x_{-2}v_{-1}$, a final use of (7) with $j = -1$ yields $x_{-2}v_{-1} = (H - 3)\, v_{-3}$. The upshot is that the following relations hold for $i = 0, \pm 1, \pm 2$:

$$x_iv_j = (1 + i + j)\, v_{i+j} \qquad (j \neq -1),$$

(25)

$$x_iv_{-1} = (i + i(i + 1)\alpha)\, v_{i-1},$$

where H has been replaced by $2\alpha + 1$.

Now it is a routine verification that the equations (25), together with $Cv_j = 0$ for all j, define a representation $A(\alpha)$ of V. This representation coincides with the given representation for the elements $x_{\pm 1}$ and $x_{\pm 2}$, which generate V. Hence $A(\alpha)$ coincides with the given representation on all of V.

For every value of the parameter $\alpha$, the representation $A(\alpha)$ is indecomposable. However, it is reducible, for $\sum_{j \neq -1} \mathbb{C}v_j$ is an invariant subspace.

IV. $a = 0$. Equation (24) is not available. Equations (21), (22), and (23) take the form $DF = 0$, $EG = -1$, $HF = 0$. If $F \neq 0$, D and H vanish, so does E by (19), a contradiction. So $F = 0$. The residual degree of freedom is used here by normalizing G to be 1.

Then $E = -1$, $H$ is again at liberty, and $D = 3H + 1$. The element $x_{-2}v_2$ is evaluated as $(H + 3)\, v_0$ by putting $j = 1$ in (7). For $i = 0, \pm 1, \pm 2$ we have verified

$$x_i v_j = j v_{i+j} \qquad (j \neq -i),$$

(26)

$$x_i v_{-i} = (-i - i(i+1)\beta) v_0,$$

where $H$ this time has been replaced by $-2\beta - 1$. The remarks made in Case III can now be repeated, except that the reducibility of the representation $B(\beta)$ given by (26) is apparent from the fact that $v_0$ is annihilated by V.

__Remark__. Although we have treated both of the cases $a = 1$ and $a = 0$ in full, the discussion could in fact have been cut in half by making use of what is called the adjoint module in [2]. For present purposes it is convenient to define the adjoint module as follows: if $x_i v_j = f(i,j)\, v_{i+j}$ in a given module, the adjoint is given by $x_i v_j = -f(i,-j - i - 1)\, v_{i+j}$. (The term $-1$ is inserted here so as to maintain the normalization $x_{-1}v_0 = 0$.) The effect of this is to replace the parameter $a$ by $1 - a$ and thus interchange the cases $a = 0$ and $a = 1$.

We summarize the discussion in the following theorem.

__Theorem__. Let V be the Virasoro algebra and $\mathfrak{h}_0$ its Cartan subalgebra. Let A be an indecomposable Harish-Chandra module over $(V, \mathfrak{h}_0)$ with the property that the constituents $A_{a+j}$ in the decomposition of A relative to $\mathfrak{h}_0$ are all one-dimensional, where $j$ ranges over all of $\mathbf{Z}$. Then A is isomorphic to one of the modules $A_{a,b}$, $A(\alpha)$, or $B(\beta)$.

## §8. __Concluding remarks__

(a) The basic setup in [5] is a little different from that used in this paper, in that it is assumed at the start that the given

V-module A is $\mathbb{Z}$-graded (i.e. graded by the integers) with the natural relation between the $\mathbb{Z}$-gradings on V and A. However, it is easy to pass from the setup of [5] to the present one. Here are the details.

A has a basis $v_j$ such that $x_iv_j$ is a scalar multiple of $v_{i+j}$ for all i and j. It is assumed that A is graded-indecomposable. The missing ingredient is that we do not yet know that the characteristic subspaces relative to $x_0$ are one-dimensional. Write $x_0v_j = s_jv_j$. Define a relation on the integers by $i \sim j$ if $s_i - s_j = i - j$. Manifestly this is an equivalence relation. If $x_iv_j \neq 0$ we deduce from

$$\begin{aligned} ix_iv_j = x_0x_i \cdot v_j &= x_0 \cdot x_iv_j - x_i \cdot x_0v_j \\ &= x_0 \cdot x_iv_j - s_jx_iv_j \end{aligned}$$

that $s_{i+j} = s_j + i$. Hence

(27) $$x_iv_j \neq 0 \text{ implies } j \sim i + j.$$

For any equivalence class I of the relation write $A_I$ for the subspace of A spanned by the $v_i$'s, $i \in I$. By (27) we see that $A_I$ is a homogeneous invariant subspace of A. One has that A is the direct sum of the $A_I$'s. Therefore (since A is graded-indecomposable) there exists I such that $A = A_I$. In particular, the $s_j$'s are distinct; indeed $s_j = s_0 + j$. Thus the characteristic subspaces relative to $x_0$ are one-dimensional, as desired.

In the reverse direction, an appropriate $\mathbb{Z}$-grading on A was available right after we assumed in §2 that all the $\alpha$'s are congruent modulo $\mathbb{Z}$.

(b) The elements $x_0$ and C clearly span a Cartan subalgebra of V. Several times above we referred to this as "the" Cartan subalgebra of V, thus suggesting uniqueness. Uniqueness does indeed hold, as follows from the following proposition.

<u>Proposition</u>. If an element p of the centerless Virasoro algebra W acts diagonally on W, then p must be a scalar multiple of $x_0$.

**Proof.** Say

$$p = \alpha_1 x_{n_1} + \alpha_2 x_{n_2} + \dots + \alpha_r x_{n_r}$$

with $n_1 < n_2 < \dots < n_r$. We can assume $n_r > 0$. There must exist $q \neq 0$ in W with pq a scalar multiple of q and with the highest $x_s$ occurring in q higher than $x_{n_r}$. But then in pq we have the term $x_{n_r+s}$ not subject to any cancellation, so that pq cannot be a scalar multiple of q.

It is an immediate corollary that the group of automorphisms of W is generated by $x_i \longrightarrow \lambda^i x_i$ and $x_i \longrightarrow -x_{-i}$.

(c) In this final remark we record some observations concerning the modules. Proofs are left to the reader.

(I) $A_{a,b}$ is irreducible if and only if $(a,b) \neq (0,0)$ or $(0,1)$.

(II) $A_{a,b}$ and $A_{c,d}$ are isomorphic if and only if $a - c \in \mathbb{Z}$ and $b = d$ or $1 - d$.

(III) $A(\alpha)$ is isomorphic to $A(\alpha')$ if and only if $\alpha = \alpha'$; the same is true for $B(\beta)$.

(IV) Duality works out as follows: $A_{a,b}$ has $A_{-a,b}$ as its inverted module, $A_{-a,1-b}$ as its adjoint, and $A_{a,1-b}$ as its contragredient; $A(\alpha)$ has $A(-\alpha)$ as its inverted module, $B(-\alpha)$ as its adjoint, and $B(\alpha)$ as its contragredient.

(V) $A(0)$ and $B(0)$ are isomorphic to $A_{0,0}$ and $A_{0,1}$, as noted above at the end of §1. There are no other isomorphisms.

(VI) $A(\alpha)$ has a simple submodule of codimension one, namely the module of Laurent series with vanishing constant term. The quotient module is the one-dimensional trivial module. Dually, $B(\beta)$ has the trivial submodule of dimension one as a submodule, with quotient the same module of Laurent series.

## References

[1] J. Dixmier, Enveloping Algebras, North-Holland, 1977.

[2] B. L. Feigin and D. B. Fuks, Invariant skew-symmetric differential operators on the line and Verma modules over the Virasoro algebra, Funct. Anal. Appl. 16(1982), no. 2, 47–63; English translation 114–126.

[3] I. M. Gelfand and D. B. Fuks, Cohomologies of the Lie algebra of vector fields on the circle, Funct. Anal. Appl. 2(1968), no. 4, 92–93; English translation 342–343.

[4] V. G. Kac, Some problems on infinite dimensional Lie algebras and their representations, pp. 117–126 in Lie Algebras and Related Topics, Springer Lecture Notes no. 933, 1982.

[5] I. Kaplansky, The Virasoro algebra, Comm. Math. Phys. 86(1982), 49–54.

[6] J. Schwartz, Dual resonance theory, Physics Reports 8c(1973), 269–335.

Added in proof:

Some results of the present work have been given a cohomological interpretation in

[7] A. Meurman and L. J. Santharoubane, Cohomology and Harish-chandra modules over the Virasoro algebra. Preprint, MSRI.

# RATIONAL HOMOTOPY THEORY OF FLAG VARIETIES ASSOCIATED TO KAC-MOODY GROUPS

By

Shrawan Kumar
Mathematical Sciences Research Institute,
Berkeley, CA
and
Tata Institute of Fundamental Research,
Colaba, BOMBAY (INDIA)

## Introduction

This paper is a sequel to my earlier paper "Geometry of Schubert cells and cohomology of Kac-Moody Lie-algebras". It uses many results from the paper, just mentioned, in an essential manner.

Let $\mathfrak{g}$ be a Kac-Moody Lie-algebra and let $\mathfrak{p}_X$ be a parabolic subalgebra of finite type. Let G be the algebraic group (in general infinite dimensional), in the sense of Šafarevič, associated with $\mathfrak{g}$ (called a Kac-Moody algebraic group) and let $P_X$ be the parabolic subgroup (of finite type) of G, associated with $\mathfrak{p}_X$. One of the principal aims of this paper is to study the rational homotopy theory of the flag varieties $G/P_X$. We prove that $G/P_X$ is a "formal" space in the sense of rational homotopy theory. Further, we explicitly determine the minimal models of the flag varieties G/B. We also prove that the Lie-algebra cohomology, with trivial coefficients, $H^*(\mathfrak{g}^1)$ (resp. $H^*(\mathfrak{g},\mathfrak{r}_X)$) is isomorphic, as graded algebras, with singular cohomology $H^*(G,\mathbb{C})$ (resp. $H^*(G/P_X,\mathbb{C})$) and the isomorphism is explicitly given by an integration map. ($\mathfrak{g}^1$ denotes the commutator subalgebra of $\mathfrak{g}$ and $\mathfrak{r}_X$ is the reductive part of $\mathfrak{p}_X$.)

Now we describe the contents of this paper in more detail.

**Chapter (0)** is devoted to recalling various definitions and well known elementary facts from Kac-Moody theory. We fix notations to be used throughout the paper.

**Chapter (1).** Main result of this section is theorem (1.6). This states that $H^*(\mathfrak{g},\mathfrak{r}_X)$ (resp. $H^*(\mathfrak{g}^1)$) is isomorphic with $H^*(G/P_X,\mathbb{C})$ (resp. $H^*(G,\mathbb{C})$), as graded algebras and moreover the isomorphism is explicitly given by an integration map. In particular, this gives a "complete" description of the cohomology algebra of the loop algebra $\mathfrak{g}_0 \otimes \mathbb{C}[t,t^{-1}]$ and its central extension (the affine algebra), for any finite dimensional semi-simple Lie-algebra $\mathfrak{g}_0$. Kac-Peterson also claim to have proved that $H^*(\mathfrak{g}^1)$ is isomorphic with $H^*(G,\mathbb{C})$. Their proofs have not yet appeared, but presumably, it is very different from ours. As more or less immediate corollaries (corollaries (1.9)) we deduce that $H^*(\mathfrak{g})$ and $H^*(\mathfrak{g}^1)$ are both Hopf algebras; for a finite dimensional simple Lie-algebra $\mathfrak{g}_0$, $H^2(\mathfrak{g}_0 \otimes \mathbb{C}[t,t^{-1}])$ is one dimensional; $H^2(\mathfrak{g}^1)$ is always 0 for any symmetrizable Kac-Moody Lie-algebra and hence, in particular, the standard map $\mathfrak{g}^1 \longrightarrow \mathfrak{g}_0 \otimes \mathbb{C}[t,t^{-1}]$ (where $\mathfrak{g}$ is the affine Lie-algebra associated with the finite dimensional simple Lie-algebra $\mathfrak{g}_0$) is a universal central extension. A similar result is true in the twisted affine case. Universality of this central extension is originally due to H. Garland, R. Wilson and V. Chari.

<u>**Chapter 2**</u>. One of the main results of this section is theorem (2.2), which states that the DGA (differential graded algebra) $C(\mathfrak{g},\mathfrak{r}_X)$ is formal (in the sense of rational homotopy theory). Our proof of this is similar to one of the proofs given by Deligne-Griffiths-Morgan and Sullivan for the formality of compact Kähler manifolds, but there is one essential difference in that the usual Hodge decomposition for Kähler manifolds is replaced by the "Hodge decomposition" with respect to the disjoint operators $d$ and $\partial$ developed in [Ku$_1$]. This theorem, coupled with a technical lemma (lemma 2.6), gives rise to theorem (2.7) which states that $G/P_X$ is a formal space (where $P_X$ is any standard parabolic of $G$ of finite type). So that, complete rational homotopy information of $G/P_X$ can be derived from the cohomology algebra $H^*(G/P_X)$. Also, in particular, all the Massey products of any order are zero over $\mathbb{Q}$. As a second application of theorem (2.2), we prove that the Leray-Serre spectral sequence in cohomology corresponding to the fibration $G \longrightarrow G/B$ degenerates at $E_3$ over $\mathbb{Q}$. In fact, recently, Kac-Peterson have proved a far reaching result that this spectral sequence degenerates

at $E_3$ even over $\mathbb{Z}/p\,\mathbb{Z}$, for any prime p.

In Chapter 3, we explicitly determine the minimal models for the flag varieties G/B (for any symmetrizable Kac-Moody group G). We also determine the Lie-algebra structure (under Whitehead product) on $\pi_*(G/B) \otimes_{\mathbb{Z}} \mathbb{Q}$. See theorem (3.8) for the complete description.

After this work was done, I learnt from Victor Kac that theorem (2.7) was observed by P. Deligne (using the machinery of $\ell$-adic cohomology) in a private communication to him. My very sincere thanks are due to Dale Peterson for many helpful conversations. I thank Heisuke Hironaka, Victor Kac, James R. Munkres, Leslie D. Saper and Pradeep Shukla for some helpful conversations.

## 0. Preliminaries and Notations

**(0.1) Definitions.**

(a) A *symmetrizable generalized Cartan matrix* $A = (a_{ij})_{1\leq i,j\leq \ell}$ is a matrix of integers satisfying $a_{ii} = 2$ for all i, $a_{ij} \leq 0$ if $i \neq j$, DA is symmetric for some diagonal matrix $D = \text{diag.}\ (q_1,\ldots,q_\ell)$ with $q_i > 0 \in \mathbb{Q}$.

(b) Choose a triple $(\mathfrak{h},\pi,\pi^V)$, unique up to isomorphism, where $\mathfrak{h}$ is a vector space over $\mathbb{C}$ of dim$\ell$+co-rank A, $\pi = \{\alpha_i\}_{1\leq i\leq \ell} \subset \mathfrak{h}^*$ and $\pi^V = \{h_i\}_{1\leq i\leq \ell} \subset \mathfrak{h}$ are linearly independent indexed sets satisfying $\alpha_j(h_i) = a_{ij}$. The *Kac-Moody algebra* $\mathfrak{g} = \mathfrak{g}(A)$ is the Lie-algebra over $\mathbb{C}$, generated by $\mathfrak{h}$ and the symbols $e_i$ and $f_i$ $(1\leq i\leq \ell)$ with the defining relations $[\mathfrak{h},\mathfrak{h}] = 0$; $[h,e_i] = \alpha_i(h)e_i$, $[h,f_i] = -\alpha_i(h)f_i$ for $h\in\mathfrak{h}$ and all $1\leq i\leq \ell$; $[e_i,f_j] = \delta_{ij}h_j$ for all $1\leq i,j\leq \ell$; $(\text{ad } e_i)^{1-a_{ij}}(e_j) = 0 = (\text{ad } f_i)^{1-a_{ij}}(f_j)$ for all $1\leq i\neq j\leq \ell$.

$\mathfrak{h}$ is canonically embedded in $\mathfrak{g}$.

**(0.2) Root space decomposition** $[K_1]$. There is available the root space decomposition $\mathfrak{g} = \mathfrak{h} \oplus \sum_{\alpha\in\Delta\subset\mathfrak{h}^*} \mathfrak{g}_\alpha$, where $\mathfrak{g}_\alpha = \{x \in \mathfrak{g}: [h,x] = \alpha(h)x,$ for all $h \in \mathfrak{h}\}$ and $\Delta = \{\alpha \in \mathfrak{h}^* - (0)$ such that $\mathfrak{g}_\alpha \neq 0\}$. Moreover $\Delta = \Delta_+ \cup \Delta_-$, where $\Delta_+ \subset \{\sum_{i=1}^{\ell} n_i\alpha_i: n_i \in \mathbb{Z}_+$ (= the non-negative integers) for all $i\}$ and $\Delta_- = -\Delta_+$. Elements of $\Delta_+$ (resp. $\Delta_-$) are called positive (resp. negative) roots.

**(0.3) Parabolics.** We fix a subset X (including $X = \emptyset$) of $\{1,\ldots,\ell\}$ of finite type, i.e., the submatrix $A_X = (a_{ij})_{i,j\in X}$ is a classical Cartan matrix of finite type. There is a natural injection $\mathfrak{g}_X = \mathfrak{g}(A_X) \hookrightarrow \mathfrak{g}(A)$. Define $\Delta_+^X$ (resp. $\Delta_-^X$) $= \Delta_+ \cap \{\sum_{i\in X} \mathbb{Z}\ \alpha_i\}$ (resp. $\Delta_- \cap \{\sum_{i\in X} \mathbb{Z}\ \alpha_i\}$), then $\mathfrak{g}_X = \mathfrak{h}_X \oplus \sum_{\alpha\in\Delta_+^X} \mathfrak{g}_\alpha \oplus \sum_{\alpha\in\Delta_-^X} \mathfrak{g}_\alpha$, where $\mathfrak{h}_X$ = linear span of $\{h_i\}_{i\in X}$.

Define the following Lie-subalgebras. $\mathfrak{n} = \sum_{\alpha\in\Delta_+} \mathfrak{g}_\alpha$; $\mathfrak{u} = \mathfrak{u}_X = \sum_{\alpha\in\Delta_+\setminus\Delta_+^X} \mathfrak{g}_\alpha$; $\mathfrak{r} = \mathfrak{r}_X = \mathfrak{g}_X + \mathfrak{h}$ and $\mathfrak{p} = \mathfrak{p}_X = \mathfrak{r} + \mathfrak{u}$. Of course $\mathfrak{r}$ is a reductive algebra. $\mathfrak{p}$ is called the *F-parabolic subalgebra* (F for finite dimensionality of $g_X$) defined by X. If $X = \emptyset$, the associated parabolic $\mathfrak{p}$ (= $\mathfrak{h} + \mathfrak{n}$) is the *Borel subalgebra*. If A itself is of finite type (i.e. A is a classical Cartan matrix), then the F-parabolic subalgebras are precisely the parabolic subalgebras of $g$ containing the Borel subalgebra $\mathfrak{h} \oplus \mathfrak{n}$.

**(0.4)** <u>**Weyl group**</u> $[K_1]$. There is a *Weyl group* $W \subset \mathrm{Aut}(\mathfrak{h}^*)$ generated by the reflections $\{r_i\}_{1\leq i\leq \ell}$ $(r_i(\beta) = \beta - \beta(h_i)\alpha_i)$, associated to the Lie-algebra $g$. $(W, \{r_i\}_{1\leq i\leq \ell})$ is a Coxeter system, hence we can talk of the lengths of elements of W.

W preserves $\Delta$. $\Delta^{re}$ is defined to be $W\cdot\pi$ and $\Delta^{im} = \Delta\setminus\Delta^{re}$. For $\alpha \in \Delta^{re}$, $\dim g_\alpha = 1$ and $\Delta \cap \mathbb{Z}\alpha = \{\alpha,-\alpha\}$.

Given a subset X of finite type, as in §(0.3), there is defined a subset $W_X^1$, of the Weyl group W, by

$$W_X^1 = \{w \in W: \Delta_+ \cap w\Delta_- \subset \Delta_+\setminus\Delta_+^X\}.$$

**(0.5)** <u>**Cartan involution**</u>. There is a ($\mathbb{C}$-linear) unique involution $\omega$ of $g$ defined by $\omega(f_i) = -e_i$ for all $1\leq i\leq \ell$ and $\omega(h) = -h$, for all $h \in \mathfrak{h}$. It is easy to see that $\omega$ leaves $g(\mathbb{R})$ (= "real points" of $g$) stable.

Further, there is a unique *conjugate linear* involution $\omega_0$ of $g$ which coincides with $\omega$ on $g(\mathbb{R})$.

**(0.6)** <u>**Algebraic group associated to a Kac-Moody Lie-algebra**</u> $g$ $[KP_1]$, $[KP_2]$ and $[T]$. A $g^1$ (= $[g,g]$) module $(V,\theta)$ ($\theta: g^1 \longrightarrow \mathrm{End}\, V$) is called integrable, if $\theta(e)$ is locally nilpotent whenever $e \in g_\alpha$, for $\alpha \in \Delta^{re}$. Let $G^*$ be the free product of the additive groups $\{g_\alpha\}_{\alpha\in\Delta^{re}}$, with canonical inclusions $i_\alpha: g_\alpha \longrightarrow G^*$. For any integrable $g^1$-module $(V,\theta)$, define a homomorphism $\theta^*: G^* \longrightarrow \mathrm{Aut}_{\mathbb{C}}V$ by $\theta^*(i_\alpha(e)) = \exp(\theta(e))$ for $e \in g_\alpha$. Let $N^*$ be the intersection of all $\mathrm{Ker}\,\theta^*$. Put $G = G^*/N^*$. Let $q$ be the canonical homomorphism: $G^* \longrightarrow G$. For $e \in g_\alpha$ ($\alpha \in \Delta^{re}$), put $\exp e = q(i_\alpha e)$, so that $U_\alpha = \exp g_\alpha$ is an additive one parameter subgroup of G. Denote by U the subgroup of G generated by the $U_\alpha$'s with $\alpha \in \Delta_+^{re}$.

Choose $\Lambda_i \in \mathfrak{h}^*(1 \leq i \leq \ell)$, satisfying $\Lambda_i(h_j) = \delta_{ij}$ for all $1 \leq j \leq \ell$. There is an embedding [KP$_2$; page 162-163]

$$i: G \longrightarrow \mathbb{A} = [\bigoplus_{i=1}^{\ell} L(\Lambda_i)] \oplus [\bigoplus_{i=1}^{\ell} L^*(\Lambda_i)]$$

defined by $i(g) = g(\sum_{i=1}^{\ell} v_{\Lambda_i}) + g(\sum_{i=1}^{\ell} v^*_{\Lambda_i})$.

Here $(L(\Lambda_i), \pi(\Lambda_i))$ is the integrable highest weight module with highest weight $\Lambda_i$. $L^*(\Lambda_1)$ is the vector space $L(\Lambda_i)$ regarded as a $\mathfrak{g}$-module under $\pi^*(\Lambda_i) = \pi(\Lambda_i) \circ \omega$; $v_{\Lambda_i}$ is a highest weight vector in $L(\Lambda_i)$ and $v_{\Lambda_i}$ is denoted $v^*_{\Lambda_i}$ regarded as an element in $L^*(\Lambda_i)$.

By "differentiating" i, we get an embedding $\bar{i}: \mathfrak{g}^1 \longrightarrow \mathbb{A}$. More explicitly $\bar{i}(x) = x(\sum_{i=1}^{\ell} v_{\Lambda_i}) + x(\sum_{i=1}^{\ell} v^*_{\Lambda_i})$, for $x \in \mathfrak{g}^1$.

$\mathbb{A}$ is endowed with a Hausdorff topology defined as follows. A set $V \subset \mathbb{A}$ is open if and only if $V \cap F$ is open in F, for all the finite dimensional vector sub-spaces F of $\mathbb{A}$. Now, put the subspace (through i) topology on G. G may be viewed as a, possibly infinite dimensional, affine algebraic group in the sense of Śafarević [Sa] with Lie-algebra $\mathfrak{g}^1$. For a proof, see [KP$_2$; §4]. In [KP$_2$; §4(G)], (a priori) a different topology is put on G but it can be seen that these two topologies, on G, actually coincide.

**(0.7)** Recall, from §(0.5), the conjugate linear involution $\omega_0$ of $\mathfrak{g}$. On "integration" this gives rise to an involution $\tilde{\omega}_0$ of G. Let K denote the fixed point set of this involution.

**(0.8)** The subgroup of $\mathrm{Aut}_{\mathbb{C}}(\mathfrak{h})$ generated by the reflections $\{\bar{r}_i\}_{1 \leq i \leq \ell}$ (resp. $\{\bar{r}_i\}_{i \in X}$) is denoted by $\bar{W}$ (resp. $\bar{W}_X$), where $\bar{r}_i(h) = h - \alpha_i(h)h_i$, for all $h \in \mathfrak{h}$. It is easy to see that, under the canonical identification $\chi$: Aut $\mathfrak{h} \longrightarrow$ Aut $(\mathfrak{h}^*)$ (given by $(\chi f)\beta(h) = \beta(f^{-1}h)$, for $f \in$ Aut $\mathfrak{h}$; $\beta \in \mathfrak{h}^*$ and $h \in \mathfrak{h}$), $\bar{W}$ corresponds with W, in fact $\chi(\bar{r}_i) = r_i$ for all $1 \leq i \leq \ell$. From now on, we would identify $\bar{W}$ with W (under $\chi$) and use the same symbol W for both.

For each $1 \leq i \leq \ell$, there exists a unique homomorphism $\beta_i$: $SL_2(\mathbb{C}) \longrightarrow G$ satisfying $\beta_i\begin{pmatrix} 1 & t \\ 0 & 1 \end{pmatrix} = \exp(te_i)$ and $\beta_i\begin{pmatrix} 1 & 0 \\ t & 1 \end{pmatrix} = \exp(tf_i)$ (for all $t \in \mathbb{C}$). Define $H_i = \beta_i\left\{\begin{pmatrix} t & 0 \\ 0 & t^{-1} \end{pmatrix} : t \in \mathbb{C}^*\right\}$; $G_i =$

$\mathcal{B}_i(SL_2(\mathbb{C}))$; $N_i$ = Normalizer of $H_i$ in $G_i$; H = the subgroup (of G) generated by all $H_i$; N = the subgroup (of G) generated by all $N_i$. There is an isomorphism $\gamma$: W $\longrightarrow$ N/H, such that $\gamma(r_i)$ is the coset $N_iH\setminus H$ mod H. See [$KP_1$; §2]. We would, sometimes, identify W with N/H under $\gamma$.

Put B = HU (U is defined in §(0.6)) and P = $P_X$ = $BW_XB$. Denote by $K_X$ the subgroup K $\cap$ $P_X$. It is easy to see that the canonical inclusion $K/K_X \longrightarrow G/P_X$ is a (surjective) homeomorphism. Use [$KP_2$; Theorem 4(d)]. (K $\subset$ G is given the subspace topology and topology on G is described in §(0.6)).

**(0.9)** <u>**Bruhat decomposition**</u> [$KP_1$]; [$KP_2$] and [T]. Recall the definition of $W_X^1$ from §(0.4). $W_X^1$ can be characterized as the set of elements of minimal length in the cosets $W_Xw$ (w $\in$ W) (each such coset contains a unique element of minimal length).

G can be written as disjoint union $G = \bigcup_{w \in W_X^1} (U\, a(w)^{-1}\, P_X)$, so that $G/P_X = \bigcup_{w \in W_X^1} (U\, a(w)^{-1}\, P_X/P_X)$.

(a(w) is an element of N satisfying a(w) mod H = $\gamma$(w). In fact, we will choose a(w) $\in$ N $\cap$ K, which is possible because KH $\supset$ N.)

$G/P_X$ is a C-W complex with cells $\{V_w = U\, a(w)^{-1}\, P_X/P_X\}_{w \in W_X^1}$ and $\dim_{\mathbb{R}} V_w$ = 2 length w. (To interchange right and left cosets we have, in the expression of $V_w$, $a(w)^{-1}$ instead of a(w) as in [$KP_2$].)

**(0.10)** <u>**Notations**</u>. Throughout the paper, unless otherwise specifically stated, all the vector spaces will be over $\mathbb{C}$ and linear maps would be $\mathbb{C}$-linear maps. For a vector space V, $\Lambda(V)$ denotes the exterior algebra and S(V) denotes the symmetric algebra.

For a Lie-algebra pair ($\mathfrak{g}$,$\mathfrak{r}$), C($\mathfrak{g}$,$\mathfrak{r}$) denotes the standard co-chain complex associated to the pair ($\mathfrak{g}$,$\mathfrak{r}$). See, e.g., [HS; §1]. For a topological space X, C(X,$\mathbb{C}$) will denote the (usual) singular co-chain complex of X with coefficients in $\mathbb{C}$. Unless otherwise stated, *cohomologies would be with complex coefficients.*

*The symmetrizability assumption on the Kac-Moody Lie-algebras* g(A) (*i.e. A is symmetrizable*) *would be implicitly assumed throughout the paper.* By a *Kac-Moody algebraic group,* we mean a group G (as defined in §(0.6)), associated to some Kac-Moody Lie algebra g. The subgroup K (defined in §(0.7)) would be called the *standard compact real* form of G (though it is non-compact, in general!). By a *standard parabolic of G,* we would mean $P_X$ (defined in §(0.8)) for some $X \subset \{1,...,\ell\}$. If, in addition, X is of finite type $P_X$ would be called a *standard parabolic of finite type.* When $X = \emptyset$, so that $P_X = B$, it is called the *standard Borel subgroup of G.*

## 1. An Analogue of Cartan-deRham Theorem for Kac-Moody Groups

(1.1) Let $\mathfrak{g} = \mathfrak{g}(A)$ be a Kac-Moody Lie-algebra associated to a generalized Cartan matrix $A = (a_{ij})_{1 \leq i,j \leq \ell}$ and let $X \subset \{1,\ldots,\ell\}$ be a subset of finite type. There is associated a group G, its standard compact real form K and a standard parabolic subgroup P $= P_X$ as described in §(0.10).

(1.2) **Definitions.**

(a) We recall the definition of a smooth map from a finite dimensional smooth manifold M to K or $K/K_X$ from [$Ku_1$; §(4.3)] ($K_X = K \cap P_X$).

Let f: M $\longrightarrow$ K be a continuous map. Consider the composite of the maps

$$M \xrightarrow{f} K \hookrightarrow G \xrightarrow{i} \mathbb{A} \qquad \text{(i is defined in §(0.6)).}$$

Since $i \circ f$: M $\longrightarrow \mathbb{A}$ is continuous, given any $x_0 \in M$, there exists an open neighborhood $N(x_0)$ of $x_0$ in M such that $i \circ f(N(x_0)) \subset F$, for some finite dimensional vector subspace F of $\mathbb{A}$. We say that f *is smooth at* $x_0$ if the restricted map $i \circ f\big|_{N(x_0)}: N(x_0) \longrightarrow F$ is smooth ($= C^\infty$) in the usual sense. *The map* f *itself is said to be smooth* if f is smooth at all $x_0 \in M$.

A map f: M $\longrightarrow K/K_X$ *is said to be smooth* if for any $x_0 \in M$, there exists an open neighborhood $N(x_0)$ (of $x_0$ in M) and a smooth lift $\tilde{f}: N(x_0) \longrightarrow K$ (i.e. $\tilde{f}$ is smooth and $\pi \circ \tilde{f} = f\big|_{N(x_0)}$, where $\pi$ is the canonical projection: $K \longrightarrow K/K_X$).

(b) By *a smooth singular* n-*simplex in* K (*resp.* $K/K_X$), we mean a continuous map f: $\Delta^n = \{(t_1,\ldots,t_n) \in \mathbb{R}^n$: $t_i \geq 0$ and $\Sigma\, t_i \leq 1\} \longrightarrow K$ (resp. f: $\Delta^n \longrightarrow K/K_X$) such that there exists an open neighborhood N of $\Delta^n$ in $\mathbb{R}^n$ and a smooth map $f_{ext}$: N $\longrightarrow$ K (resp. $f_{ext}$: N $\longrightarrow K/K_X$) extending f.

Let us denote by $\Delta^n_{C^\infty}(K)$ (resp. $\Delta^n_{C^\infty}(K/K_X)$), the free abelian group on the set of all the smooth singular n-simplexes f in K (resp. in $K/K_X$).

Finally, denote $\sum_{n \geq 0} \mathrm{Hom}_{\mathbb{Z}}(\Delta^n_{C^\infty}(K), \mathbb{C})$ (resp.

$\sum_{n\geq 0} \mathrm{Hom}_{\mathbb{Z}}(\Delta^n_{C^\infty}(K/K_X),\mathbb{C}))$ by $C_{C^\infty}(K,\mathbb{C})$ (resp. $C_{C^\infty}(K/K_X,\mathbb{C})$).

(c) Let M be a finite dimensional smooth manifold with a smooth map $f: M \longrightarrow K$ (resp. $f: M \longrightarrow K/K_X$). Given a $u \in C^n(g^1)$ (resp. $u \in C^n(g^1,r^1)$, $r = r_X$ is defined in §(0.3) and $r^1 = r \cap g^1$), we construct a smooth n-form $f^*(u)$ on M as follows.

Fix a $x_0 \in M$. Choose a local smooth lift $\tilde{f}: N(x_0) \longrightarrow K$. (When $f: M \longrightarrow K$, $\tilde{f}$ is, of course, f itself.) Consider the map $i\circ L_{\tilde{f}(x_0)^{-1}}\circ\tilde{f}: N(x_0) \longrightarrow \mathbf{A}$, where $L_{\tilde{f}(x_0)^{-1}}$ is the left translation (by $\tilde{f}(x_0)^{-1}$): $K \longrightarrow K$. Define $(f^*u)_{x_0} = (i\circ L_{\tilde{f}(x_0)^{-1}}\circ\tilde{f})^*\tilde{u}$, where $\tilde{u}$ is any translation invariant n-form on $\mathbf{A}$ (so that $\tilde{u}$ is given by $\tilde{u}_0 \in \mathrm{Hom}_{\mathbb{C}}(\Lambda^n(\mathbf{A}),\mathbb{C}))$ satisfying $\tilde{u}_0\big|_{\Lambda^n(g^1)} = u$. ($g^1$ is identified as a subspace of $\mathbf{A}$ via $\bar{i}$, see §(0.6).)

It is a routine checking, using the following facts, that $f^*(u)$ is well defined, i.e., $(f^*u)_{x_0}$ does not depend upon the particular choices of $\tilde{f}$; $\tilde{u}$ and further $(f^*u)$ is a smooth n-form on M.

Let M be a (finite dim.) smooth manifold and $m_0 \in M$. Given two smooth maps $f,f'$: $(M,m_0) \longrightarrow (G,e)$ (i.e. $\bar{f} = i\circ f: M \longrightarrow \mathbf{A}$ is smooth and so is $\bar{f}'$), then the following are true.

(1) The map $ff'^{-1}$: $(M,m_0) \longrightarrow (G,e)$, defined by $ff'^{-1}(m) = f(m)\cdot(f'(m))^{-1}$ for all $m \in M$, is smooth and $d(\overline{ff'^{-1}})_{m_0} = (d\bar{f})_{m_0} - (d\bar{f}')_{m_0}$.

(2) Fix any $a \in \mathbf{A}$, then the map $f_a$: $M \longrightarrow \mathbf{A}$, defined by $f_a(m) = f(m)\cdot a$ is smooth.

(3) $(d\bar{f})_{m_0}(T_{m_0}(M)) \subset \bar{i}(g^1)$.

(4) Fix a $g \in G$, then the map $gfg^{-1}$: $(M,m_0) \longrightarrow (G,e)$, defined by

$(gfg^{-1})m = gf(m)g^{-1}$, is smooth and for any $v \in T_{m_0}(M)$,

$\overline{d(gfg^{-1})_{m_0}v = \bar{i}((\mathrm{Ad}\ g)x(v))}$, where $x(v) \in g^1$ is the element satisfying $(d\bar{f})_{m_0}v = \bar{i}(x(v))$. (Ad: $G \longrightarrow \mathrm{Aut}(g^1)$ is defined in [$KP_1$; §2].)

(1) and (2) are easy in view of [$KP_2$; §4]. Dale Peterson showed me proofs of (3) and (4).

(1.3) Integration map. We describe an "integration" map $\int: C(g^1) \longrightarrow C_{C^\infty}(K,\mathbb{C})$ as follows.

$$(\int u)f = \int_{\Delta^n} (f^* u), \text{ for } u \in C^n(g^1) \text{ and } f: \Delta^n \longrightarrow K \text{ a}$$

simplex $\in \Delta^n_{C^\infty}(K)$.

Exactly similarly, we can define an integration map $\int: C(g^1,r^1) \longrightarrow C_{C^\infty}(K/K_X,\mathbb{C})$.

We have the following two technical lemmas.

(1.4) Lemma. The integration maps $\int: C(g^1) \longrightarrow C_{C^\infty}(K,\mathbb{C})$ and $\int: C(g^1,r^1) \longrightarrow C_{C^\infty}(K/K_X,\mathbb{C})$ are both co-chain maps. Further they induce algebra homomorphisms in cohomology.

Proof. We would prove that $\int: C(g^1) \longrightarrow C_{C^\infty}(K,\mathbb{C})$ is a co-chain map, which induces algebra homomorphism in cohomology. The proof of the analogous statement for $K/K_X$ is similar.

To prove that $\int$ is a co-chain map, in view of Stokes' theorem, it suffices to show that for any (finite dimensional) smooth manifold M and a smooth map $f: M \longrightarrow K$, we have, for any $u \in C^n(g^1)$, $d(f^* u) = f^*(du)$.

Extend u arbitrarily to an element $u_0$ of $\mathrm{Hom}_{\mathbb{C}}(\Lambda^n_{\mathbb{C}}(\mathbf{A}),\mathbb{C})$. ($g^1$ is canonically embedded in $\mathbf{A}$ via $\bar{i}$. See §(0.6).) The embedding $i|_K: K \longrightarrow \mathbf{A}$ is K-equivariant (K acting on K by left multiplication and of course $\mathbf{A}$ is a representation space for K). Extend $u_0$ to a K-invariant form $\hat{u}_0$ on $\mathbf{A}$, though defined only on

$i(K)$. Since the representation map: $G \times \mathbb{A} \longrightarrow \mathbb{A}$ is regular (see [KP₂; §4]), $\hat{u}_0$ can further be extended to a smooth (in the obvious sense) n-form $\bar{u}_0$ defined on whole of $\mathbb{A}$. Of course, $(i\circ f)^*(d\bar{u}_0) = d((i\circ f)^*\bar{u}_0)$. Further, $(i\circ f)^*\bar{u}_0$ can be easily seen to be the form $f^*(u)$. So, in view of K-invariance of $\hat{u}_0$ on $i(K)$, it is enough to show that

$$(d\bar{u}_0)_{i(e)}(\bar{i}x_0,\dots,\bar{i}x_n) = du(x_0,\dots,x_n), \text{ for all } x_0,\dots,x_n \in g^1.$$

Fix any ad locally-finite elements $x_0,\dots,x_n \in g^1$. Consider the 1-parameter group of diffeomorphisms $\phi(x_i)$: $\mathbb{R} \times \mathbb{A} \longrightarrow \mathbb{A}$, defined by $\phi(x_i)(t,a) = \exp(tx_i)a$. It can be easily seen that the corresponding vector field $\bar{x}_i$ on $\mathbb{A}$ is given by $\bar{x}_i(a) = x_i a$. Now (we would write e for i(e)),

$$(d\bar{u}_0)_e(\bar{x}_0,\dots,\bar{x}_n) = \sum_{i=0}^{n} (-1)^i(\bar{x}_i)_e(\bar{u}_0(\bar{x}_0,\dots,\hat{\bar{x}}_i,\dots,\bar{x}_n)) +$$

$$\sum_{i<j} (-1)^{i+j}(\bar{u}_0)_e([\bar{x}_i,\bar{x}_j],\bar{x}_0,\dots,\hat{\bar{x}}_i,\dots,\hat{\bar{x}}_j,\dots,\bar{x}_n)$$

$$(*) = \sum_{i=0}^{n} (-1)^i \operatorname*{Lt.}_{t\to 0} \frac{1}{t}[(\bar{u}_0)_e(\exp(-tx_i)x_0 \exp(tx_i)e,\dots,\underset{i\text{-th place}}{\wedge}, \dots,\exp(-tx_i)x_n \exp(tx_i)e) -(\bar{u}_0)_e(x_0e,\dots,\hat{x_i e},\dots,x_ne)]$$

$$+ \sum_{i<j} (-1)^{i+j}(\bar{u}_0)_e([\bar{x}_i,\bar{x}_j],\bar{x}_0,\dots,\hat{\bar{x}}_i,\dots,\hat{\bar{x}}_j,\dots,\bar{x}_n)$$

But since, for all $0 \leqslant j \leqslant n$,

$$\operatorname*{Lt.}_{t\to 0} \frac{\exp(-tx_i)x_j \exp(tx_i)e-x_je}{t}$$

$$= \operatorname*{Lt.}_{t\to 0} \frac{(\mathrm{Ad}(\exp(-tx_i))x_j)e-x_je}{t}$$

$$= \operatorname*{Lt.}_{t\to 0} \frac{((\exp(\mathrm{ad}(-tx_i)))x_j)e-x_je}{t}$$

$$= - [x_i,x_j]e$$

Also, $[\bar{x}_i,\bar{x}_j]_e = \operatorname*{Lt.}_{t\to 0} \dfrac{\exp(-tx_i)x_j \exp(tx_i)e-x_je}{t}$

$$= -[x_i,x_j]\,e.$$

Putting these in (*) we get

$$(d\bar{u}_0)_e(\bar{x}_0,\ldots,\bar{x}_n) = \sum_{i=0}^{n} (-1)^{i+1} \sum_{j\neq i} (\bar{u}_0)_e(x_0e,\ldots,[x_i,x_j]\,e,\ldots,$$
$$\hat{x_i}e,\ldots,x_ne)$$

$$+ \sum_{i<j} (-1)^{i+j+1}(\bar{u}_0)_e([x_i,x_j]\,e,x_0e,\ldots,\hat{x_i}e,\ldots,$$
$$\hat{x_j}e,\ldots,x_ne)$$

$$= \sum_{j<i} (-1)^{i+j+1}(\bar{u}_0)_e([x_i,x_j]\,e,x_0e,\ldots,\hat{x_j}e,\ldots,$$
$$\hat{x_i}e,\ldots,x_ne)$$

$$+ \sum_{j>i} (-1)^{i+j}(\bar{u}_0)_e([x_i,x_j]\,e,x_0e,\ldots,\hat{x_i}e,\ldots,$$
$$\hat{x_j}e,\ldots,x_ne)$$

$$+ \sum_{i<j} (-1)^{i+j+1}(\bar{u}_0)_e([x_i,x_j]\,e,x_0e,\ldots,\hat{x_i}e,\ldots,$$
$$\hat{x_j}e,\ldots,x_ne)$$

$$= \sum_{j<i} (-1)^{i+j+1}(\bar{u}_0)_e([x_i,x_j]\,e,x_0e,\ldots,\hat{x_j}e,\ldots,$$
$$\hat{x_i}e,\ldots,x_ne)$$

(since the last two expressions cancel each other)

$$= \sum_{i<j} (-1)^{i+j}(\bar{u}_0)_e([x_i,x_j]\,e,x_0e,\ldots,\hat{x_i}e,\ldots,$$
$$\hat{x_j}e,\ldots,x_ne)$$

(interchanging i and j)

$$= du\,(x_0,\ldots,x_n).$$

Since ad locally-finite elements in $g^1$ span $g^1$, this proves that $\int$ is a co-chain map.

Now we prove that $\int$ induces algebra homomorphism in cohomology.

Let $\mathrm{Sing}_{C^\infty}(K)$ (resp. Sing(K)) denote the simplicial set $n \to \mathrm{Sing}^n_{C^\infty}(K)$ (resp. $\mathrm{Sing}^n(K)$), where $\mathrm{Sing}^n_{C^\infty}(K)$ (resp. $\mathrm{Sing}^n(K)$) denotes the set of all the smooth (resp. continuous) singular n-simplexes in K with the standard face and degeneracy maps. Let $\Omega_{p\cdot dR}(\mathrm{Sing}_{C^\infty}(K)) = \sum_{p \geq 0} \Omega^p_{p\cdot dR}(\mathrm{Sing}_{C^\infty}(K))$ denote the piece-wise smooth de-Rham complex associated to the simplicial set $\mathrm{Sing}_{C^\infty}(K)$, where an element of $\Omega^p_{p\cdot dR}(\mathrm{Sing}_{C^\infty}(K))$ is, by definition, a function $\theta$ which assigns to each element of $\mathrm{Sing}^n_{C^\infty}(K)$ ($n = 0,1,2,\ldots$) a complex valued smooth p-form on $\Delta^n$ (i.e. a p-form on $\Delta^n \subset \mathbb{R}^n$, which extends to a smooth p-form on an open neighborhood U of $\Delta^n$), such that $\theta$ commutes with the face and degeneracy operators. $\Omega_{p\cdot dR}$ is made into a DGA (DGA is defined in §(2.1)(a)) under pointwise addition, multiplication and the usual differential of forms. Define a DGA morphism $\eta$: $C(g^1) \longrightarrow \Omega_{p\cdot dR}(\mathrm{Sing}_{C^\infty}(K))$, by $(\eta w)s = s^*w$, for $w \in C^p(g^1)$ and for any smooth singular n-simplex s: $\Delta^n \longrightarrow K$.

There is a canonical integration map $\tilde{\int}$: $\Omega_{p\cdot dR}(\mathrm{Sing}_{C^\infty}(K)) \longrightarrow C_{C^\infty}(K)$, defined by

$$(\tilde{\int}\theta)s = \int_{\Delta^p} \theta(s), \text{ for } \theta \in \Omega^p_{p\cdot dR}(\mathrm{Sing}_{C^\infty}(K))$$

and for any smooth singular simplex s: $\Delta^p \longrightarrow K$. (We denote the integration map here by $\tilde{\int}$ to distinguish it from our earlier integration map $\int$.)

By Stokes' theorem $\tilde{\int}$ is a co-chain map. Further, it is known (see [$S_2$; §7] and our next lemma (1.5)) that $\tilde{\int}$: $\Omega_{p \cdot dR}(\mathrm{Sing}_{C^\infty}(K)) \longrightarrow C_{C^\infty}(K)$ induces algebra isomorphism in cohomology. Of course, by definition, $\tilde{\int} \circ \eta = \int$ and hence the assertion, that $\int$ induces algebra homomorphism in cohomology, follows.

(1.5) **Lemma**. The restriction map $\gamma$: $C(K,\mathbb{C}) \longrightarrow C_{C^\infty}(K,\mathbb{C})$ induces isomorphism in cohomology, where $C(K,\mathbb{C})$ is the usual (continuous) singular co-chain complex with complex coefficients. A similar statement holds good with K replaced by $K/K_X$ throughout.

**Proof**. For any $n \geq 0$, let $\mathcal{C}^n_\infty$ be the sheaf on K associated with the presheaf (for any open set U in K) $U \longmapsto \mathrm{Hom}_{\mathbf{Z}}(\Delta^n_{C^\infty}(U),\mathbb{C})$, where $\Delta^n_{C^\infty}(U)$ denotes the free abelian group on the set of all the smooth singular simplexes $\phi$: $\Delta^n \longrightarrow U$. There is clearly a sheaf sequence

$$(S) \ldots \quad 0 \longrightarrow \mathbb{C} \longrightarrow \mathcal{C}^0_\infty \xrightarrow{d} \mathcal{C}^1_\infty \xrightarrow{d} \mathcal{C}^2_\infty \xrightarrow{d} \cdots$$

($\mathbb{C}$ denotes the constant sheaf on K).

To prove the lemma, it suffices (see [Wa; Chapter 5]) to show that the above sequence (S) is exact and all the sheaves $\mathcal{C}^n_\infty$ are fine sheaves.

(a) $\mathcal{C}^n_\infty$ <u>are fine sheaves</u>. Choose a locally finite open cover $\{U_\alpha\}$ of K (K being a paracompact space, this is possible). Choose a (discontinuous) partition of unity $\{\phi_\alpha\}$ subordinate to the cover $\{U_\alpha\}$ in which the functions $\phi_\alpha$ take values 0 or 1 only. For each $\alpha$ and n, define an endomorphism $e_\alpha$ of $\mathcal{C}^n_\infty$ by setting $(e_\alpha f)\sigma = \phi_\alpha(\sigma(0))f(\sigma)$, for $\sigma$ a smooth singular n-simplex in U and $f \in \mathrm{Hom}_{\mathbf{Z}}(\Delta^n_{C^\infty}(U),\mathbb{C})$. This provides a partition of unity for all the sheaves $\mathcal{C}^n_\infty$, concluding that they are fine.

(b) <u>The sequence (S) is exact</u>. We need to prove a Poincaré type lemma. Write any element $a \in \mathbf{A}$ as

$$a = \sum_{i=1}^{l} \sum_{\lambda_i \in \text{Weights of } L(\Lambda_i)} v^i_{\lambda_i}(a)$$

$$+ \sum_{i=1}^{l} \sum_{\mu_i \in \text{Weights of } L^*(\Lambda_i)} w^i_{\mu_i}(a),$$

where $v^i_{\lambda_i}$ (resp. $w^i_{\mu_i}$) denotes $\lambda_i$ (resp. $\mu_i$) weight vector $\in L(\Lambda_i)$ (resp. $L^*(\Lambda_i)$). Let $N = \{a \in \mathbb{A}: v^i_{\Lambda_i}(a) \notin \mathbb{R}^+(-v_{\Lambda_i})$ and $w^i_{-\Lambda_i}(a) \notin \mathbb{R}^+(-v^*_{\Lambda_i})$ for any $i\}$. Fix a smooth function $\varphi: \mathbb{R} \longrightarrow [0,1]$ satisfying $\varphi(t) = 0$ for all $t \leqslant 3/4$ and $\varphi(t) = 1$ for all $t \geqslant 1$. Define a contraction $H: \mathbb{R} \times N \longrightarrow N$ by

$$H(t,a) = \sum_{i=1}^{l} \frac{\sum_{\lambda_i} \varphi(t)^{ht\cdot(\Lambda_i-\lambda_i)} v^i_{\lambda_i}(a)}{\|\sum_{\lambda_i} \varphi(t)^{ht\cdot(\Lambda_i-\lambda_i)} v^i_{\lambda_i}(a)\|}$$

$$+ \sum_{i=1}^{l} \frac{\sum_{\mu_i} \varphi(t)^{ht\cdot(\Lambda_i+\mu_i)} w^i_{\mu_i}(a)}{\|\sum_{\mu_i} \varphi(t)^{ht\cdot(\Lambda_i+\mu_i)} w^i_{\mu_i}(a)\|}, \text{ for } t \geqslant 1/2 \text{ and}$$

$$= \sum_{i=1}^{l} \frac{\varphi(2t)\alpha_i(a)+1-\varphi(2t)}{|\varphi(2t)\alpha_i(a)+1-\varphi(2t)|} v_{\Lambda_i} +$$

$$+ \sum_{i=1}^{l} \frac{\varphi(2t)\beta_i(a)+1-\varphi(2t)}{|\varphi(2t)\beta_i(a)+1-\varphi(2t)|} v^*_{\Lambda_i}, \text{ for } t \leqslant 3/4,$$

where $\dfrac{v^i_{\Lambda_i}(a)}{\|v^i_{\Lambda_i}(a)\|} = \alpha_i(a)v_{\Lambda_i}$ and $\dfrac{w^i_{-\Lambda_i}(a)}{\|w^i_{-\Lambda_i}(a)\|} = \beta_i(a)v^*_{\Lambda_i}$ (If $\lambda_i = \Lambda_i - \sum_j n_j\alpha_j$ (resp. $\mu_i = -\Lambda_i + \sum_j n_j\alpha_j$), then $ht\cdot\Lambda_i-\lambda_i$ (resp.

$\mu_i+\Lambda_i$) is defined to be $\Sigma\, n_j$ · $\varphi(t)^0$ is defined to be $\equiv 1$.)

H has the following properties.

(1) H is smooth (in the obvious sense).

(2) $H(\mathbb{R} \times (N \cap i(K))) \subset N \cap i(K)$.

(3) $H(t,a) = a$, for all $t \geqslant 1$ and $a \in N \cap i(K)$.

(4) $H(t,a) = i(e)$, for all $t \leqslant 1/4$ and all $a \in N$.

Now we are ready to show that

$$0 \longrightarrow \mathbb{C} \longrightarrow C^0_{C^\infty}(K \cap i^{-1}(N)) \xrightarrow{d} C^1_{C^\infty}(K \cap i^{-1}(N)) \xrightarrow{d} \cdots$$

is exact. It suffices to find a homotopy operator, i.e., a linear map $h_p$: $C^p_{C^\infty}(K \cap i^{-1}(N)) \longrightarrow C^{p-1}_{C^\infty}(K \cap i^{-1}(N))$, for all $p \geqslant 1$, satisfying

(*) $$d\circ h_p + h_{p+1}\circ d = \mathrm{Id}.$$

For a smooth singular simplex $\sigma$: $\Delta^{p-1} \longrightarrow K \cap i^{-1}(N)$, define a smooth singular simplex $\tilde{h}_p\sigma$: $\Delta^p \longrightarrow K \cap i^{-1}(N)$ by,

$$(\tilde{h}_p\sigma)(t_1,\ldots,t_p) = i^{-1}H(t_1+\ldots+t_p,\ i\sigma(\frac{t_2}{t_1+\ldots+t_p},\ldots,\frac{t_p}{t_1+\ldots+t_p}))$$

$$\text{for } t_1+\ldots+t_p>0$$

$$= e \text{ for } t_1+\ldots+t_p<1/4.$$

Now, define $(h_pf)\sigma = f(\tilde{h}_p\sigma)$, for $f \in C^p_{C^\infty}(K \cap i^{-1}(N))$.

It is easy to see that (*) is satisfied.

Since K is homogeneous under left multiplication and also that, we can choose a co-final system $\{N^\epsilon\}$ of open neighborhoods of i(e) in N such that $H(\mathbb{R} \times (N^\epsilon \cap i(K))) \subset N^\epsilon \cap i(K)$, we get that the sheaf sequence (S) is exact.

The case of $K/K_X$ is similar. (We can define a similar smooth contraction of the open set $\bigcap_{w \in W_X} a(w)\, U^- \, a(w)^{-1} \subset G/P_X$.)

**Remark.** A contraction, similar to H above, has earlier been used by Kac–Peterson to prove contractibility of U.

We come to the main theorem of this section.

**(1.6) Theorem.** Let $\mathfrak{g} = \mathfrak{g}(A)$ be the Kac–Moody Lie-algebra associated to a symmetrizable generalized Cartan matrix $A = (a_{ij})_{1 \leq i,j \leq \ell}$ and let $X \subset \{1,\ldots,\ell\}$ be a subset of finite type. Then the integration maps (defined in §(1.3))

(a) $$\int : C(\mathfrak{g}^1, \mathfrak{r}_X^1) \longrightarrow C_{C^\infty}(K/K_X, \mathbb{C}) \text{ and}$$

(b) $$\int : C(\mathfrak{g}^1) \longrightarrow C_{C^\infty}(K, \mathbb{C})$$

both induce algebra isomorphisms in cohomology.

Recall that $\mathfrak{g}^1 = [\mathfrak{g},\mathfrak{g}]$; $\mathfrak{r}_X^1 = \mathfrak{g}^1 \cap \mathfrak{r}_X$ ($\mathfrak{r}_X$ is defined in §(0.3)) K is the standard compact real form of the Kac–Moody algebraic group G associated to $\mathfrak{g}$ and $K_X = K \cap P_X$ ($P_X$ is the standard parabolic subgroup of G). See §(0.10).

In particular, in view of lemma (1.5), the Lie-algebra cohomology $H^*(\mathfrak{g}^1, \mathfrak{r}_X^1)$ (resp. $H^*(\mathfrak{g}^1)$) is algebra isomorphic with the singular cohomology $H^*(K/K_X, \mathbb{C})$ (resp. $H^*(K, \mathbb{C})$). Also, by [L; §6], the canonical inclusion $C(\mathfrak{g}, \mathfrak{r}_X) \hookrightarrow C(\mathfrak{g}^1, \mathfrak{r}_X^1)$ induces isomorphism in cohomology.

**Proof.**

(a) By lemma (1.5) and the Bruhat decomposition §(0.9), $\dim H^n(C_{C^\infty}(K/K_X, \mathbb{C})) = \dim H^n(K/K_X, \mathbb{C}) = \#\{$elements of length $n/2$ in $W_X^1\}$. ($W_X^1$ is defined in §(0.4).) Also, by [L; §6] (see also [Ku$_1$; §(3.3)]), $\dim H^n(\mathfrak{g}, \mathfrak{r}_X) = \#\{$elements of length $n/2$ in $W_X^1\}$. Hence $\dim H^n(K/K_X, \mathbb{C}) = \dim H^n(\mathfrak{g}, \mathfrak{r}_X)$. Since, by lemma (1.4), $\int$ induces algebra homomorphism, it suffices to show (for dimensional considerations) that the induced map $H(\int): H^{2n}(\mathfrak{g}, \mathfrak{r}_X) \longrightarrow$

$H^{2n}(C_{C^\infty}(K/K_X,\mathbb{C}))$ is injective for all $n \geq 0$.

For any $\omega \in W_X^1$ of length n, $U\, a(\omega)^{-1}\, P_X/P_X$ is an open cell (of real dim 2n) in $G/P_X$ (i.e. homeomorphic with $\mathbb{C}^n$). See §(0.9). Further, this extends to a smooth singular simplex $\sigma_\omega$: $\Delta^{2n} \longrightarrow G/P_X \in \Delta^{2n}_{C^\infty}(G/P_X)$, so that $\partial(\sigma_\omega)$ is a 2n-1 dim cycle in $(G/P_X)^{2n-2}$. But since $H_{2n-1}((G/P_X)^{2n-2}) = 0$, there exists a 2n-dim chain $b_\omega$ in $(G/P_X)^{2n-2}$ such that $\partial(b_\omega) = \partial(\sigma_\omega)$. In fact, we can further choose $b_\omega \in \Delta_{C^\infty}(G/P_X)$.

By [Ku$_1$; Theorem 4.5], there are "d;$\partial$ harmonic" forms $\{s_o^\omega\}_{\omega\in W_X^1}$ such that $\int_{\sigma_\eta} s_o^\omega = \int_{B\, a(\eta)^{-1} P_X/P_X} s_o^\omega = \delta_{\omega,\eta}$, for $\omega;\eta \in W_X^1$ with $\ell(\omega) = \ell(\eta) = n$. So $\int_{\sigma_\eta - b_\eta} s_o^\omega = \int_{\sigma_\eta} s_o^\omega - \int_{b_\eta} s_o^\omega = \delta_{\omega,\eta}$, since $\int_{b_\eta} s_o^\omega = 0$. (Actually the integrand itself is 0, as $s_o^\omega$ is a 2n-form and $b_\eta$ is a chain in $(G/P_X)^{2n-2}$.) Since $\{s_o^\omega\}_{w\in W_X^1 \text{ with } l(w)=n}$ is a $\mathbb{C}$-basis of $H^{2n}(\mathfrak{g},r_X)$, by [Ku$_1$; §3], this immediately gives injectivity of $H(\int)$. Hence (a) follows.

(b) There is a Hochschild-Serre filtration $\tilde{F} = \{\tilde{F}_p\}_{p\geq 0}$ of $C(\mathfrak{g}^1)$ with respect to the subalgebra $\mathfrak{h}^1$, defined as follows. $\tilde{F}_p^n = \{u \in C^n(\mathfrak{g}^1)$: $u(r_1,\dots,r_n) = 0$, whenever n-p+1 of the arguments $r_j$ belong to $\mathfrak{h}^1\}$ and $\tilde{F}_p = \sum_{n\geq 0} \tilde{F}_p^n$.

Also there is a Leray-Serre filtration $G = \{G_p\}$ of $C_{C^\infty}(K)$, associated with the fibration $\pi$: $K \longrightarrow K/T$ (where $T = B \cap K$), defined by $G_p = \{c \in C_{C^\infty}(K)$: $c\big|_{\Delta(K^{p-1})\cap\Delta_{C^\infty}(K)} = 0$, where $K^{p-1}$ denotes $\pi^{-1}((K/T)^{p-1})$ and $\Delta(K^{p-1})$ denotes the usual (continuous) singular chain complex of $K^{p-1}$. ($(K/T)^{p-1}$, as earlier, denotes (p-1)-th skeleton of $K/T \approx G/B$ under the Bruhat decomposition.)

It is fairly easy to see that $\int(\tilde{F}_p) \subset G_p$ for all p.

Let $E_r(\tilde{F})$ and $E_r(G)$ be the spectral sequences associated with the filtrations $\tilde{F}$ and $G$ respectively. Since $\int$ preserves filtrations, it induces a map $E_r(\int)$: $E_r(\tilde{F}) \longrightarrow E_r(G)$ for all r.

By [HS; §6], $E_2^{p,q}(\tilde{F}) \cong H^p(g^1,h^1) \otimes H^q(h^1)$ and converging to the cohomology $H^*(g^1)$. (Although, in [HS], this is proved under the assumption that $g^1$ is finite dimensional, it can easily be adopted to our situation since $h^1$ acts reductively on $g^1$.)

Further, we can suitably modify the proof of lemma (1.5) to give the following generalization (of lemma (1.5)).

For any $p \geq 0$, the restriction map: $\mathrm{Hom}_{\mathbb{Z}}(\Delta(K^{p-1}),\mathbb{C}) \longrightarrow \mathrm{Hom}_{\mathbb{Z}}(\Delta_{C^\infty}(K) \cap \Delta(K^{p-1}),\mathbb{C})$ induces isomorphism in cohomology.

Using the five lemma, this gives that the filtration $G = \{G_p\}$ is regular (and hence strongly convergent) in the sense of [CE; page 324] and also (by Leray-Serre) $E_2^{p,q}(G) \cong H^p(K/T) \otimes H^q(T)$.

By part (a) of this theorem $H(\int)$: $H^p(g^1,h^1) \xrightarrow{\sim} H^p(K/T)$, for all $p \geq 0$. From this it is fairly easy to see that $E_2^{p,q}(\tilde{F}) \xrightarrow{\sim} E_2^{p,q}(G)$ for all p and q. Hence $E_r^{p,q}(\tilde{F}) \xrightarrow{\sim} E_r^{p,q}(G)$ for all $r \geq 2$ and all $p,q \geq 0$.

This completes the proof of part (b) as well.

(1.7) <u>Remark</u>. Kac-Peterson also claim to have proved that $H^*(g^1)$ is isomorphic with $H^*(K,\mathbb{C})$, although their proofs have not yet appeared.

The following lemma is trivial to verify.

(1.8) <u>Lemma</u>. For any Lie-algebra $\mathfrak{g}$ and a subalgebra $\mathfrak{s}$, $H^*(\mathfrak{g},\mathfrak{s}) \approx H^*(\mathfrak{g}/\mathfrak{z},\mathfrak{s}/\mathfrak{z})$, for a central subalgebra $\mathfrak{z}$ of $\mathfrak{g}$ such that $\mathfrak{z} \subset \mathfrak{s}$.

**<u>Proof</u>.** In fact the co-chain complex $C(\mathfrak{g},\mathfrak{s})$ itself is isomorphic with $C(\mathfrak{g}/\mathfrak{z},\mathfrak{s}/\mathfrak{z})$,

(1.9) <u>Corollaries</u>.

(a) For any Kac-Moody Lie-algebra $g$, $H^*(g^1)$ and $H^*(g)$ are both Hopf algebras.

(b) Let $g_0$ be a finite dimensional simple Lie-algebra and let $\theta$ be an automorphism of $g_0$, of order k, induced by an automorphism of the Dynkin diagram (so that k = 1, 2 or 3). Then

$H^2(\tilde{g}^{(k)})$ is one dimensional, where $\tilde{g}^{(k)} = \sum_{m=-\infty}^{\infty} g_m \otimes t^m$ $(g_m = \{x \in g: \theta(x) = e^{2\pi(-1)^{1/2}m/k} \cdot x\})$.

(c) Let $g$ be any Kac-Moody Lie-algebra then $H^2(g^1) = 0$. In particular, let $g$ be the affine Lie-algebra associated to a finite dim. simple Lie-algebra $g_o$ and an automorphism $\theta$ (of $g_o$), of order k, as in (b). Then the one dimensional central extension $0 \longrightarrow \mathfrak{z} \longrightarrow g^1 \longrightarrow \tilde{g}^{(k)} \longrightarrow 0$ (see [$W_i$; page 210], $g^1$ is nothing but $\hat{g}^{(k)}$ in the notation of [$W_i$]) is universal.

(d) $H^2(g) \approx \Lambda^2(g/g^1)$ and $H^3(g) \approx H^3(K) \oplus \Lambda^3(g/g^1)$, for any $g$.

<u>Proof</u>.

(a) Since $H^*(g^1) \approx H^*(K,\mathbb{C})$; K is a topological group and $H^*(g) \approx H^*(g^1) \otimes \Lambda(g/g^1)$ by [$Ku_2$; Proposition 1.9], (a) follows.

(b) We prove (b) in the special case $\theta = 1$. The general case is exactly similar. Specializing theorem (1.6) (a) to the affine Lie-algebra $g$ associated with $g_o$ and choosing an appropriate maximal parabolic $p_X$, we get that $H^2(g^1,h^1+g_o)$ is one dimensional, since, from Bruhat decomposition, $K/K_X$ can be easily seen to have only one cell in dim 2.

By lemma (1.8), taking $s = h^1 + g_o$ and $\mathfrak{z}$ = centre of $g^1$, we get $H^*(g^1,h^1+g_o) \approx H^*(g_o \otimes \mathbb{C}[t,t^{-1}],g_o)$. Now using [HS; Corollary in §6] (since $g_o$ is acting reductively on $g_o \otimes \mathbb{C}[t,t^{-1}]$, this is available) and the fact that $H^1(g_o) = H^2(g_o) = 0$, we get (b).

(c) By theorem (1.6) (b), we have $H^2(g^1) \approx H^2(K)$. But, by [$KP_2$; Theorem 4], K is simply connected. Further, using the long exact homotopy sequence for the fibration $K \longrightarrow K/T$, we get

$$0 \longrightarrow \pi_2(K) \longrightarrow \pi_2(K/T) \longrightarrow \pi_1(T) \longrightarrow 0 .$$

Since $\pi_2(K/T)$ ($\approx H_2(K/T)$) and $\pi_1(T)$ are free abelian groups with equal ranks, we get $\pi_2(K) = 0$. This gives $H^2(g^1) = 0$.

Now universality of the central extension follows immediately from (1.9) (b) together with standard facts on central extensions. See

[G; §1].

(d) This follows easily from (c) and [$Ku_2$; Proposition 1.9].

**(1.10) Remarks.** (1.9) (b) is due to the referee of [G]. See [G; §2]. (1.9) (c) in the affine case is, independently, due to Garland [G, Theorem (3.14)] and Vyjayanthi Chari (unpublished) and the twisted affine case is due to Wilson [$W_1$]. (1.9) (d) is strengthening of some results due to Berman [B].

**(1.11) Remark.** Using mixture of topological and geometric arguments, we show that, in general, the inclusion of the space of bi-invariant forms $C(\mathfrak{g}^1)^{\mathfrak{g}^1} \longrightarrow C(\mathfrak{g}^1)$ *does not* induce isomorphism in cohomology. The counterexample exists in any irreducible Kac-Moody Lie-algebra except in the case when it is a finite dimensional Lie-algebra or $\widehat{s\ell}(2)$.

## 2. Formality of Flag Varieties Associated to Kac-Moody Groups

We recall some, fairly known, definitions from rational homotopy theory. See, e.g., [DGMS]; [GM]; [Q]; [$S_1$]; [$S_2$].

**(2.1) Definitions.**

(a) A *differential graded algebra/$\mathbb{C}$* (*abbreviated to DGA*) is a graded algebra (over $\mathbb{C}$) $A = \bigoplus_{p\geq 0} A^p$ with a differential (i.e. $d^2 = 0$) $d: A \longrightarrow A$ of degree +1, such that

(1) A is graded commutative, i.e.,

$$a\cdot b = (-1)^{pq} b\cdot a \text{ for } a \in A^p \text{ and } b \in A^q.$$

(2) d is a derivation, i.e.,

$$d(a\cdot b) = (da)\cdot b + (-1)^p a\cdot db \text{ for } a \in A^p.$$

A is said to be *connected* if $H^0(A)$ is the ground field $\mathbb{C}$ and A is *one-connected* if, in addition, $H^1(A) = 0$.

(b) A DGA $\mu$ is a *minimal differential algebra*, if

(1) d is decomposable, i.e., $d(\mu^+) \subset \mu^+ \cdot \mu^+$ ($\mu^+$ denotes the augmentation ideal $\sum_{p>0} \mu^p$).

(2) $\mu$ may be written as an increasing union of sub DGA's $\mu_0 = \mathbb{C} \subset \mu_1 \subset \mu_2 \subset \ldots$, $\bigcup_{i\geq 0} \mu_i = \mu$ with $\mu_i \subset \mu_{i+1}$ an elementary extension for all $i \geqslant 0$, i.e., $\mu_{i+1}$ is a graded algebra of the form $\mu_i \otimes F(V_{d_i})$, for some $d_i > 0$ ($F(V_{d_i})$ denotes the symmetric (resp. exterior) algebra on $V_{d_i}$ if $d_i$ is even (resp. odd). We assign grade degree $d_i$ to elements of $V_{d_i}$) and such that $d_{\mu_{i+1}}|_{\mu_i} = d_{\mu_i}$ and $d_{\mu_{i+1}}(V_{d_i}) \subset \mu_i$.

(c) A *minimal model* for a DGA A is a minimal

differential algebra $\mu_A$ together with a DGA homomorphism $\rho: \mu_A \longrightarrow A$ such that $\rho$ induces isomorphism in cohomology.

An important fact is that every one connected DGA A, such that $H^i(A)$ is finite dimensional for all i, has a minimal model unique up to isomorphism. See [DGMS; Theorem 1.1 (a)].

In this paper, we would only consider one-connected DGA's A with the additional assumption that $H^i(A)$ is finite dimensional for all i. *From now on, this would be our implicit assumption on DGA's.*

(d) A minimal differential algebra $\mu$ is said to be *formal* if there is a DGA homomorphism $\psi: \mu \longrightarrow H^*(\mu)$ inducing the identity on cohomology. ($H^*(\mu)$ is equipped with identically zero differential.)

(e) *The homotopy type of a DGA* A *is a formal consequence of its cohomology* if its minimal model is formal.

Now we can state one of the main theorems of this section.

**(2.2) Theorem.** Let $g = g(A)$ be the Kac-Moody Lie-algebra associated to a symmetrizable generalized Cartan matrix $A = (a_{ij})_{1\leq i,j\leq \ell}$ and let $X \subset \{1,\dots,\ell\}$ be a subset of finite type.

Then, the homotopy type of the DGA $C(g,r)$ is formal consequence of its cohomology, where $r = r_X$ and $C(g,r)$ are defined in §(0.3) and §(0.10) respectively.

**Proof.** Our proof of this theorem is similar to the first proof of formality of Kähler manifolds, given by Deligne-Griffiths-Morgan-Sullivan [DGMS; §6]. One essential difference, however, is that the Hodge decomposition with respect to the operators $d;d^*$ ($d^*$ is the adjoint of d) is replaced by the 'Hodge decomposition' proved in [$Ku_1$], for the "disjoint" operators d and $\partial$.

We need the following $dd^c$ lemma.

**(2.3) Lemma.** Recall the definition of the operators d;d';d": $C(g,r) \longrightarrow C(g,r)$ from [$Ku_1$; §3]. As in [DGMS], define the operator $d^c = i(d'' - d')$ acting on $C(g,r)$. Then, we have

(1) $\operatorname{Im} d \cap \operatorname{Ker} d^c \subset \operatorname{Im}(dd^c)$ and

(2) $\text{Im } d^c \cap \text{Ker } d \subset \text{Im } (dd^c)$.

**Proof**. Let $\omega \in \text{Im } d \cap \text{Ker } d^c$. Since $d = d' + d''$

$$(I_1)\ldots \qquad d'\omega = 0 = d''\omega$$

From the 'Hodge type decomposition' [$Ku_1$; Theorem 3.13 and Remark 3.14] and disjointness of $(d',\partial')$ [$Ku_1$; Proposition 3.7], we get $\omega \in \text{Im } d' \oplus \text{Ker } S'$. Further, again by using [$Ku_1$; Lemma (3.8), Theorem (3.13), Remark (3.14) and Lemma (3.5)], we get $\text{Im } d' \subset \text{Im } S = \text{Im } d \oplus \text{Im } \partial$. Since, by assumption, $\omega \in \text{Im } d$, we get $\omega \in \text{Im } d'$. Write $\omega = d'\eta$, for some $\eta \in C(g,r)$. Express $\eta = d''\eta_1 + \partial''\eta_2 + \eta_3$, for some $\eta_1, \eta_2 \in C(g,r)$ and $\eta_3 \in \text{Ker } S'' = \text{Ker } S'$. This gives, on taking $d'$,

$$(I_2)\ldots \qquad \omega = d'\eta = d'd''\eta_1 + d'\partial''\eta_2 \quad (\text{since } \eta_3 \in \text{Ker } S').$$

Using $d'\partial'' + \partial''d' = 0$ and $d'd'' + d''d' = 0$ (see [$Ku_1$; Lemma (3.1) and identity $(I_{18})$]), we get $d'\eta = -d''d'\eta_1 - \partial''d'\eta_2$. So $d''d'\eta = -d''\partial''\, d'\eta_2 = 0$ (since $d''d'\eta = d''\omega = 0$, by $(I_1)$). By disjointness of the pair $(d'',\partial'')$ [$Ku_1$; Proposition 3.7], $\partial''d'\eta_2 = 0$. Putting this in $(I_2)$, we get the first part of this lemma. The second part follows exactly similarly.

(2.4) **Proof of Theorem (2.2)**. Denote by $H_{d^c}(g,r)$ the cohomology of the complex $C(g,r)$ under $d^c$ and by $Z_{d^c}(g,r)$ the $d^c$ closed forms in $C(g,r)$. Consider the diagram

$$C(g,r) \overset{i}{\hookleftarrow} Z_{d^c}(g,r) \xrightarrow{\alpha} H_{d^c}(g,r),$$

where i is the canonical inclusion and $\alpha$ the canonical projection.

Since $dd^c = -d^c d$, $Z_{d^c}(g,r)$ is stable under d. Moreover, by the previous lemma, the differential induced by d on $H_{d^c}(g,r)$ is zero.

We prove that i and $\alpha$ both induce isomorphism in cohomology, if we consider $C(g,r)$; $Z_{d^c}(g,r)$ and $H_{d^c}(g,r)$ as co-chain complexes under d.

(1) $\alpha^*$ *is surjective*: Given $\omega \in Z_{d^c}(g,r)$, we need to show that there exists a $\eta \in C(g,r)$ such that $\omega + d^c\eta$ is d closed. By $dd^c$-lemma, $d\omega = -dd^c\eta$, so $d(\omega + d^c\eta) = 0$.

(2) $\alpha^*$ *is injective*: We need to show that $\mathrm{Im}\, d^c \cap \mathrm{Ker}\, d \subset d(Z_{d^c}(g,r))$, which is immediate from $dd^c$-lemma.

(3) $i^*$ *is injective*: We need to prove that $\mathrm{Im}\, d \cap \mathrm{Ker}\, d^c \subset d(Z_{d^c}(g,r))$. Use $dd^c$-lemma.

(4) $i^*$ *is surjective*: We need to show that

$$\mathrm{Im}\, d + (\mathrm{Ker}\, d^c \cap \mathrm{Ker}\, d) = \mathrm{Ker}\, d.$$

By [$Ku_1$; Theorem 3.13, Remark 3.14 and Lemma (3.5)], $\mathrm{Ker}\, S \subset \mathrm{Ker}\, d^c \cap \mathrm{Ker}\, d$ and $\mathrm{Ker}\, S + \mathrm{Im}\, d = \mathrm{Ker}\, d$. This gives surjectivity of $i^*$.

Theorem follows, now, by choosing a minimal model $\mu$ for the DGA $Z_{d^c}(g,r)$. (Observe that $C(g,r)$, hence $Z_{d^c}(g,r)$, is one-connected and $H^i(C(g,r))$ is finite dimensional for all i, by [L; §6] or [$Ku_1$; Theorem 3.15].) □

We recall the following

(2.5) __Definition__ [$S_2$]; [DGMS]. A *polyhedron* Y (we assume, for simplicity, that Y is simply connected and $H^i(Y,\mathbb{Q})$ is finite dimensional for all i) *is said to be a formal consequence of its cohomology over* $\mathbb{Q}$ *(or a formal space over* $\mathbb{Q}$*)*, if the homotopy type of the DGA of $\mathbb{Q}$-polynomial forms $E_Y^*$ (see [DGMS; §2] for the definition of $E_Y^*$) is a formal consequence of its cohomology.

The formality of Y does not depend upon particular choice of simplicial structure on Y, in fact let $\tilde{E}_Y^*$ denote the DGA of $\mathbb{Q}$-polynomial forms on Y with respect to some other triangulation of Y then the minimal models of $E_Y \otimes_{\mathbb{Q}} \mathbb{C}$ and $\tilde{E}_Y \otimes_{\mathbb{Q}} \mathbb{C}$ are isomorphic. This can be easily seen by taking a common subdivision. Now using [HaS; Theorem 6.8] or [$S_2$; §12], we see that the minimal models of $E_Y^*$ and $\tilde{E}_Y^*$ are themselves isomorphic.

**(2.6) Lemma.** Minimal models of the DGA's $C(g,r)$ and $E^*_{G/P_X} \otimes_{\mathbb{Q}} \mathbb{C}$ are isomorphic.

**Proof.** We have described the DGA $\Omega_{p\cdot dR}(\mathrm{Sing}_{C^\infty}(K/K_X))$ of piece-wise smooth forms associated to the simplicial set $\mathrm{Sing}_{C^\infty}(K/K_X)$ during the proof of lemma (1.4). Further, we described an integration map $\tilde{\int}: \Omega_{p\cdot dR}(\mathrm{Sing}_{C^\infty}(K/K_X)) \longrightarrow C_{C^\infty}(K/K_X)$. Exactly similarly, we can define $\Omega_{p\cdot dR}(\mathrm{Sing}\ K/K_X)$ associated to the simplicial set $\mathrm{Sing}(K/K_X)$ and also $\Omega_{PL}(\mathrm{Sing}(K/K_X)) \hookrightarrow \Omega_{p\cdot dR}(\mathrm{Sing}(K/K_X))$, where $\Omega_{PL}$ consists of only polynomial forms $/\mathbb{C}$ (with respect to the Barycentric co-ordinates on $\Delta^n$). We have the following commutative diagram

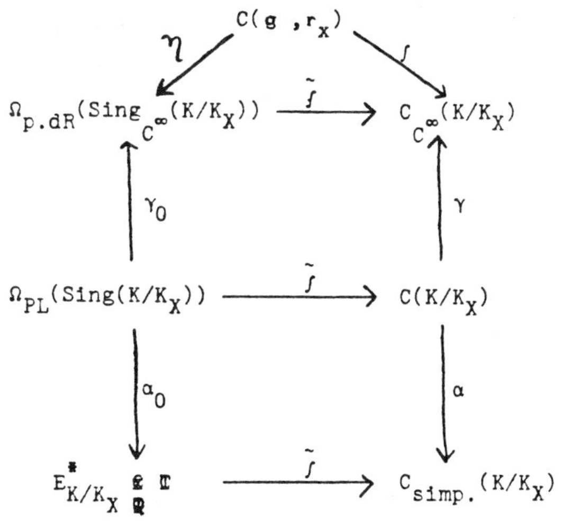

, where $E^*_{K/K_X}$ is the space of $\mathbb{Q}$-polynomial forms on $K/K_X$ with respect to some (fixed) triangulation of $K/K_X$, $C_{simp.}(K/K_X)$ is the simplicial co-chain complex of $K/K_X$, the maps $\alpha$, $\alpha_0$, $\gamma$, $\gamma_0$ are the canonical restrictions and the map $\eta$ is defined during the proof of lemma (1.4).

All the three horizontal maps induce algebra isomorphisms in cohomology.* (See [$S_2$; §7] and our lemma (1.5).) By lemma (1.5) (resp. theorem (1.6)) $\gamma$ (resp. $\eta$) induces algebra isomorphism in cohomology. $\alpha$, of course, induces algebra isomorphism in cohomology. Hence $\alpha_0$ and $\gamma_0$ both induce isomorphisms in cohomology, which proves the lemma.

As an immediate corollary of theorem (2.2), lemma (2.6) and [HaS; Corollary 6.9], we get the following.

**(2.7) Theorem.** Let G be a Kac-Moody algebraic group and let $P = P_X$ be a standard parabolic (of G) of finite type (see §(0.10) for terminologies).

Then the space G/P is a formal space over $\mathbb{Q}$.

So, complete rational homotopy information of G/P can be derived from the cohomology algebra $H^*(G/P,\mathbb{Q})$. In particular, the rational homotopy groups $\pi_*(G/P) \otimes_{\mathbb{Z}} \mathbb{Q}$, viewed as a graded Lie-algebra under Whitehead product, depends only on the cohomology ring $H^*(G/P,\mathbb{Q})$. Moreover, all Massey products of any order are zero over $\mathbb{Q}$.

**(2.8) Remarks.**

(a) Compare the above theorem with formality of Kähler manifolds proved in [DGMS].

---

*A more detailed proof can be found in Chapter 12 of "Lectures on Minimal models by S. Halperin, Publications de L' U.E.R. Mathematiques pures et Appliquees".

(b) Since $H^*(G,\mathbb{Q})$ is a Hopf algebra, the minimal model $\mu_G$ of G (i.e. the minimal model of DGA of $\mathbb{Q}$-polynomial forms $E_G^*$) is $H^*(G,\mathbb{Q})$, so that $H^*(G,\mathbb{Q}) \approx S(\pi_{even}(G) \underset{\mathbb{Z}}{\otimes} \mathbb{Q}) \otimes \Lambda(\pi_{odd}(G) \underset{\mathbb{Z}}{\otimes} \mathbb{Q})$ as graded algebras, where $\pi_{even}(G)$ (resp. $\pi_{odd}(G)$) denotes $\sum_{n=0}^{\infty} \pi_{2n}(G)$ (resp. $\sum_{n=0}^{\infty} \pi_{2n+1}(G)$).

(c) In the next section, we would specifically determine the minimal model of G/B and the Lie-algebra $\pi_*(G/B) \underset{\mathbb{Z}}{\otimes} \mathbb{Q}$ under Whitehead product.

As an application of our theorem (2.2), we prove degeneracy of the Leray-Serre fiber spectral sequence/$\mathbb{Q}$ corresponding to the fibration $K \longrightarrow K/T$.

Recently, Kac-Peterson have proved an important result that this spectral sequence degenerates at $E_3$ even over any finite field.

**(2.9)** <u>**Proposition**</u>. Let K be the standard compact real form of a Kac-Moody algebraic group G and let B be the standard Borel subgroup of G. (See §(0.10) for the notations.)

Leray-Serre spectral sequence in cohomology/$\mathbb{Q}$ corresponding to the fibration $K \longrightarrow K/T$ where $T = B \cap K$ degenerates at $E_3$, i.e., $E_3^{p,q} \approx E_\infty^{p,q}$ for all p and q.

<u>**Proof**</u>. <u>**Step I**</u>. Let A be a one connected DGA and let $G_o$ be a finite dimensional connected Lie-group. Given a linear map $\theta$: $P \longrightarrow Z(A)$ of degree +1 (where $P \subset H^*(G,\mathbb{C})$ is the linear subspace generated by primitive elements and $Z(A) = \{a \in A: da = 0\}$), we put a *twisted differential* $D = D_\theta$ on the tensor product of graded algebras $A \underset{\mathbb{C}}{\otimes} H^*(G_o)$, to make it a DGA, as follows.

$$D\big|_A = \text{differential of } A \text{ and}$$

$$Dx = \theta(x), \text{ for all } x \in P.$$

Denote the DGA, thus obtained, by $A_\theta$.

There is a filtration $F = \{F_p\}$, of the co-chain complex

$A_\theta$, defined by $F_p = \sum_{\ell \geq p} A^\ell \otimes H^*(G_0)$. Clearly, $F_p$ is D-stable. Further, it is easy to see that the corresponding spectral sequence has $E_2^{p,q} \approx H^p(A) \otimes H^q(G_0)$ and converges to $H^*(A_\theta)$.

The above construction is motivated by Hirsch lemma. Also the only property of $H^*(G_0)$, which we are using is that it is free (in the graded sense) graded algebra on P.

__Step II.__ In Step I, if we assume that the homotopy type of A is formal consequence of its cohomology and $G_0$ is a torus T then the above spectral sequence degenerates at $E_3$.

To prove this, fix a minimal model $\rho\colon \mu \longrightarrow A$ and a DGA morphism, inducing the identity at cohomology, $\psi\colon \mu \longrightarrow H^*(\mu)$. There exist linear maps $\tilde{\theta}\colon H^1(T) \longrightarrow \mu^2$ and $y\colon H^1(T) \longrightarrow A^1$, such that $\rho\circ\tilde{\theta}(x)-\theta(x) = dy(x)$ for all $x \in H^1(T)$. Further, there exists a DGA isomorphism $\xi\colon A_{\rho\circ\tilde{\theta}} \longrightarrow A_\theta$ defined by $\xi|_A = \mathrm{Id}\cdot$ and $\xi(x)$

$= 1\otimes x + y(x)\otimes 1$, for $x \in H^1(T)$. We have the following DGA morphisms

$$H^*(\mu)_{\psi\circ\tilde{\theta}} \xleftarrow{\psi\otimes \mathrm{Id}\cdot} \mu_{\tilde{\theta}} \xrightarrow{\rho\otimes \mathrm{Id}\cdot} A_{\rho\circ\tilde{\theta}} \xrightarrow{\xi} A_\theta$$

All of these morphisms preserve filtrations and induce isomorphisms at $E_2$ level. Hence degeneracy of the spectral sequence for $A_\theta$ at $E_3$ is equivalent to the degeneracy of the spectral sequence for $H^*(\mu)_{\psi\circ\tilde{\theta}}$ at $E_3$.

We come to prove the degeneracy of the spectral sequence for $H^*(\mu)_{\psi\circ\tilde{\theta}}$ at $E_3$. By definition (see, e.g., [GH; page 441]) $E_s^p = Z_s^p/(Z_{s-1}^{p+1} + D(Z_{s-1}^{p-s+1}))$, where $Z_s^p = \{a \in F_p\colon Da \in F_{p+s}\}$

and the differential $d_s\colon E_s^p \longrightarrow E_s^{p+s}$ is $\bar{a} \longrightarrow \overline{Da}$ for $a \in Z_s^p$. So, it suffices to show that $D(Z_s^p) \subset DZ_{s-1}^{p+1}$, for all $s \geq 3$. Let $a = \sum_{t \geq p} a_t \in Z_s^p$, where $a_t \in H^t(\mu) \otimes H^*(T)$. By definition of D,

$Da_t \in H^{t+2}(\mu) \otimes H^*(T)$. Since $a \in Z_s^p$, $Da \in F_{p+s}$; in particular $Da_p = 0$ and hence $Da = D(\sum_{t\geq p+1} a_t) \in DZ_{s-1}^{p+1}$.

__Step III.__ Consider the DGA $C(\mathfrak{g}^1,\mathfrak{h}^1)$ and a degree +1 (transgression)

map $\theta(f) = d_{g^1}(\tilde{f})$, for all $f \in (h^1)^*$, where $\tilde{f}|_{h^1} = f$ and $\tilde{f}|_{g_\alpha} = 0$ for root spaces $g_\alpha$ corresponding to all the (nonzero) roots $\alpha$.

($d_{g^1}$ denotes the usual co-chain map of $C(g^1)$. It is easy to see that $d_{g^1}(\tilde{f})$ is, in fact, an element of $C(g^1,h^1)$.) As in Step I, $\theta$ gives rise to a DGA $C(g^1,h^1)_\theta = C(g^1,h^1) \underset{\mathbb{C}}{\otimes} \Lambda(h^{1*})$. There is a DGA morphism $\psi$: $C(g^1,h^1)_\theta \longrightarrow C(g^1)$, defined by $\psi|_{C(g^1,h^1)} = i$ ($i$ is the canonical inclusion: $C(g^1,h^1) \longrightarrow C(g^1)$) and $\psi(f) = \tilde{f}$ for $f \in h^{1*}$.

In §(1.3), we have defined a co-chain map $\int: C(g^1) \longrightarrow C_{C^\infty}(K)$. Composing with $\psi$, we get a co-chain map $\int \circ \psi$: $C(g^1,h^1)_\theta \longrightarrow C_{C^\infty}(K)$. We have described a filtration $F = \{F_p\}$ of $C(g^1,h^1)_\theta$ in Step I. Also $C_{C^\infty}(K)$ has a Leray-Serre filtration $G = \{G_p\}$, described in §(1.6). It is fairly easy to see that $\int \circ \psi(F_p) \subset G_p$ for all p. Further, by Step I, $E_2^{p,q}(F) \simeq H^p(g^1,h^1) \otimes \Lambda^q(h^{1*})$. In view of theorem (1.6) (a) (applied in the special case $X = \emptyset$), we get that $\int \circ \psi$ induces isomorphism: $E_2^{p,q}(F) \longrightarrow E_2^{p,q}(G)$ for all p and q and hence degeneracy of the spectral sequence, corresponding to the filtration G, at $E_3$ is equivalent to the degeneracy corresponding to the filtration F, which, in turn, follows from Step II and theorem (2.2). This establishes the proposition.

**(2.10)** Remark. The proof of proposition (2.9) can be modified to give the following generalization of (2.9).

Let Y be a simply connected space such that $H^i(Y,\mathbb{Q})$ is finite dimensional, for all i. Assume further that Y is a formal space/$\mathbb{Q}$ and let $E \longrightarrow Y$ be any principal T bundle (T is a torus), then the corresponding Leray-Serre spectral sequence in cohomology/$\mathbb{Q}$ degenerates at $E_3$.

## 3. Determination of Minimal Model for G/B.

(3.1) From the proof of proposition (2.9), we know that $H^*(K,\mathbb{C})$, as a graded algebra, is isomorphic with the cohomology of the DGA $H^*(K/T)_\beta = H^*(K/T,\mathbb{C}) \underset{\mathbb{C}}{\otimes} \Lambda(\mathfrak{h}^{1*})$, where the notation $H^*(K/T)_\beta$ is as in Step I of the proof of proposition (2.9) and $\beta: \mathfrak{h}^{1*} \longrightarrow H^2(K/T)$ is the map defined by $\beta(f) = [\int (d_{\mathfrak{g}^1}\tilde{f})]$, for all $f \in \mathfrak{h}^{1*}$. ($\tilde{f}$ is, as in Step III of the proof of proposition (2.9), an element of $C^1(\mathfrak{g}^1)$ satisfying $\tilde{f}|_{\mathfrak{h}^1} = f$ and $\tilde{f}|_{\mathfrak{g}^\alpha} = 0$, for root spaces $\mathfrak{g}^\alpha$ corresponding to all the roots $\alpha$. $\int$ is the integration map, defined in §(1.3), from $C(\mathfrak{g}^1,\mathfrak{h}^1)$ to $C_{C^\infty}(K/T)$ and [ ] denotes the cohomology class.) Extend $\beta$ (again denoted by $\beta$ itself) to an algebra homomorphism (called the Borel homomorphism) from $S(\mathfrak{h}^{1*}) \longrightarrow H^*(K/T)$. $H^*(K/T)$ becomes a $S(\mathfrak{h}^{1*})$-module under $\beta$.

It is fairly easy to see that the DGA $H^*(K/T)_\beta$ can be identified with the standard chain complex $\Lambda(\mathfrak{h}^{1*},H^*(K/T))$, corresponding to the abelian Lie-algebra $\mathfrak{h}^{1*}$ with coefficients in $H^*(K/T)$ (considered as $\mathfrak{h}^{1*}$-module under $\beta$). So $H^*(K,\mathbb{C})$, which is isomorphic with the cohomology of the DGA $H^*(K/T)_\beta$, is isomorphic (as a graded algebra) with $H_*(\mathfrak{h}^{1*},H^*(K/T))$.

By [$K_2$; page 4, assertion 4] (in fact it is valid even over $\mathbb{Z}/p\mathbb{Z}$), $H^*(K/T,\mathbb{C})$ is free as $\bar{S}(\mathfrak{h}^{1*}) = S(\mathfrak{h}^{1*})/\mathrm{Ker}\ \beta$-module (Ker $\beta$ denotes the kernel of $\beta$: $S(\mathfrak{h}^{1*}) \longrightarrow H^*(K/T,\mathbb{C})$). Hence

$$
\begin{aligned}
H^*(K,\mathbb{C}) &\approx H_*(\mathfrak{h}^{1*},H^*(K/T)) \\
&\approx H_*(\mathfrak{h}^{1*},\bar{S}(\mathfrak{h}^{1*}) \otimes_{\bar{S}(\mathfrak{h}^{1*})} H^*(K/T)) \\
&\approx H_*(\mathfrak{h}^{1*},\bar{S}(\mathfrak{h}^{1*})) \otimes_{\bar{S}(\mathfrak{h}^{1*})} H^*(K/T)
\end{aligned}
$$

$$
(I_3)\ldots \quad H^*(K,\mathbb{C}) \approx H_*(\mathfrak{h}^{1*},\bar{S}(\mathfrak{h}^{1*})) \underset{\mathbb{C}}{\otimes} [H^*(K/T)/\langle\bar{S}^+(\mathfrak{h}^{1*})\rangle]
$$

as graded algebras. (Since $H_*(\mathfrak{h}^{1*},\bar{S}(\mathfrak{h}^{1*}))$ is trivial $\mathfrak{h}^{1*}$-module.)

$\langle\bar{S}^+(\mathfrak{h}^{1*})\rangle$ denotes the ideal, in $H^*(K/T)$, generated by $\sum_{i\geqslant 1} \bar{S}^i(\mathfrak{h}^{1*})$.

**(3.2)** <u>**Definition**</u>. A *graded algebra A is said to be free* if A is isomorphic (as graded algebra) with $S(W_0) \otimes \Lambda(W_1)$, where $W_0$ (resp. $W_1$) is evenly > 0 (resp. oddly) graded vector space.

**(3.3)** <u>**Lemma**</u>. Let A be a free graded algebra and let B and C be two graded subalgebras of A such that A, as a graded algebra, is isomorphic with $B \otimes C$ then B and C are free algebras.

<u>**Proof**</u>. Choose a graded algebra isomorphism $\phi: B \otimes C \longrightarrow A$. It is fairly easy to see that $\phi(V' \otimes 1 \oplus 1 \otimes V'') \oplus A^+ \cdot A^+ = A^+$, where $V' \subset B^+$ (resp. $V'' \subset C^+$) is any graded vector space such that $V' \oplus B^+ \cdot B^+ = B^+$ (resp. $V'' \oplus C^+ \cdot C^+ = C^+$) and $A^+$ denotes $\sum_{i>0} A^i$. Further, for a free graded algebra D and any graded vector space $W \subset D^+$ such that $W \oplus D^+ \cdot D^+ = D^+$, F(W) is isomorphic as graded algebras, with D. (Where F(W) denotes $S(W_0) \otimes \Lambda(W_1)$; $W_0$ (resp. $W_1$) is linear span of evenly (resp. oddly) graded elements in W.)

In particular $F(V' \otimes 1 \oplus 1 \otimes V'') \xrightarrow{\theta} A$ is an isomorphism, where $\theta$ is the graded algebra homomorphism with $\theta|_{V'\otimes 1\oplus 1\otimes V''} = \phi|_{V'\otimes 1\oplus 1\otimes V''}$. Clearly $\theta(F(V' \otimes 1)) \subset \phi(B)$ and $\theta(F(1\otimes V'')) \subset \phi(C)$. But since $\theta$ is an (surjective) isomorphism, we get $\theta(F(V' \otimes 1)) = \phi(B)$ and $\theta(F(1 \otimes V'')) = \phi(C)$. This proves the lemma.

**(3.4)** We return to the situation of §(3.1). K being a topological group, $H^*(K,\mathbb{C})$ is a free graded algebra. Write

$(I_4)$...
$$H^*(K,\mathbb{C}) \approx \Lambda(W_1) \otimes S(W_0),$$

where $W_0$ (resp. $W_1$) is an evenly (resp. oddly) graded vector space.

Since, clearly, all the elements of $H_*(\mathfrak{h}^{1*},\bar{S}(\mathfrak{h}^{1*}))$ of positive degree are nilpotent and $H^*(K/T)$ consists of evenly graded elements only, we get from $(I_3)$ and lemma (3.3)

$(I_5)$...
$$H_*(\mathfrak{h}^{1*},\bar{S}(\mathfrak{h}^{1*})) \approx \Lambda(W_1) \text{ and}$$

$(I_6)$...
$$H^*(K/T)/\langle\bar{S}^+(\mathfrak{h}^{1*})\rangle \approx S(W_0) \text{ as graded algebras.}$$

We prove the following.

**(3.5)** **Lemma.** $H^*(K/T) \approx \bar{S}(\mathfrak{h}^{1*}) \otimes S(W_0)$ as graded algebras.

**Proof.** Consider the graded algebra homomorphism $p: \bar{S}(\mathfrak{h}^{1*}) \underset{\mathbb{C}}{\otimes} S(W_0) \longrightarrow H^*(K/T)$ defined by $p(\bar{a} \otimes b) = \beta(a) \cdot \theta(b)$, for $a \in S(\mathfrak{h}^{1*})$ and $b \in S(W_0)$. ($\beta$: $S(\mathfrak{h}^{1*}) \longrightarrow H^*(K/T)$ is the Borel homomorphism defined in §(3.1); $\bar{a}$ denotes a mod Ker $\beta$ and $\theta$ is any graded algebra homomorphism: $S(W_0) \longrightarrow H^*(K/T)$ such that $\pi \circ \theta$ is an isomorphism as in $(I_6)$, where $\pi$: $H^*(K/T) \longrightarrow H^*(K/T)/\langle\bar{S}^+(\mathfrak{h}^{1*})\rangle$ is the canonical projection.) From $(I_6)$ it is fairly easy to see that p is surjective. We assert that p is injective as well.

Let J be the kernel of p, so there is an exact sequence of $\bar{S}(\mathfrak{h}^{1*})$-modules ($\bar{S}(\mathfrak{h}^{1*})$ acts on $H^*(K/T)$ via $\beta$ and it acts on $\bar{S}(\mathfrak{h}^{1*}) \otimes S(W_0)$ by multiplication on the first factor.)

$$0 \longrightarrow J \longrightarrow \bar{S}(\mathfrak{h}^{1*}) \underset{\mathbb{C}}{\otimes} S(W_0) \longrightarrow H^*(K/T) \longrightarrow 0$$

considering $\mathbb{C} = \bar{S}(\mathfrak{h}^{1*})/\bar{S}^+(\mathfrak{h}^{1*})$ as $\bar{S}(\mathfrak{h}^{1*})$-module by multiplication, we get an exact sequence.

$$(I_7)\ldots \quad \mathrm{Tor}_1^{\bar{S}(\mathfrak{h}^{1*})}(\mathbb{C},H^*(K/T)) \longrightarrow J/\bar{S}^+(\mathfrak{h}^{1*})\cdot J \longrightarrow$$

$$(\bar{S}(\mathfrak{h}^{1*})\underset{\mathbb{C}}{\otimes}S(W_0))/(\bar{S}^+(\mathfrak{h}^{1*})\underset{\mathbb{C}}{\otimes}S(W_0)) \longrightarrow H^*(K/T)/\langle\bar{S}^+(\mathfrak{h}^{1*})\rangle \longrightarrow 0$$

(Since $\mathbb{C} \underset{\bar{S}(\mathfrak{h}^{1*})}{\otimes} M \cong M/\bar{S}^+(\mathfrak{h}^{1*})\cdot M$, for any $\bar{S}(\mathfrak{h}^{1*})$-module M.)

By $(I_6)$, $(\bar{S}(\mathfrak{h}^{1*}) \underset{\mathbb{C}}{\otimes} S(W_0))/(\bar{S}^+(\mathfrak{h}^{1*}) \underset{\mathbb{C}}{\otimes} S(W_0)) \approx S(W_0) \longrightarrow H^*(K/T)/\langle\bar{S}^+(\mathfrak{h}^{1*})\rangle$ is an isomorphism. Also, $H^*(K/T)$ is $\bar{S}(\mathfrak{h}^{1*})$-free module and hence $\mathrm{Tor}_1^{\bar{S}(\mathfrak{h}^{1*})}(\mathbb{C},H^*(K/T)) = 0$. Putting these in $(I_7)$, we get $J/\bar{S}^+(\mathfrak{h}^{1*})\cdot J = 0$, i.e.,

$$(I_8)\ldots \qquad J = \bar{S}^+(\mathfrak{h}^{1*})\cdot J$$

Assume, if possible, that $J \neq 0$. Pick a homogeneous element $a \neq 0 \in J$ of minimal degree. By $(I_8)$, a can be written as $a = \sum_i \lambda_i a_i$, for some homogeneous elements $\lambda_i \in \bar{S}^+(\mathfrak{h}^{1*})$ and $a_i \in J$. Since $\bar{S}^+(\mathfrak{h}^{1*})$ has no elements of degree 0, we have $\deg a_i < \deg a$, contradicting the minimality of deg a. This proves the lemma.

**(3.6)** **Determination of minimal model for G/B.** Since by theorem (2.7) G/B is a formal space over $\mathbb{Q}$, in view of the Lemma (3.5), it suffices to determine the minimal model for the DGA $\bar{S}(\mathfrak{h}^{1*})$ (with $d \equiv 0$).

Denote by I the graded ideal Ker $\beta$. Choose a $\mathbb{C}$-linear graded splitting s of the canonical projection: $I \longrightarrow I/S^+(\mathfrak{h}^{1*})\cdot I$ and let $\{f_1,\dots,f_{m_0}\}$ be a homogeneous $\mathbb{C}$-basis of $s(I/S^+(\mathfrak{h}^{1*})\cdot I)$ with $f_i$ of degree $\ell(i)$ (assigning deg 1 to the elements of $\mathfrak{h}^{1*}$). By [K$_2$], $\{f_1,\dots,f_{m_0}\}$ is a $S(\mathfrak{h}^{1*})$-regular sequence. (Since $S(\mathfrak{h}^{1*})$ is Noetherian, $m_0$ is finite.) As $\beta: \mathfrak{h}^{1*} \longrightarrow H^2(G/B)$ is an isomorphism, $\ell(i) \geq 2$ for all $1 \leq i \leq m_0$.

Define a minimal differential algebra $\mu_0 = S(\mathfrak{h}^{1*}) \otimes [\bigotimes_{i=1}^{m_0} \Lambda(x^i_{2\ell(i)-1})]$ ($\Lambda(x^i_{2\ell(i)-1})$ denotes the exterior algebra on 1 dim vector space in grade degree $2\ell(i)-1$ and the elements of $\mathfrak{h}^{1*}$ are assigned grade degree 2) with $d|_{S(\mathfrak{h}^{1*})} \equiv 0$ and $d(x^i_{2\ell(i)-1}) = f_i$.

Define a DGA homomorphism $\theta: \mu_0 \longrightarrow S(\mathfrak{h}^{1*})/\mathrm{Ker}\ \beta$ by $\theta|_{S(\mathfrak{h}^{1*})}$ is the canonical projection and $\theta(x^i_{2\ell(i)-1}) = 0$. (Since $f_i \in \mathrm{Ker}\ \beta$, $\theta$ is a co-chain map.)

**(3.7)** **Lemma.** $\theta$ induces isomorphism in cohomology.

**Proof.** $H^*(\mu_0)$ can be, easily, identified with $\mathrm{Tor}^{\mathbb{C}[y_1,\dots,y_{m_0}]}(\mathbb{C}, S(\mathfrak{h}^{1*}))$, where $\mathbb{C}$ is trivial $\mathbb{C}[y_1,\dots,y_{m_0}]$ module and $S(\mathfrak{h}^{1*})$ is $\mathbb{C}[y_1,\ldots,y_{m_0}]$ module under $y_i\cdot f = f_i\cdot f$ for all $1 \leqslant i \leqslant m_0$ and $f \in S(\mathfrak{h}^{1*})$.

Further $\mathrm{Tor}_i^{\mathbb{C}[y_1,\ldots,y_{m_0}]}(\mathbb{C},S(\mathfrak{h}^{1*})) = 0$ for all $i \geqslant 1$. This follows from [Se; Proposition 2 on page IV–4], $\{f_1,\ldots,f_{m_0}\}$ is a $S(\mathfrak{h}^{1*})$-regular sequence. Of course $\mathrm{Tor}_o^{\mathbb{C}[y_1,\ldots,y_{m_0}]}(\mathbb{C},S(\mathfrak{h}^{1*})) = S(\mathfrak{h}^{1*})/\langle f_1,\ldots,f_{m_0}\rangle = S(\mathfrak{h}^{1*})/\mathrm{Ker}\ \beta$. (Since the ideal $\langle f_1,\ldots,f_{m_0}\rangle$, generated by $f_1,\ldots,f_{m_0}$, is equal to Ker $\beta$.) This easily gives that $\theta$ induces isomorphism in cohomology.

We summarize all this in the following

(3.8) **Theorem**. Let G be a Kac–Moody algebraic group and B the standard Borel subgroup of G. (See §(0.10).) Then

(1) Let $\{f_1,\ldots,f_{m_0}\} \subset \mathrm{Ker}\ \beta$ be a homogeneous $\mathbb{C}$-basis of Ker $\beta$ modulo $S^+(\mathfrak{h}^{1*})\cdot\mathrm{Ker}\ \beta$, with $f_i$ of degree $\ell(i)$ (assigning degree 1 to the elements of $\mathfrak{h}^{1*}$) ($\beta$ is the Borel map defined in §(3.1)).

Then the minimal model of the space G/B (this is defined to be the minimal model of the DGA $E^*_{G/B} \otimes_{\mathbb{Q}} \mathbb{C}$, with respect to some triangulation of G/B. See §(2.5)) is of the form

$$\mu_{G/B} = S(W_0) \otimes S(\mathfrak{h}^{1*}) \otimes [\bigotimes_{i=1}^{m_0} \Lambda(x^i_{2\ell(i)-1})],$$

where $W_0$ is an evenly graded vector space which is isomorphic (as graded vector spaces/$\mathbb{C}$) with $\sum_{n\geqslant 1} \pi_{2n}(G) \otimes_{\mathbb{Z}} \mathbb{C}$ and $\Lambda(x^i_{2\ell(i)-1})$ is the exterior algebra on a 1 dim. vector space in grade degree $2\ell(i)-1$. Further the differential d on $\mu_{G/B}$ is described as follows

$$d\big|_{S(W_0)} \equiv 0$$

$$d\big|_{S(\mathfrak{h}^{1*})} \equiv 0$$

$$d(x^i_{2\ell(i)-1}) = f_i$$

In particular, $\sum_{n\geq 1} \pi_{2n-1}(G) \otimes_{\mathbb{Z}} \mathbb{C} \approx \sum_{n\geq 1} \pi_{2n-1}(G/B) \otimes_{\mathbb{Z}} \mathbb{C}$ is finite dimensional and $\dim_{\mathbb{C}}(\pi_{2n-1}(G/B) \otimes_{\mathbb{Z}} \mathbb{C}) = \#\{f_j: \deg f_j = n\}$.

(2) The map: $H^*(G/B,\mathbb{C}) \longrightarrow H^*(G,\mathbb{C})$ (induced by the canonical projection: $G \longrightarrow G/B$) has the kernel precisely equal to the ideal generated by $H^2(G/B)$ and the image of $H^*(G/B,\mathbb{C})$ in $H^*(G,\mathbb{C})$ is isomorphic (as a graded algebra) with $S(W_0)$.

(3) <u>Determination of Whitehead product in</u> $\pi_*(G/B) \otimes \mathbb{C}$. The Whitehead product map $[\ ,\ ]: (\pi_n(G/B) \otimes_{\mathbb{Z}} \mathbb{C}) \otimes (\pi_m(G/B) \otimes_{\mathbb{Z}} \mathbb{C}) \longrightarrow \pi_{n+m-1}(G/B) \otimes_{\mathbb{Z}} \mathbb{C}$ is given by

(a) $[\alpha,\beta] = 0$ for $\alpha \in \pi_n(G/B) \otimes_{\mathbb{Z}} \mathbb{C}$ and $\beta \in \pi_m(G/B) \otimes_{\mathbb{Z}} \mathbb{C}$ unless $n = m = 2$

(b) $(\pi_2(G/B) \otimes_{\mathbb{Z}} \mathbb{C}) \otimes (\pi_2(G/B) \otimes_{\mathbb{Z}} \mathbb{C}) \longrightarrow \pi_3(G/B) \otimes_{\mathbb{Z}} \mathbb{C}$ is surjective.

<u>Proof</u>. (1) follows easily from theorem (2.7); $(I_4)$; lemma (3.5); §(3.6) and lemma (3.7) coupled with [DGMS; Theorem (3.3)(a)].

From §(3.1), it is easy to see that the map: $H^*(G/B,\mathbb{C}) \longrightarrow H^*(G,\mathbb{C})$ has kernel precisely equal to $<\bar{S}^+(\mathfrak{h}^{1*})>$. Hence, by $(I_6)$, (2) follows.

To prove (3), observe that the Whitehead product $[\ ,\ ]: (\pi_n(G) \otimes_{\mathbb{Z}} \mathbb{C}) \otimes (\pi_m(G) \otimes_{\mathbb{Z}} \mathbb{C}) \longrightarrow \pi_{n+m-1}(G) \otimes_{\mathbb{Z}} \mathbb{C}$ is zero ($G$ being a group). From the homotopy exact sequence, corresponding to the fibration $G \longrightarrow G/B$, $\pi_n(G) \approx \pi_n(G/B)$ for $n \geq 3$. Hence $[\ ,\ ]: (\pi_n(G/B) \otimes_{\mathbb{Z}} \mathbb{C}) \otimes (\pi_m(G/B) \otimes_{\mathbb{Z}} \mathbb{C}) \longrightarrow \pi_{n+m-1}(G/B) \otimes_{\mathbb{Z}} \mathbb{C}$ is zero unless one of m and n is equal to 2. From first part of this theorem and [DGMS; Theorem (3.3)(a)], it is fairly easy to see that

$[\ ,\ ]: (\pi_2(G/B) \underset{\mathbb{Z}}{\otimes} \mathbb{C}) \otimes (\pi_m(G/B) \underset{\mathbb{Z}}{\otimes} \mathbb{C}) \longrightarrow \pi_{m+1}(G/B) \underset{\mathbb{Z}}{\otimes} \mathbb{C}$ is also zero for $m \geq 3$. Finally the map $d\colon \pi^3_{G/B} = \sum_{\substack{\text{those } i \\ \text{s.t. } \ell(i)=2}} \mathbb{C}x^i_{2\ell(i)-1} \longrightarrow \mu^2_{G/B} \otimes \mu^2_{G/B} = \mathfrak{h}^{1*} \otimes \mathfrak{h}^{1*}$ can be easily seen (using its definition) to be injective.

This completes the proof of the theorem.

## References


[B] Berman, S.: On the low dimensional cohomology of some infinite dimensional simple Lie-algebras. Pacific Journal of Mathematics 83 (1979), 27-36.

[CE] Cartan, H. and Eilenberg, S.: Homological Algebra. Princeton University Press, Princeton (1956).

[DGMS] Deligne, P.; Griffiths, P.; Morgan, J. and Sullivan, D.: Real homotopy theory of Kähler manifolds. Inventiones Math. 29 (1975), 245-274.

[G] Garland, H.: The arithmetic theory of loop groups. Publications Math. I.H.E.S. 52 (1980), 181-312.

[GH] Griffiths, P. and Harris, J.: Principles of Alg. Geometry. Wiley-Interscience, New York (1978).

[GM] Griffiths, P.A. and Morgan, J.W.: Rational homotopy theory and differential forms. Progress in Mathematics Vol. 16. Birkhäuser (1981).

[HaS] Halperin, S. and Stasheff, J.: Obstructions to

homotopy equivalences. Adv. in Math. 32 (1979), 233-279.

[Hi] Hironaka, H.: Triangulations of Algebraic sets. Proc. of Symposia in Pure Mathematics 29 (1975), 165-185.

[HS] Hochschild, G. and Serre, J.P.: Cohomology of Lie-algebras. Annals of Math. 57 (1953), 591-603.

[$K_1$] Kac, V.G.: Infinite dimensional Lie algebras. Progress in Mathematics Vol. 44. Birkhäuser (1983).

[$K_2$] Kac, V.G.: Torsion in cohomology of compact Lie groups. MSRI preprint no. 023-84-7 (1984).

[$KP_1$] Kac, V.G. and Peterson, D.H.: Infinite flag varieties and conjugacy theorems. Proc. Nat. Acad. Sci. USA 80 (1983), 1778-1782.

[$KP_2$] Kac, V.G. and Peterson, D.H.: Regular functions on certain infinite dimensional groups. Arithmetic and Geometry, ed. Artin, M. and Tate, J. (Birkhäuser, Boston) 1983, 141-166.

[$Ku_1$] Kumar, S.: Geometry of Schubert cells and cohomology of Kac-Moody Lie-algebras. MSRI preprint no. 012-84 (1984). (To appear in Journal of Diff. Geometry.)

[$Ku_2$] Kumar, S.: Homology of Kac-Moody Lie algebras with arbitrary coefficients. MSRI preprint no. 033-84 (1984). (To appear in Journal of Algebra.)

[L] Lepowsky, J.: Generalized Verma modules, loop

space cohomology and Macdonald-type identities. Ann. Scien. Éc. Norm. Sup. 12 (1979), 169-234.

[M] Moody, R.V.: A new class of Lie-algebras. J. Algebra 10 (1968), 211-230.

[Q] Quillen, D.: Rational homotopy theory. Ann. of Math. 90 (1969), 205-295.

[$S_1$] Sullivan, D.: Genetics of homotopy theory and the Adams conjecture. Ann. of Math. 100 (1974), 1-79.

[$S_2$] Sullivan, D.: Infinitesimal computations in topology. Publ. Math. I.H.E.S. 47 (1978), 269-331.

[Sa] Śafarević, I.R.: On some infinite dimensional groups II. Math. USSR Izvestija 18 (1982), 185-194.

[Se] Serre, J.P.: Algèbre Locale-Multiplicités. Lecture notes in Mathematics, Springer-Verlag (1965).

[T] Tits, J.: Resumé de cours. College de France, Paris (1981-82).

[Wa] Warner, F.W.: Foundations of Differentiable manifolds and Lie-groups. Scott, Foresman & Co. (1971).

[$Wh_1$] Whitney, H.: Geometric Integration theory. Princeton University Press (1957).

[$Wh_2$] Whitney, H.: On products in a Complex. Ann. of Math. 39 (1938), 397-432.

[Wi] Wilson, R.: Euclidean Lie-algebras are universal central extensions. Lecture notes in Math. 933 (1982), 210-213.

# THE TWO-SIDED CELLS OF THE AFFINE WEYL GROUP OF TYPE $\tilde{A}_n$

By

George Lusztig
Department of Mathematics
M.I.T., Cambridge, MA 02139

Recently, J. Y. Shi [4] determined the left cells (in the sense of [1]) of the affine Weyl group of type $\tilde{A}_n$; he has also obtained some partial results on the two-sided cells, which imply, in particular, that their number is at most equal to the number of partitions of n. In this paper we shall complete the results of Shi by proving the conjecture in [2,3.6] on the two-sided cells which implies in particular that the number of two-sided cells is exactly the number of partitions of n. The new ingredient in the proof is the function a(w) on an affine Weyl group which has been introduced in [3]. In the first three sections of this paper we shall recall the definition of a(w) and describe its connection with Gelfand-Kirillov dimension. In Section 4, we describe, following Vogan, an analogue of the notion of left cell, which we call left V-cell; in Section 5 we prove a finiteness theorem for the left V-cells of affine Weyl groups. Sections 6, 7 are concerned with affine Weyl groups of type $\tilde{A}_n$.

I wish to thank Roger Carter for explaining to me (in April 1983) the results of Shi.

---

Supported in part by the National Science Foundation.
This text is based on a lecture given at the Conference on Infinite-Dimensional Lie Groups, MSRI, Berkeley, CA, May 1984.

1. Let (W,S) be the Weyl group of a simple complex Lie algebra g with respect to a fixed Cartan subalgebra h and a Borel subalgebra $b \supset h$. For each $w \in W$, we denote by $L_{(w)}$ the irreducible g-module with highest weight $w\rho-\rho$, where $\rho$ is half the sum of the positive roots of g with respect to $b \supset h$. Let A(w) be the Gelfand-Kirillov dimension of the g-module $L_w$ and let a(w) be defined by $a(w) = \nu - A(ww_0)$ where $w_0$ is the longest element of W and $\nu = \ell(w_0)$ is the length of $w_0$. It is known that

$$(1.1) \qquad \ell(w) \leqslant A(w) \leqslant \nu, \qquad 0 \leqslant a(w) \leqslant \ell(w)$$

for all $w \in W$, and A, a, are constant on each two-sided cell of W.

Let us now replace g by a Kac-Moody Lie algebra; then, (W,S) becomes a possibly infinite Coxeter group. The g-module $L_{(w)}$ is still defined. However, there is no reasonable definition of the function A(w) in this case. Indeed, if such a function would exist and satisfy (1.1) it would sometime have infinite values since a two-sided cell of an affine Weyl group may have elements of arbitrarily large length. One can still hope that the function a(w) can be defined in general.

2. We shall now give a heuristic definition of the function a(w) in the general case. Let G be a group over the finite field $F_p$ associated with our Kac-Moody Lie algebra and let $B \subset G$ be a Borel subgroup. Let $\mathcal{F}$ be the space of all functions on G/B with complex values and with finite support. Then $\mathcal{F}$ is naturally a G-module and an $H_0$-module where $H_0$ is the Hecke algebra of B-biinvariant functions on G, with support on finitely many double cosets. Let $(C_w)_{w \in W}$ be the basis of H defined in [1], and let $\mathcal{F}_w$ be the image of the linear map of $H_0$ into itself defined as multiplication by $C_w$. When W is finite, then $\mathcal{F}_w$ is finite dimensional and $p^{a(w)}$ is the largest power of p dividing the integer dim $(\mathcal{F}_w)$, (if $p \geqslant 7$). In the general case, one can hope that dim $(\mathcal{F}_w)$ is still well defined as a p-adic integer; if this is so, one can gain define a(w) as the largest integer such that $p^{a(w)}$ divides dim $(\mathcal{F}_w)$, (for p large enough). For example, for w = e, we can take dim $(\mathcal{F}_w) = \sum_{w \in W} p^{\ell(w)}$; this is a

p-adic unit, hence $a(w) = 0$.

3. Let $(W,S)$ be any Coxeter group such that $S$ is finite. For each subset $I \subset S$ which generates a finite subgroup $W_I$, we denote by $\nu_I$ the length of the longest element of $W_I$ and we define $\nu$ to be the largest of the numbers $\nu_I$. We shall consider the (formal) Hecke algebra $H$ associated to $(W,S)$ over the ring $\mathbf{Z}[q^{1/2},q^{-1/2}]$ ($q^{1/2}$ is an indeterminate); it has two natural basis: $(T_w)_{w \in W}$ and $(C_w)_{w \in W}$, (see [1]). Let $\tilde{T}_w = q^{-\ell(w)/2}T_w$. It seems likely that, for all $x, y \in W$, we have

$$(3.1) \quad q^{\nu/2}\tilde{T}_x\tilde{T}_y \in \sum_w \mathbf{Z}[q^{1/2}]\tilde{T}_w.$$

At any rate, this is trivial for finite $W$ and true for $W$ an affine Weyl group [3].

Assume now that (3.1) holds for $(W,S)$. We have

$\tilde{T}_w \in \sum_w \mathbf{Z}[q^{1/2}]C_{w'}$ for all $w$, hence from (3.1) it follows that

$$(3.2) \quad q^{\nu/2}\tilde{T}_x\tilde{T}_y \in \sum_w \mathbf{Z}[q^{1/2}]C_{w'},$$

for all $x,y \in W$.

Let us now fix $w \in W$. Following [3] we define $a(w)$ to be the smallest integer $\geqslant 0$ such that the coefficient of $C_w$ in the expansion of $q^{a(w)/2}\tilde{T}_x\tilde{T}_y$ (in the $C_{w'}$-basis) is in $\mathbf{Z}[q^{1/2}]$ for all $x,y \in W$. From (3.2) we see that $a(w)$ is well defined and $a(w) \leqslant \nu$. Thus the a-function is defined, at least for finite or affine Weyl groups. For finite Weyl groups it coincides with the function considered in Section 1, (see [3]).

We shall need the following properties of the a function (see [3]):

(3.3) $a(w)$ is constant on the two-sided cells of $W$.

(3.4) $w' \underset{L}{\leqslant} w \Rightarrow a(w) \leqslant a(w')$.

(3.5) $w' \underset{L}{\leqslant} w,\ a(w) = a(w') \Rightarrow \mathcal{R}(w') = \mathcal{R}(w).$

(3.6) $a(w) = a(w^{-1}).$

(Here, $\underset{L}{\leqslant}$ is the preorder defined in [1], and $\mathcal{R}(w) = \{s \in S \mid ws < w\}$.)

4. The function a is very useful in the study of left cells and two-sided cells of W. We shall now consider a variant (due to Vogan) of the notion of cell, which is more easily computable.

Let $s \neq t$ be two elements of S such that $(st)^3 = 1$. If $x \in W$ satisfies $\mathcal{R}(x) \cap \{s,t\} = \{s\}$, then exactly one of the elements $xs, xt$ (say $x^*$) satisfies $\mathcal{R}(x^*) \cap \{s,t\} = \{t\}$.

We now define, following Vogan [5], a sequence of equivalence relations $\underset{i}{\sim}$ on W, $(i \geqslant 0)$, where W is an arbitrary Coxeter group which is assumed to have the following property: for any $s \neq t$ in S, the product st has order 2 or 3. By definition, $x \underset{0}{\sim} y \Leftrightarrow \mathcal{R}(x) = \mathcal{R}(y)$. Assume now that $i \geqslant 1$ and that $\underset{j}{\sim}$ is already defined for $0 \leqslant j < i$. Given $x, y \in W$, we say that $x \underset{i}{\sim} y$ if the following two conditions are satisfied:

(a) $x \underset{i-1}{\sim} y$

(b) for any $s \neq t$ in S such that $(st)^3 = 1$ and such that $\mathcal{R}(x) \cap \{s,t\} = \mathcal{R}(y) \cap \{s,t\} = \{s\}$, we have $\mathcal{R}(x^*) = \mathcal{R}(y^*)$, where $x^*, y^*$ are defined as above in terms of s,t.

Finally, we say that $x \sim y$ if $x \underset{i}{\sim} y$ for all $i \geqslant 0$. The equivalence classes on W for $\sim$ are called left V-cells.

Any left V-cell is a union of left cells, (see [1]). We now define the two-sided V-cells of W. Two elements $x, y \in W$ are said to be in the same two-sided V-cell if there exists a sequence $x = x_0, x_1, \ldots, x_n = y$ in W such that for each $i \in [1,n]$ we have $x_{i-1} \sim x_i$ or $x_{i-1}^{-1} \sim x_i^{-1}$. Any two-sided V-cell is a union of left V-cells.

5. We now state the following result.

**Theorem**. *Let* (W,S) *be an affine Weyl group of type* $\tilde{A}_n$, $\tilde{D}_n$ *or* $\tilde{E}_n$. *Then the number of left* V*-cells of* W *is finite; hence, the number of two-sided* V*-cells of* W *is finite.*

(The analogous result for left cells is only known in type $\tilde{A}_n$ [4] or $\tilde{B}_2, \tilde{G}_2$ [3].)

Let $E_{\geqslant i}$ be the $\mathbb{Q}$-subspace of the group algebra $\mathbb{Q}[W]$ with basis $C_w$ ($w \in W$, $a(w) = i$). (We regard $\mathbb{Q}[W]$ as the specialization $q^{1/2} \to 1$ of the Hecke algebra H and we write $C_w$ for the specialization of $C_w \in H$.) From (3.4) it follows that $E_{\geqslant i}$ is a left ideal of $\mathbb{Q}[W]$, for any i. We set $E_i = E_{\geqslant i}/E_{\geqslant i+1}$ and we denote by $e_w$ the image of $C_w \in E_{\geqslant i}$ under the natural map $E_{\geqslant i} \to E_i$. For each left V-cell M we denote $[M]_i$ the $\mathbb{Q}$-vector space generated in $E_i$ by the elements $e_w$, ($w \in M$, $a(w) = i$). We have the following result:

(5.1) $[M]_i$ is a left $\mathbb{Q}[W]$ submodule of $E_i$.

We shall now prove the theorem assuming (5.1). First note that the $\mathbb{Q}$-subspaces $[M]_i$ for the various left V-cells M form a direct sum decomposition of the vector space $E_i$. According to (5.1), they also form a direct sum decomposition of $E_i$ as a left $\mathbb{Q}[W]$-module, and hence also as a $\mathbb{Q}[T]$-module, where T is the group of translations in W. Clearly, $E_{\geqslant 0}$ is a free left $\mathbb{Q}[T]$-module of finite rank. Since $\mathbb{Q}[T]$ is a noetherian ring and $E_i$ is a subquotient of $E_{\geqslant 0}$, it follows that $E_i$ is a finitely generated $\mathbb{Q}[T]$-module. Hence all but finitely many summands $[M]_i$ must be zero. Hence, for each i, there are only finitely many left V-cells which meet the set $\{w \in W \mid a(w) = i\}$. Since a(w) can only take finitely many values ($0 \leqslant a(w) \leqslant \nu$), it follows that there are only finitely many left V-cells in W. It remains to prove (5.1). We must show that $\sigma e_w \in [M]_i$ for any $\sigma \in S$ and any $w \in M$ such that $a(w) = i$. If $\sigma w < w$, then $\sigma e_w = -e_w$. Thus, we may assume that $\sigma w > w$.

Then $\sigma C_w$ is a linear combination of elements $C_{w'}$, $(w' \underset{L}{\leq} w)$, and we see that it is enough to prove:

(5.2) $\quad w' \underset{L}{\leq} w,\ a(w) = a(w') \Rightarrow w \underset{i}{\sim} w'$ for all $i \geq 0$.

The conclusion of (5.2) for $i = 0$ follows from (3.5). Assume now that $i \geq 1$ and that, with the assumptions of (5.2) we know already that $w \underset{j}{\sim} w'$ for all $j$, $0 \leq j < i$. We now show that $w \underset{i}{\sim} w'$. Let $s \neq t$ be two elements of S such that $(st)^3 = 1$ and such that $\mathcal{R}(w) \cap \{s,\}$ $= \mathcal{R}(w') \cap \{s,t\} = \{s\}$ and let $w^*, w'^*$ be defined in terms of s,t as in Section 4. We must show that $\mathcal{R}(w^*) = \mathcal{R}(w'^*)$. Since $w, w^*$ are in the same right cell, we have $a(w) = a(w^*)$, see (3.3). Similarly, we have $a(w') = a(w'^*)$. Moreover, from $w' \underset{L}{\leq} w$, $\mathcal{R}(w') = \mathcal{R}(w)$, it follows that $w'^* \underset{L}{\leq} w^*$, (see the proof of Corollary 4.3 in [1]). Hence we have $w'^* \underset{L}{\leq} w^*$, $a(w^*) = a(w'^*)$. Using again (3.5), we see that $\mathcal{R}(w^*) = \mathcal{R}(w'^*)$. Thus, we have proved by induction that $w \underset{i}{\sim} w'$ for all $i \geq 0$. This completes the proof of (5.2) and hence also that of the theorem.

6. In the rest of this paper, we assume that (W,S) is an affine Weyl group of type $\tilde{A}_n$. As in [2,3.6], we shall identify W with the group of all permutations $\sigma: \mathbb{Z} \to \mathbb{Z}$ such that $\sigma(i + n) = \sigma(i) + n$ for all $i \in \mathbb{Z}$ and such that $\sum_{i=1}^{n} (\sigma(i) - i) = 0$. The simple reflections are $\sigma_0, \sigma_1, \ldots, \sigma_{n-1}$ where $\sigma_i(j) = j+1$ for $j \equiv i \pmod n$, $\sigma_i(j) = j - 1$ for $j \equiv i + 1 \pmod n$, $\sigma_i(j) = j$ for $j \not\equiv i, i + 1 \pmod n$. Following [2,3.6] we associate with each $\sigma \in W$ a sequence of integers $d_1, d_2, \ldots, d_n = n$ by the requirement that $d_k$ is the maximum cardinal of a subset of $\mathbb{Z}$ whose elements are non-congruent to each other modulo n and which is a disjoint union of k subsets each of which has its natural order reversed by $\sigma$. From a general theorem of C. Greene it follows that $d_1 \geq d_2 - d_1 \geq d_3 - d_2 \geq \ldots \geq d_n - d_{n-1}$ (a partition of n). The following result was conjectured in [2,3.6].

__Theorem.__ *Two elements* $\sigma, \sigma' \in W$ *are in the same two-sided cell if and only if they give rise to the same partition of* n.

The proof is based on the following result, which is a reformulation of the work of Shi [4].

**Theorem** (Shi). *For* (W,S) *of type* $\tilde{A}_n$, *the left* V-*cells are precisely the left cells, and the two-sided* V-*cells are precisely the fibres of the map* W→{*partitions of* n} *described above.*

We now prove our theorem. By the second assertion in Shi's theorem, we see that we have to prove that the two-sided V-cells of W are precisely the two-sided cells of W. Let $x,y \in W$ be two elements in the same two-sided V-cell; we must show that they are in the same two-sided cell. We are immediately reduced to the case where x,y are in the same left V-cell. By the first assertion in Shi's theorem, it follows that x,y are in the same left cell. Hence they are in the same two-sided cell.

Conversely, assume that $x,y \in W$ are in the same two-sided cell. Then, by (3.3), we have a(x) = a(y), and there exists a sequence $x = x_0, x_1, \ldots, x_m = y$ such that for each $i \in [1,n]$ we have $x_{i-1} \underset{L}{\leqslant} x_i$ or $x_{i-1}^{-1} \underset{L}{\leqslant} x_i^{-1}$. We then have $a(x_{i-1}) \geqslant a(x_i)$ (by (3.4) and (3.6)); since $a(x_0) = a(x_n)$, we see that $a(x_0) = a(x_1) = \ldots = a(x_n)$. From $x_{i-1} \underset{L}{\leqslant} x_i$ and $a(x_{i-1}) = a(x_i)$ it follows (see (5.2)) that $x_{i-1}, x_i$ are in the same left V-cell. Similarly, from $x_{i-1}^{-1} \leqslant x_i^{-1}$ and $a(x_{i-1}^{-1}) = a(x_i^{-1})$ it follows that $x_{i-1}^{-1}, x_i^{-1}$ are in the same left V-cell. It follows that x,y are in the same two-sided V-cell. This completes the proof of our theorem.

7. Let $S_0$ be the subset of S consisting of $\sigma_1, \sigma_2, \ldots, \sigma_n$ and let $W_0$ be the subgroup of W generated by $S_0$. If in Shi's theorem we replace (W,S) by $(W_0,S_0)$ and the map W→{partitions of n} by its restriction to $W_0$, we get again a true statement: its first

assertion is contained in [1, §5] and its second assertion follows from the work of C. Greene. The arguments in Section 6 can then be applied without change to $W_0$. We deduce that:

(7.1) Two elements in $W_0$ (= symmetric group in n letters) are in the same two-sided cell of $W_0$ if and only if they give rise to the same partition of n. We deduce the following:

__Corollary.__ *Any two-sided cell of* W *contains a unique two-sided cell of* $W_0$*; hence the two-sided cells of* W *and* $W_0$ *are in a natural* 1-1 *correspondence.*

8. With the notations in Section 1, we set $\tilde{g} = g \otimes \mathbb{C}[t,t^{-1}] \subset \hat{g} = g \otimes \overline{\mathbb{C}(t)}$, ($\overline{\mathbb{C}(t)}$ = algebraic closure of $\mathbb{C}(t)$), $g_\alpha$ = root subspace of g with respect to h, corresponding to a root $\alpha$. We say that $\alpha > 0$ if $g_\alpha \subset b$, and $\alpha < 0$ otherwise. Let $\tilde{u}^+$ be the $\mathbb{C}$-subspace of $\tilde{g}$ spanned by $g_\alpha \otimes t^i$, ($\alpha > 0$, $i \geqslant 0$), $g_\alpha \otimes t^i$, ($\alpha < 0$, $i > 0$), $h \otimes t^i$, ($i > 0$) and let $\tilde{u}^-$ be the $\mathbb{C}$-subspace of $\tilde{g}$ spanned by $g_\alpha \otimes t^i$, ($\alpha < 0$, $i \leqslant 0$), $g_\alpha \otimes t^i$, ($\alpha > 0$, $i < 0$), $h \otimes t^i$, ($i < 0$). Let $\tilde{W}$ be the group of automorphisms of $\tilde{g}$ generated by the transformations (a) $x \mapsto w(x)$, where w is in the Weyl group of g with respect to h and (b) $x \otimes t^i \mapsto x \otimes t^{i+\chi(\alpha)}$, ($x \in g_\alpha$), $x \otimes t^i \mapsto x \otimes t^i$, ($x \in h$), where $\chi$ is an element in the lattice of coroots. Then $\tilde{W}$ is an affine Weyl group with length function given by $\ell(w) = \dim \tilde{u}^+ \cap w(\tilde{u}^-)$, ($w \in \tilde{W}$). It is easy to check that $\tilde{u}^+ \cap w(\tilde{u}^-)$, ($w \in \tilde{W}$), is a finite dimensional $\mathbb{C}$-vector subspace of $\tilde{g}$ contained in the nilpotent radical of some Borel subalgebra $\hat{b} \supset h$ of $\hat{g}$; thus, it consists entirely of elements which are nilpotent in the Lie algebra $\hat{g}$, hence it has an open dense subset which is contained in a single nilpotent orbit in $\hat{g}$. This gives a canonical map $\tilde{W} \to \{$nilpotent orbits in $\hat{g}\}$. (An analogous map from finite Weyl groups to nilpotent orbits has been studied by Steinberg and Spaltenstein.) The reader can perhaps check that in case of the affine Weyl group of type $\tilde{A}_n$, this map coincides with the map into the set of partitions of n described in Section 6. For general affine Weyl groups the fibres of our map from $\tilde{W}$ to nilpotent orbits do not coincide with the two-sided cells of $\tilde{W}$.

## References.

[1] D. Kazhdan, G. Lusztig, Representations of Coxeter groups and Hecke algebras, Invent. Math. 53(1979), 165-184.

[2] G. Lusztig, Some examples of square integrable representations of semisimple p-adic groups, Trans. Amer. Math. Soc. 277(1983), 623-653.

[3] G. Lusztig, Cells in affine Weyl roups, to appear in Proceedings of the International Symposion on Algebraic Groups, Katata(Japan), 1983.

[4] J. Y. Shi, to appear.

[5] D. Vogan, A generalized $\tau$-invariant for the primitive spectrum of a semisimple Lie algebra, Math. Ann. 242(1979), 209-224.

# LOOP GROUPS, GRASSMANNIANS AND KdV EQUATIONS

by

Andrew Pressley

This is a report on joint work of Graeme Segal and George Wilson. Its aim is to describe a construction of a large class of solutions of the KdV equations (and, more generally, the KP equations) in terms of the geometry of a certain infinite dimensional Grassmannian Gr. The solutions obtained include all the "quasi-periodic" solutions arising from the algebro-geometric construction of Krichever, and in particular the rational, soliton and multi-soliton solutions.

This is closely related to the work of M. and Y. Sato [5]. They describe the solutions of the KdV equations in terms of the Plücker coordinates on their "Universal Grassmann Manifold." Our description involves the action on Gr of the group of nowhere vanishing holomorphic functions on the disc, and is essentially coordinate-free.

In the exposition, we shall concentrate on giving the main ideas. Proofs may be found in [7] and in the forthcoming book [4].

## §1. Introduction

The classical Korteweg-de Vries (KdV) equation is

$$(1.1)\quad \frac{\partial u}{\partial t} = \frac{1}{4}\frac{\partial^3 u}{\partial x^3} + \frac{3}{2} u \frac{\partial u}{\partial x} ;$$

we shall regard it as describing the time (t) evolution of a function u of one space variable (x). It therefore gives rise to a flow on a certain space $\mathcal{C}^{(2)}$ of functions of x (the reason for the notation will appear later). The

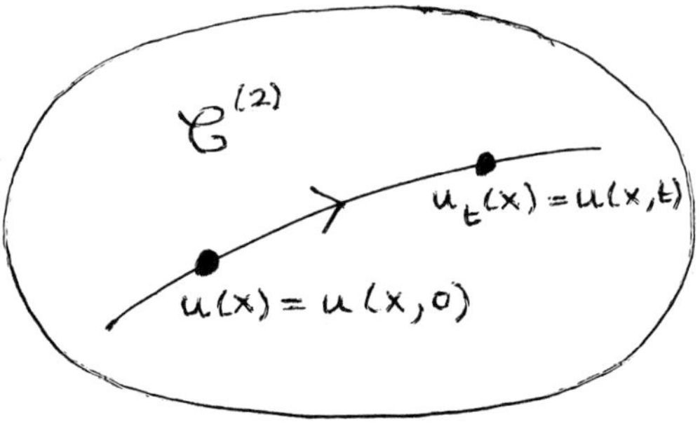


specification of precisely which functions lie in $\mathscr{C}^{(2)}$ is one of the more difficult problems of the theory, and will be ignored for the moment. We shall return to it in §5.

The starting point for the solution of (1.1) is to rewrite it in Lax form:

$$\frac{\partial L}{\partial t} = [P_3, L] \,, \tag{1.2}$$

where L and $P_3$ are differential operators acting on functions of x given by

$$L = D^2 + u, \qquad P_3 = D^3 + \frac{3}{2}\, u\, D + \frac{3}{4}\frac{\partial u}{\partial x}\,, \qquad D = \frac{\partial}{\partial x}\,.$$

We can interpret (1.2) as saying that the KdV flow consists of isospectral deformations of the Schrödinger operator $D^2 + u$.

It is now natural to consider all equations of the form

$$\frac{\partial L}{\partial t} = [P, L] \tag{1.3}$$

where P is any differential operator. If P has order m then the commutator [P,L] will in general have order m + 1, and since $\frac{\partial L}{\partial t}$ is simply multiplication by $\frac{\partial u}{\partial t}$, it is clear that P must be very special for (1.3) to have any chance of being solvable. In fact, it turns out that there is a canonical sequence of operators $\{P_r\}_{r\geq 1}$ such that the operators P for which (1.3) makes sense are precisely the constant linear combinations of the $P_r$ (we shall describe how to construct the $P_r$ in §2). This gives an infinite hierarchy of equations

(1.4) $\frac{\partial L}{\partial t_r} = [P_r, L]$ .

We use $t_r$ for the time variable in the $r^{th}$ equation to distinguish the different flows. For $r = 3$ we recover (1.2), and for all $r$ (1.4) is equivalent to an evolution equation of the form

$$\frac{\partial u}{\partial t_r} = \left\{\begin{array}{l}\text{universal polynomial in } u\\ \text{and its } x\text{-derivatives.}\end{array}\right\}$$

It is a fundamental fact of the theory that all the KdV flows commute. This means that if we move from some initial point $u \in \mathcal{C}^{(2)}$ for a time $t_r$ along the $r^{th}$ flow, and then for a time $t_s$ along the $s^{th}$ flow, the result is the same as if we had moved first along the $s^{th}$ flow for time $t_s$, and then along the $r^{th}$ flow for time $t_r$:

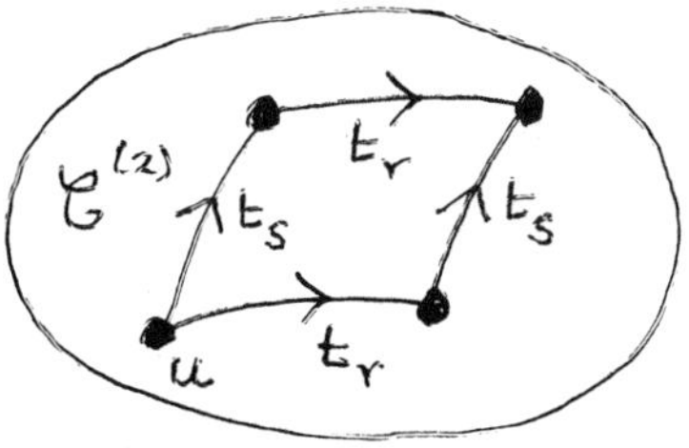


This makes it possible to write the solution of the KdV hierarchy in the form $u(x, t_1, t_2, t_3, \ldots)$, this being the result of flowing for a time $t_k$ along the $k^{th}$ flow for all $k$. In fact, some of these variables are redundant. If $r$ is even, then $P_r = L^{r/2}$ commutes with $L$, so the $r^{th}$ flow is stationary. And $P_1 = D$, so the $t_1$ flow is uniform translation in $x$, so it is usual to identify $t_1$ with $x$.

The geometrical construction of solutions of the hierarchy (1.4) is based on an infinite dimensional Grassmannian $Gr^{(2)}$, defined as follows. Let $H$ be the Hilbert space $L^2(S^1, \mathbb{C})$, where we think of the circle $S^1$ as $\{z \in \mathbb{C} \mid |z| = 1\}$. Multiplication by $z$ is a unitary operator on $H$, which we denote simply by $z$. Let $H_+$ be the subspace of $H$ consisting of boundary values of holomorphic functions in the disc $|z| < 1$. Then

$$(1.5) \quad Gr^{(2)} = \left\{ \begin{array}{c|l} \text{closed} & 1)\ z^2W \subset W, \text{ and} \\ \text{subspaces} & 2)\ W \text{ is "comparable"} \\ W \subset H & \quad \text{with } H_+ \end{array} \right\}.$$

The meaning of "comparable" will be explained in §3, but $Gr^{(2)}$ should be thought of as a completion of the space of subspaces for which ($z^2W \subset W$ and) $W \cap H_+$ has finite codimension in both $W$ and $H_+$.

The relation between this Grassmannian and the KdV equations is expressed by giving a projection map $\pi: Gr^{(2)} \to \mathcal{C}^{(2)}$. In fact, $\pi$ is simply the quotient map corresponding to the action of an infinite dimensional group $\Gamma_-$ on $Gr^{(2)}$. $\Gamma_-$ is the group of holomorphic maps $g: D_\infty \to \mathbb{C}^*$ which preserve base points, so $g(\infty) = 1$ ($D_\infty$ is the disc $|z| \geq 1$ in the Riemann sphere). $\Gamma_-$ acts on $H$ by multiplication operators, and obviously commutes with $z$, so it acts on $Gr^{(2)}$. In fact, the action is free and $\pi$ can be made into a smooth fibre bundle.

The KdV flows correspond to an action of the group $\Gamma_+$ of based holomorphic maps $g: D_0 \to \mathbb{C}^*$, where $D_0$ is the disc $|z| \leq 1$ and "based" now means $g(0) = 1$. The action of $\Gamma_+$ on $Gr^{(2)}$ obviously commutes with that of $\Gamma_-$, so induces an action on $\mathcal{C}^{(2)}$. The main result is that the action on $\mathcal{C}^{(2)}$ of the group element $e^{-t_r z^r} \in \Gamma_+$ corresponds precisely to flowing for a time $t_r$ along the $r^{th}$ KdV flow.

Notice that since $\Gamma_+$ is abelian, it is now obvious that the KdV flows commute. Moreover, since $z^2W \subset W$, it follows that

$$\exp\Big( -\sum_{k \text{ even}} t_k z^k \Big) \cdot W = W$$

for any $W \in Gr^{(2)}$, so the even flows are stationary.

The connection with loop groups arises because, as we shall explain in §3, $Gr^{(2)}$ can be identified with the group $\Omega U_2$ of based maps $f: S^1 \to U_2$, where "based" means $f(1) = 1$ (we postpone the discussion of exactly what class of maps f we allow). But the corresponding actions of the groups $\Gamma_{\pm}$ on $\Omega U_2$ are not easy to describe directly, so it is usually more convenient to work with the Grassmannian model. However, this does suggest that there ought to be KdV hierarchies corresponding to $\Omega U_n$ for all n, and more generally still, to $\Omega G$ for any compact Lie group G. In fact, the generalization to $\Omega U_n$ is obvious, and we shall always work in this generality from now on. The $n^{th}$ KdV hierarchy is a collection of infinitely many commuting flows on a space $\mathcal{C}^{(n)}$, a point of which is a sequence of n-1 functions $(u_0(x),\ldots,u_{n-2}(x))$. The hierarchies corresponding to an arbitrary group G have been described by Drinfel'd and Sokolov [1], though it is not yet completely clear how these should be accommodated within our geometric framework. This will be the subject of a future work by George Wilson, but we shall make a few remarks on it in §5.

## §2. Formal Theory of the KdV Equation: The Lax Form

We shall now describe the standard construction of the KdV hierarchies corresponding to the loop groups $\Omega U_n$. This is due to Gel'fand and Dickii [2], though some of the ideas go back at least to Schur [6].

The $n^{th}$ KdV hierarchy is a system of Lax equations

(2.1) $$\frac{\partial L}{\partial t_r} = [P_r,L] , \qquad r = 1, 2, 3, \ldots,$$

where L is the $n^{th}$ order operator

(2.2) $$L = D^n + u_{n-2}(x,t)\, D^{n-2} + \ldots + u_1(x,t)\, D + u_0(x,t) .$$

To construct the differential operators $P_r$, we introduce an algebra of <u>formal pseudo-differential operators</u> Psd. An element of Psd is an expression

$$\sum_{i=-\infty}^{N} r_i(x)\, D^{-i}$$

for some N, where the $r_i$ are formal power series in x. To multiply two elements of Psd we need to know how to move a power of D past a "function" u(x). The rule for this is determined (for both positive and negative powers of D) by requiring associativity and the relation

$$Du = uD + \frac{\partial u}{\partial x} .$$

For example,

$$D^{-1}u = uD^{-1} - \frac{\partial u}{\partial x} D^{-2} + \frac{\partial^2 u}{\partial x^2} D^{-3} - \dots .$$

__Lemma__ __2.3__. **(Schur)** *L has a unique* $n^{th}$ *root in* Psd *of the form*

$$L^{1/n} = D + \sum_{i=1}^{\infty} q_i(x)D^{-i} .$$

The proof is trivial: writing out the equation $(L^{1/n})^n = L$ and equating coefficients we obtain conditions of the form

$$u_{n-i-1} = nq_i + (\text{polynomial in } q_1,\dots,q_{i-1} \text{ and their x-derivatives})$$

(we put $u_j = 0$ for $j < 0$); these can be solved inductively for the $q_i$.

We now define $P_r = (L^{r/n})_+$, where the + means that we discard from $L^{r/n}$ all terms involving negative powers of D. We call

$$(2.4) \quad \frac{\partial L}{\partial t_r} = [(L^{r/n})_+, L]$$

__the__ __$r^{th}$__ __equation__ __of__ __the__ __$n^{th}$__ __KdV__ __hierarchy__. It is equivalent to a system of evolution equations of the form

$$\frac{\partial u_i}{\partial t_r} = \left\{\begin{matrix}\text{a universal polynomial in the} \\ u_j \text{ and their x-derivatives.}\end{matrix}\right\}$$

for i = 0, 1, ..., n-2.

<u>**Examples**</u>

1. For $r = 3$, $n = 2$ we obtain the KdV equation

$$\frac{\partial u}{\partial t} = \frac{1}{4}\frac{\partial^3 u}{\partial x^3} + \frac{3}{2} u \frac{\partial u}{\partial x}$$

(we have put $u = u_0$, $t = t_3$).

2. Obviously $P_1 = D$ for every n, so just as for the classical KdV hierarchy, the $t_1$ flow is just uniform translation in the x-variable. So we identify $t_1$ with x.

3. If n divides r, $P_r = L^{r/n}$ commutes with L so the $r^{th}$ flow is stationary.

We write the solution of (2.4) as a vector $(u_0(x,t), \ldots, u_{n-2}(x,t))$ where the $u_i(x,t)$ are formal power series in x and $t = (t_2,t_3,\ldots)$.

The basic idea for solving (2.4) is to follow the time evolution of L by comparing its eigenfunctions with those of its highest order part $D^n$. This is effected by means of the following result.

<u>**Lemma**</u> <u>**2.5**</u>. *There is an element* $K \in Psd$ *of the form*

$$K = 1 + \sum_{i=1}^{\infty} a_i(x) D^{-i}$$

*such that* $K^{-1}LK = D^n$. *Moreover,* K *is unique up to right multiplication by a constant coefficient operator* $1 + C_1D^{-1} + C_2D^{-2} + \ldots$ .

Again the proof is trivial. The equation $LK = KD^n$ reduces to a system of ordinary differential equations

$$\frac{da_i}{dx} = \begin{bmatrix} \text{polynomial in the } a_j \text{ for} \\ j<i, \text{ and their derivatives} \end{bmatrix}$$

which can be solved inductively for the $a_i$. The non-uniqueness arises from the arbitrary constant introduced at each stage of the integration.

This means that the equation

$$(2.6) \qquad L\psi = z^n\psi$$

has a formal solution of the form

$$(2.7) \qquad \psi = e^{xz}\left(1 + \sum_{i=1}^{\infty} a_i(x)z^{-i}\right)$$

unique up to multiplication by a formal series $1 + C_1z^{-1} + \ldots$. Such a $\psi$ is called a formal Baker function (or wave function) of L.

Our strategy for solving the KdV equations will be to construct a Baker function corresponding to each point of an infinite dimensional Grassmannian. We shall then be able to reverse the above construction to recover L. Of course, if we begin with a formal Baker function as above, we would only expect to obtain a formal solution of the KdV hierarchy, i.e. one given by formal power series. But the operators we shall construct are indeed genuine solutions of the KdV hierarchies. We shall discuss their regularity properties in §5.

To complete our discussion of the formal theory of the KdV hierarchies, we must explain their relationship to the Kadomtsev–Petviashvili (KP) hierarchy. This is the family of equations

$$(2.8) \qquad \frac{\partial Q}{\partial t_r} = [(Q^r)_+, Q]$$

where Q is an element of Psd of the form

$$(2.9) \qquad Q = D + \sum_{i=1}^{\infty} q_i(x,t)\, D^{-i} .$$

The KP hierarchy contains the $n^{th}$ KdV hierarchy for all n, for if Q is the $n^{th}$ root of a differential operator L, then it is easy to check that L satisfies (2.4). In fact, the assignment $L \to L^{1/n} = Q$ gives a one-to-one correspondence between solutions of the $n^{th}$ KdV hierarchy and solutions of the KP hierarchy in which $Q^n$ is a differential operator (see [9] for a simple proof).

## §3. The Infinite Dimensional Grassmannian and Its Relation to KdV

We now define the geometric object underlying the present theory.

**Definition 3.1.**

$$\mathrm{Gr} = \left\{ \begin{array}{c|c} \text{closed} & \text{(a) } pr_+ : W \to H_+ \text{ is a} \\ \text{subspaces} & \text{Fredholm operator} \\ W \subset H & \text{(b) } pr_- : W \to H_- \text{ is} \\ & \text{Hilbert-Schmidt.} \end{array} \right\}$$

The Hilbert spaces $H$ and $H_+$ were defined in §1; $H_- = H_+^{\perp}$ and $pr_\pm$ are the orthogonal projections. Conditions (a) and (b) together are what we mean by saying "$W$ is comparable with $H_+$". It is not hard to see that Gr is a smooth complex Hilbert manifold modelled on the space of Hilbert-Schmidt operators $H_+ \to H_-$ (provided with the Hilbert-Schmidt norm). The connected components of Gr are indexed by the integers, a subspace $W$ lying in the $k^{th}$ component when the Fredholm operator $pr_+ : W \to H_+$ has index $k$. Only the $k = 0$ component will play any role in the application to KdV; we denote it by $^{\circ}\mathrm{Gr}$. The set $U$ of subspaces $W$ for which $pr_+ : W \to H_+$ is an isomorphism is dense and open in $^{\circ}\mathrm{Gr}$; it is sometimes called the "big cell".

The full group GL(H) of bounded invertible operators on H does not act on Gr, as it does not preserve conditions (a) and (b) in (3.1). The subgroup of GL(H) which does act is described in the following definition.

**Definition 3.2.** The restricted general linear group $GL_{res}(H)$ is the subgroup of GL(H) consisting of operators A whose block decomposition

$$A = \begin{pmatrix} a & b \\ c & d \end{pmatrix},$$

with respect to $H = H_+ \oplus H_-$, has the off diagonal terms $b$ and $c$

Hilbert-Schmidt. The identity component $GL^{\circ}_{res}(H)$ is the part where a (and hence also d) have index zero.

A simple argument with Fourier series shows that $\Gamma_{\pm} \subset GL_{res}(H)$, so that $\Gamma_{\pm}$ act on Gr. For any $W \in {}^{\circ}Gr$ let

$$\Gamma_{+}^{W} = \{g \in \Gamma_{+} \mid g^{-1}W \in U\} .$$

This is obviously an open set in $\Gamma_{+}$, and since U is dense in ${}^{\circ}Gr$, it is plausible that $\Gamma_{+}^{W}$ is dense in $\Gamma_{+}$. In fact this is true, though not easy to prove directly. In any case, for any $g \in \Gamma_{+}^{W}$, there is a unique element $\psi_W(g) \in W$ such that

$$(3.3) \quad \psi_W(g) = g \cdot \left[1 + \sum_{i=1}^{\infty} a_i(g) z^{-i}\right] ,$$

namely, $g^{-1}\psi_W(g)$ is the inverse image of the constant $1 \in H_{+}$ under the isomorphism $pr_{+} = g^{-1}W \to H_{+}$. We call $\psi_W$ the <u>Baker function</u> of $W \in {}^{\circ}Gr$. The similarity with (2.7) can be made clearer by writing the element $g \in \Gamma_{+}$ in the form

$$g(z) = \exp\,(xz + t_2 z^2 + \ldots) .$$

Then $\psi_W$ becomes a function of x and $t = (t_2, t_3, \ldots)$.

We omitted the word "formal" from our notation for $\psi_W$ because $\psi_W$ is a genuine analytical object. For example:

(i) the series (3.3) converges in a neighborhood of $z = \infty$;

(ii) the coefficients $a_i$ are holomorphic functions on $\Gamma_{+}^{W}$, and extend meromorphically to $\Gamma_{+}$.

We shall comment in §5 on the proofs of these facts, which are not trivial.

For future reference, note that $\psi_W$ has the following equivariance property:

(3.4) $\psi_w(g_1 g_2) = g_1 \cdot \psi_{g_1^{-1} W}(g_2)$.

We shall now reverse the construction of §2 and produce from $\psi_w$ a solution of the KP hierarchy.

**Lemma 3.5.** *For each $r \geq 2$ there is a unique differential operator $P_r$ of the form*

$$P_r = D^r + P_{r2}(x,t) D^{r-2} + \ldots + P_{rr}(x,t)$$

*such that*

(3.6) $\dfrac{\partial \psi_w}{\partial t_r} = P_r \psi_w$.

As usual, the proof is by comparing coefficients. We see that there is a unique $P_r$ for which

$$\frac{\partial \psi_w}{\partial t_r} - P_r \psi_w = g \cdot (\text{expression involving } z^{-1}, z^{-2}, \ldots).$$

Since W is a linear space, the left-hand side lies in W, and the right-hand side is in $g \cdot H_-$. Since $g^{-1} W \in U$, both sides are zero.

Following the formal theory of §2, we next define

$$K = 1 + \sum_{i=1}^{\infty} a_i(x,t) D^{-i} \in \text{Psd}.$$

Then (3.6) is equivalent to

(3.7) $\dfrac{\partial K}{\partial t_r} + K D^r = P_r K$

so

$$P_r = (K D^r K^{-1})_+ = (Q^r)_+$$

where $Q = KDK^{-1} = D + \sum_{i=1}^{\infty} q_i(x) D^{-i}$, say. Then the main result is:

**Theorem 3.8.** $Q$ *satisfies the* KP *hierarchy*:

$$\frac{\partial Q}{\partial t_r} = [(Q^r)_+, Q] .$$

This is immediate from the above formulae.

The next question is: When is $Q^n$ a differential operator? The only possibility would be $Q^n = P_n$, so by (2.6) we should have

(3.9) $\quad P_r \psi_w = z^n \psi_w.$

We can easily compute that

$$P_r \psi_w - z^n \psi_w = g \cdot \sum_{i=1}^{\infty} \frac{\partial u_i}{\partial t_n} z^{-i} .$$

**If we assume that** $z^n W \subset W$, then we conclude that both sides are zero by the same argument as in (3.5). That motivates the definition

(3.10) $\quad Gr^{(n)} = \{W \in Gr \mid z^n W \subset W\} ,$

and we can now state

**Corollary 3.11.** *If* $W \in Gr^{(n)}$, *then* $L_w = P_n$ *satisfies the* $n^{th}$ KdV *hierarchy* (2.4).

Let us reformulate this slightly. Let $\mathcal{C}^{(n)}$ be the space of all "initial values" $L_w(x,0)$ for $W \in Gr^{(n)}$. Since $L_{w'} = L_w$ if and only if $W' = g \cdot W$ for some $g \in \Gamma_-$, $\mathcal{C}^{(n)}$ can be identified with the space of orbits of $\Gamma_-$ on $Gr^{(n)}$. Further, (3.4) implies

$$\psi_w(x, t_r) = e^{t_r z^r} \psi_{e^{-t_r z^r} W}(x, 0)$$

(we are confusing $t_r$ with the vector $(0,\dots,0,t_r,0,\dots)$) and hence

$$L_W(x,t_r) = L_{e^{-t_r z^r} W}(x,0) .$$

This means that the action of $e^{-t_r z^r} \in \Gamma_+$ on $Gr^{(n)}$ corresponds to translation in $\mathcal{C}^{(n)}$ along the $r^{th}$ KdV flow for a time $t_r$.

Since $\Gamma_+$ is an abelian group, it is now obvious that the flows commute. Also, since $z^n W \subset W$, we deduce that

$$\exp\left[-\sum_{k\geq 1} t_{kn} z^{kn}\right] \cdot W = W$$

so the $r^{th}$ flow is stationary whenever r is a multiple of n.

Another reason for introducing $Gr^{(n)}$ is that it can be identified with the based loop group $\Omega U_n$. The most natural action of $\Omega U_n$ is on the Hilbert space $H^{(n)} = L^2(S^1,\mathbb{C}^n)$. We may identify $H^{(n)}$ with H by making the basis element $z^k e_j$ of $H^{(n)}$ (where $e_1,\dots,e_n$ is the usual basis of $\mathbb{C}^n$) correspond to the element $z^{nk+j-1}$ of H. Then $\Omega U_n$ acts on H, and provided the loops are sufficiently smooth, we shall get an action on Gr too. Moreover, since the action on $H^{(n)}$ obviously commutes with multiplication by z, that on H commutes with $z^n$, so we get an action on $Gr^{(n)}$. In fact, if we consider maps $S^1 \to U_n$ of Sobolev class $\frac{1}{2}$, the action is transitive, and identifies $Gr^{(n)}$ with $\Omega U_n$.

The loop group interpretation makes it natural to consider various subspaces of Gr, for example

(3.12) $Gr_0 = \{W \in Gr \mid z^N H_+ \subset W \subset z^{-N} H_+ \text{ for some } N\}$

(3.13) $Gr_1 = \{W \in Gr \mid pH_+ \subset W \subset q^{-1} H_+ \text{ for some polynomials } p,q\}$.

For $Gr_0^{(n)} = Gr_0 \cap Gr^{(n)}$ corresponds exactly to the loops in $U_n$ given by trigonometric polynomials, and $Gr_1^{(n)}$ to those given by rational functions. Note that $Gr_0$ and $Gr_1$ are preserved by the action of $\Gamma_+$, but definitely not by that of $\Gamma_-$ (for example, it is not hard to show that if $W \in Gr_0$ and $g \in \Gamma_-$, then $g \cdot W$ is never in $Gr_0$ unless $g = 1$).

## §4. Examples

The simplest solutions of the KP hierarchy arise from subspaces $W \in Gr_0$. In fact, from such subspaces we always get a solution of one of the KdV hierarchies, since

$$Gr_0 = \bigcup_{n \geqslant 0} Gr_0^{(n)}$$

(this is immediate from the definition (3.12)). In other words, any subspace in $Gr_0$ comes from a loop.

By the remark at the end of §3, ${}^\circ Gr_0^{(n)}$ may be identified with a subspace of $\mathcal{E}^{(n)}$, so the corresponding flows of the $n^{th}$ KdV hierarchy may be viewed as taking place on $Gr_0^{(n)}$ itself.

The description of the flows is particularly simple in the case $n = 2$, the original KdV hierarchy. The orbits of $\Gamma_+$ on ${}^\circ Gr_0^{(2)}$ are cells, with one cell of each even dimension:

$${}^\circ Gr_0^{(2)} = \bigcup_{k \geqslant 0} C_k$$

(this is usually called the "Bruhat decomposition"). There is an explicit homeomorphism $C_k \cong \mathbb{C}^k$, for any subspace $W \in C_k$ contains a unique element of the form

$$w = z^{-k} \exp (a_1 z + a_3 z^3 + \dots + a_{2k-1} z^{2k-1})$$

where $a_1, a_3, \dots, a_{2k-1}$ are complex numbers. The "centre" $a_1 = a_3 = \dots = 0$ of the cell $C_k$ corresponds to the subspace

$$H_k = \text{span } \{z^{-k}, z^{-k+2}, \dots, z^k, z^{k+1}, z^{k+2}, \dots\} ,$$

and the corresponding loop is

$$z \mapsto \begin{bmatrix} z^k & 0 \\ 0 & z^{-k} \end{bmatrix} .$$

The initial value $u \in \mathcal{C}^{(2)}$ is

$$u(x) = -\frac{k(k+1)}{x^2} .$$

The KdV flows can be described very simply in terms of the above coordinates on $C_k$. Namely, the $r^{th}$ KdV flow is just uniform translation in the coordinate $a_r$ if r is odd and $1 \leqslant r \leqslant 2k-1$, and is stationary on $C_k$ otherwise.

It is known that the solutions obtained from $Gr_0^{(2)}$ are precisely those which are rational and vanish at $x = \pm\infty$.

More interesting solutions are available if we pass to $Gr_1$. In general, a subspace $W \in Gr_1$ will not correspond to any loop, and so will only give a solution of the KP hierarchy.

Let us consider the subspace

$$W_{p,\lambda} = \left\{ \begin{array}{l} L^2\text{-closure of the space of functions } f \\ \text{holomorphic in the disc } |z|<1, \text{ except} \\ \text{possibly for a simple pole at } z=0, \text{ and} \\ \text{satisfying } f(-p)=\lambda f(p) \end{array} \right\},$$

where $\lambda$ and $p$ are non-zero complex numbers, and $|p| < 1$. This subspace clearly lies in ${}^{\circ}Gr_1^{(2)}$ (in fact, $(z^2-p^2)H_+ \subset W \subset z^{-1}H_+$). The corresponding Baker function must be of the form

$$\psi_W = e^{\Sigma t_r z^r} \left[1 + \frac{a(x,t)}{z}\right]$$

and the condition $\psi_W(-p) = \lambda\psi_W(p)$ forces

$$a(x,t) = -p\tanh(\theta+\alpha)$$

where $\theta = \sum_{r \text{ odd}} t_r p^r$ and $e^{2\alpha} = \lambda$. The function u for which the Lax operator $L = D^2 + u$ satisfies $L\psi = z^2\psi$ is found to be

$$u = 2p^2 \operatorname{sech}^2 (\theta+\alpha).$$

If we suppress $t_r$ for $r > 3$ and write $t_1 = x, t_3 = t$, we get

$$u(x,t) = 2p^2 \operatorname{sech}^2 (p(x+p^2t)+\alpha)$$

which we recognize as a traveling "solitary wave":

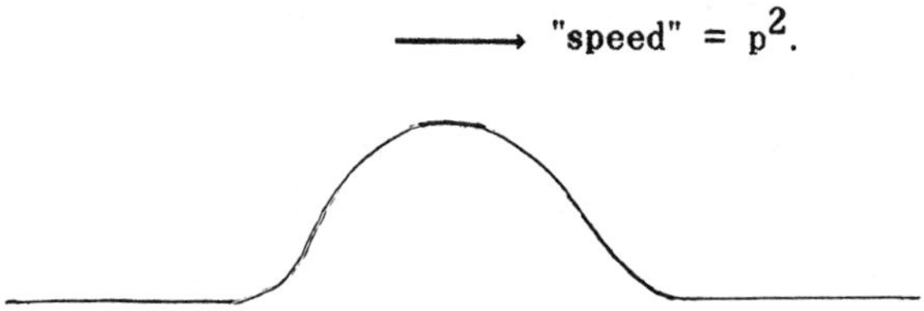


This is called the <u>1-soliton</u> solution of KdV. The n-soliton solutions can be obtained by a simple generalization of this construction.

## §5. <u>Concluding Remarks</u>

### (a) <u>The Algebro-Geometric Construction of Krichever</u>

In [3] Krichever gave a construction of certain solutions of the KdV equations, starting with a line bundle $\mathcal{L}$ on a non-singular curve X (and some other data). We shall indicate how a generalization of his construction can be accommodated within the present framework.

Our data consists of all 5-tuples $(X,\mathcal{L},x_\infty,z,\phi)$, where

$X$ = complete irreducible complex curve (possibly singular),

$\mathcal{L}$ = coherent rank 1 torsion free sheaf on X,

$x_\infty$ = non-singular point of X,

$z^{-1}$ = local parameter on X near $x_\infty$; we can assume z is an isomorphism $z:D_\infty \xrightarrow{\cong} X_\infty$, where $X_\infty$ is a neighborhood of $x_\infty$ in X,

$\phi$ = trivialization of $\mathcal{L}$ over $X_\infty$.

We write $X_0 = \overline{X \setminus X_\infty}$ (closure in the complex topology) and think of $S^1$ as embedded in X as the boundary of $X_\infty$.

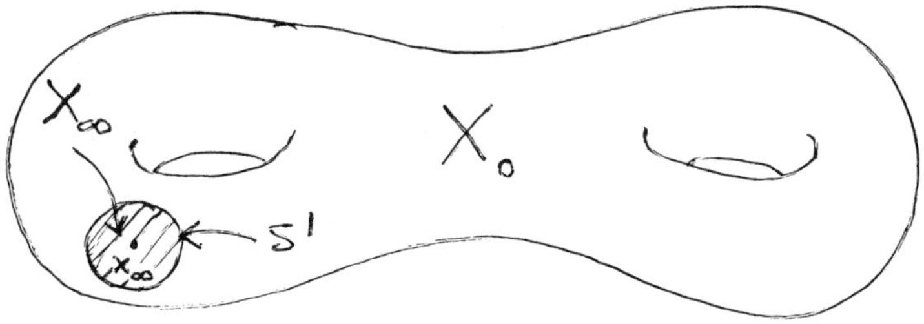


To such a 5-tuple, we associate the following subspace of H:

$$W = \left\{ \begin{array}{l} L^2\text{-closure of the set of analytic} \\ \text{functions on } S^1 \text{ which extend} \\ \text{to sections of } \mathscr{L} \text{ over } X_0 \end{array} \right\} .$$

This makes sense because the restriction to $S^1$ of any section of $\mathscr{L}$ over $X_0$ can be identified with an analytic function on $S^1$ by means of the trivialization $\phi$. It is not hard to show that $W \in Gr$; the condition that $W \in {}^\circ Gr$ is

(5.1) $\quad \dim H^\circ(X,\mathscr{L}) - \dim H^1(X,\mathscr{L}) = 1.$

When (5.1) is satisfied, the data $(X,\mathscr{L},x_\infty,z,\phi)$ therefore gives rise to a solution of the KP equations. This is essentially Krichever's construction. We obtain a solution of the $n^{th}$ KdV hierarchy precisely when $z^n$ extends to a holomorphic function on the whole of X, except for its $n^{th}$ order pole at $x_\infty$.

The action of an element $g \in \Gamma_+$ on the data is

$$g \cdot (X,\mathscr{L},x_\infty,z,\phi) = (X,\mathscr{L}\otimes\mathscr{L}_g,x_\infty,z,\phi\otimes\phi_g) ,$$

where $\mathscr{L}_g$ is the line bundle on X obtained by "clutching" trivial bundles on $X_0$ and $X_\infty$ using g, and $\phi_g$ is the natural trivialization of $\mathscr{L}_g$ over $X_\infty$. Since the map $g \mapsto \mathscr{L}_g$ is clearly a homomorphism of $\Gamma_+$ into the group of line bundles on X of degree zero, we recover the familiar statement that the KdV flows correspond to straight lines on the Jacobian of X.

(b) Generalization to Other Groups

Drinfel'd and Sokolov [1] have defined hierarchies of equations corresponding to the loop group of any compact Lie group G. The equations they define are analogues, not of the KdV equations, but of the so-called "modified KdV" (MKdV) equations. In the case $G = U_n$, these can easily be described in terms of our model. The discussion here is based on [8].

The $n^{th}$ MKdV hierarchy is a system of Lax equations

$$\frac{\partial M}{\partial t} = [P,M]$$

where M is the first order n x n matrix differential operator

$$(5.2) \qquad M = \begin{pmatrix} 0 & 0 & \cdots & 0 & D+v_n \\ D+v_1 & 0 & \cdots & 0 & 0 \\ 0 & D+v_2 & \cdots & 0 & 0 \\ \vdots & \vdots & & \vdots & \vdots \\ 0 & 0 & \cdots & D+v_{n-1} & 0 \end{pmatrix}$$

and $v_1,\ldots,v_n$ are n functions of x and t whose sum is identically zero. One checks that

$$M^n = \text{diag}\,(L_1,L_2,\ldots,L_n)$$

where

$$(5.3) \qquad L_i = (D + v_{n+i-1})(D + v_{n+i-2}) \cdots (D + v_i),$$

the indices being counted mod n. The condition $\Sigma v_i = 0$ means that each $L_i$ has no term in $D^{n-1}$, as in (2.2).

Let $Q = \text{diag}\,(L_1^{1/n}, L_2^{1/n},\ldots,L_n^{1/n})$, so that the diagonal entries of Q are elements of Psd of the form (2.9). Then the equation

$$\frac{\partial M}{\partial t_r} = [(Q^r)_+, M] \tag{5.4}$$

is the $r^{th}$ equation of the $n^{th}$ MKdV hierarchy.

From (5.3) it follows that

$$\frac{\partial}{\partial t}(M^n) = [(Q^r)_+, M^n]$$

and hence that <u>each $L_i$ satisfies the KdV equation</u> (2.4). The n transformations (5.2) from solutions of MKdV to solutions of KdV are called <u>Miura transformations</u>.

To interpret the MKdV equations, we introduce an infinite dimensional flag manifold $Fl^{(n)}$, an element of which is a sequence of subspaces $W_1, W_2, \ldots, W_n \in {}^\circ Gr^{(n)}$ such that

$$z^n W_1 \subset z^{n-1} W_n \subset \ldots \subset zW_2 \subset W_1.$$

If $\psi_i$ is the Baker function of $W_i$, then the argument preceding (3.11) shows that there are unique functions $v_1(x,t), \ldots, v_n(x,t)$ such that

$$(D + v_i)\psi_i = z\psi_{i+1}, \quad i=1, \ldots, n,$$

where $\psi_{n+1}$ means $\psi_1$. In other words, if we construct M from the $v_i$ as in (5.2), then

$$M\psi = z\psi$$

where

$$\psi = \begin{bmatrix} \psi_1 \\ \cdot \\ \cdot \\ \cdot \\ \psi_n \end{bmatrix}.$$

But from (3.6) we know

$$\frac{\partial \psi}{\partial t_r} = Q_+^r \psi$$

and this implies that M satisfies (5.3).

The Miura transformations correspond simply to the n natural projections $\pi_i : \mathrm{Fl}^{(n)} \to \mathrm{Gr}^{(n)}$ taking $(W_1, \ldots, W_n)$ to $W_i$. So we have n commutative diagrams

$$\begin{array}{ccc}
\mathrm{Fl}^{(n)} & \xrightarrow{\quad \pi_i \quad} & {}^{\circ}\mathrm{Gr}^{(n)} \\
\Big\downarrow & & \Big\downarrow \\
\left\{\begin{array}{l}\text{solutions of } n^{\text{th}} \\ \text{MKdV hierarchy}\end{array}\right\} & \xrightarrow[\text{transformation}]{i^{\text{th}} \text{ Miura}} & \left\{\begin{array}{l}\text{solutions of } n^{\text{th}} \\ \text{KdV hierarchy}\end{array}\right\}
\end{array}$$

The group $\Gamma_+$ acts on these diagrams, commuting with the maps involved, and inducing the MKdV and KdV flows.

Since both $\mathrm{Fl}^{(n)}$ and $\mathrm{Gr}^{(n)}$ have analogues corresponding to an arbitrary compact Lie group instead of $U_n$ (see [4]), we should expect that the Drinfel'd-Sokolov construction has a geometric interpretation. We hope this will be dealt with by George Wilson in a future work.

### (c) The Class of Solutions Obtained

Apart from exhibiting some explicit solutions of the KdV equations in §4, we have not so far made any comment on what sort of solutions are to be found in $\mathcal{C}^{(n)}$. This is a difficult question, and we can only give a partially satisfactory answer, as follows:

(i) *The operators* $L_W \in \mathcal{C}^{(n)}$ *extend meromorphically to the whole complex x-plane, and have only regular singular points.*

But not every operator of this type arises from a subspace in $Gr^{(n)}$. A precise characterization can, however, be given in terms of the Baker function:

(ii) *The n formal power series*

$$\psi(0,z),\ D\psi(0,z),\ \ldots,\ D^{n-1}\psi(0,z)$$

*converge in a neighborhood of* $z = \infty$.

But it is not easy to recognize which operators L arise from Baker functions with this property.

It would be interesting to obtain more direct information on the analytical character of the solutions obtained from the Grassmannian construction.

(d) Representation Theoretic Meaning of the Baker Function

In the formal development of the theory it is not convenient to work directly in terms of the Baker function. Rather, one introduces an intermediate object, the $\tau$-function. This has a simple meaning in terms of the representation theory of the loop group: it is just a matrix element in the basic representation of $\Omega U_n$. From its construction it is clearly holomorphic, and one of the main results of the theory is to give a formula for the Baker function as the quotient of two values of the $\tau$-function. This is how one proves the Baker function is meromorphic, and the facts stated in (c) above also arise from a detailed analysis of the $\tau$-function. Presumably, the Baker function itself has a simple meaning in terms of representation theory, but this is somewhat obscure at the moment.

## References

1. Drinfel'd, V. G., and Sokolov, V. V., Equations of Korteweg-de Vries type and simple Lie algebras, Soviet Math. Dokl. 23

(1981), 457-462.

2. Gel'fand, I. M., and Dickii, L. A., Fractional powers of operators and Hamiltonian systems, Funct. Anal. Appl. 10 (1976), 259-273.

3. Krichever, I. M., Integration of non-linear equations by means of algebraic geometry, Funct. Anal. Appl. 11 (1977), 12-26.

4. Pressley, A. N., and Segal, G. B., Loop groups and their representations, to be published by Oxford University Press (1985).

5. Sato, M., Soliton equations as dynamical systems on an infinite dimensional Grassmann manifold, Proc. Symp., Kyoto 1981, RIMS Kokyuroku 439 (1981), 30-46.

6. Schur, I., Über vertauschbare lineare Differentialausdrücke, Berliner Math. Ges. Sitzber. 3 (Archiv der Math. Beilage (3) 8), (1904), 2-8.

7. Segal, G. B., and Wilson, G., Loop groups and equations of KdV type, Preprint, Oxford University, (1983).

8. Wilson, G., Preliminary (sketch) version of a sequel to [6].

9. Wilson, G., Commuting flows and conservation laws for Lax equations, Math. Proc. Camb. Phil. Soc. 86 (1979), 131-143.

# AN ADJOINT QUOTIENT FOR CERTAIN GROUPS ATTACHED TO KAC-MOODY ALGEBRAS

By

Peter Slodowy
Mathematical Sciences Research Institute
Berkeley, California
and
Mathematisches Institut, Universität Bonn
Bonn, West Germany

## 0. Introduction

In this article we want to give a survey of that part of our Habilitationsschrift [16] which deals with conjugacy classes in certain groups G attached to Kac-Moody Lie algebras. These investigations were motivated on one side by the result of Brieskorn relating simple singularities and simple algebraic groups (see for instance [14]) and on the other side by recent results of Looijenga on the deformation theory of simply elliptic and cusp singularities ([9],[10]). The results in [16] show that at least to some extent there is a similar relationship between these singularities and associated Kac-Moody Lie groups as there is between simple singularities and simple algebraic groups. Here, we shall limit ourselves to the group-theoretical aspects, i.e. we give a definition (due to E. Looijenga) of an adjoint quotient for an arbitrary Kac-Moody Lie group G and we analyze the structure of its fibers. A large part of the notes will be dedicated to an explanation of Looijenga's "partial compactification" $\hat{T}/W$ of the quotient of a maximal torus T of G by the Weyl group W since this space will figure as the base of the adjoint quotient of G. Its stratification into boundary components induces a partition of G which can be described in terms of the building associated to G. We conjecture a representation-theoretic interpretation of this partition which seems to be relevant when dealing with a character-theoretic construction of the adjoint quotient. Some open problems in that direction are mentioned at the end. Detailed proofs may be found in [16]. There also the relations to singularities are explained.

## 1. Review of the Finite-Dimensional Case

Let G be a simply connected semisimple algebraic group over $\mathbb{C}$, B a Borel subgroup and $T \subset B$ a maximal torus of G of dimension $\ell$ = rank (G). We denote the normalizer of T in G by N. Then $N/T = W$ is the finite Weyl group of (G,T).

The <u>adjoint quotient</u> of G is the quotient of G by its adjoint action

$$\begin{aligned} G \times G &\longrightarrow G \\ (g,x) &\longmapsto Ad(g)x = g\, x\, g^{-1} \end{aligned}$$

in the category of algebraic varieties, i.e. it is the morphism

$$\chi\colon G \longrightarrow G/Ad(G)$$

dual to the inclusion of $\mathbb{C}$-algebras

$$\mathbb{C}[G] \hookleftarrow \mathbb{C}[G]^{Ad(G)}$$

where $\mathbb{C}[G]$ denotes the algebra of regular functions on G and $\mathbb{C}[G]^{Ad(G)}$ the finitely generated subalgebra of Ad(G)-invariants (cf. [17], [18] where one also finds details on the other topics of this section).

There are essentially three ways to explicitly define $\chi$.

1) Let $\pi_i\colon G \longrightarrow GL(V_i)$, $i=1,\dots,\ell$, denote the fundamental irreducible representations of G and $\chi_i\colon G \longrightarrow \mathbb{C}$, $\chi_i(g) = \text{trace } \pi_i(g)$, the corresponding characters. Then $\mathbb{C}[G]^{Ad(G)}$ is generated freely as a polynomial algebra by the $\chi_i$

$$\mathbb{C}[G]^{Ad(G)} = \mathbb{C}[\chi_1,\dots,\chi_\ell],$$

and $\chi$ may be realized by the map

$$G \longrightarrow \mathbb{C}^{\ell}$$
$$g \longmapsto (\chi_1(g),...,\chi_{\ell}(g)) \ .$$

By the exponential invariant theory ([8]), the restriction of $\chi$ to T induces an isomorphism

$$T/W \simeq \mathbb{C}^{\ell} = G/Ad(G)$$

resp.

$$\mathbb{C}[T]^W \simeq \mathbb{C}[\chi_1,...,\chi_{\ell}] = \mathbb{C}[G]^{Ad(G)} \ .$$

2) Identifying G/Ad(G) with T/W we may alternatively define $\chi$: G $\longrightarrow$ T/W as follows. Any element g $\in$ G splits in a unique way as the product

$$g = s.u = u.s$$

of a semisimple element s $\in$ G and a unipotent element u $\in$ G and any semisimple element s $\in$ G is conjugate in G to an element t $\in$ T. If we denote the W-orbit of t in T by $\bar{t}$, we may define

$$\chi(g) = \bar{t} \in T/W.$$

To see that this map is well-defined it is sufficient to prove that two elements in T are conjugate in G exactly when they are conjugate under W. This is an easy consequence of the Bruhat decomposition for G.

3) The last definition of $\chi$ uses the least amount of technical preparation. Let g $\in$ G. Then g is conjugate into the Borel subgroup B which splits as the semidirect product T $\ltimes$ U of T with the unipotent radical U of B. Let b = t.u be an element conjugate to g and put

$$\chi(g) = \bar{t} \in T/W .$$

The fact that $\chi$ is well-defined that way and, in particular, is invariant under conjugation follows again from the Bruhat decomposition.

The analysis of the fibers of $\chi$ is due to Steinberg ([17], [18]) who followed the model given by Kostant in the situation of the Lie algebra ([7]). The essential tool here is the Jordan decomposition in G mentioned in 2) above. Each fiber of $\chi$ has the form $\chi^{-1}(\chi(t))$ for some $t \in T$. It contains a unique closed Ad(G)-orbit which at the same time is the unique semisimple conjugacy class of $\chi^{-1}(\chi(t))$ and thus equal to the orbit $G/Z_G(t)$ of t. Let Uni(t) denote the set of unipotent elements in the reductive subgroup $Z_G(t)$. Then the map

$$g * u \longmapsto gtug^{-1}$$

induces a G-equivariant isomorphism

$$G \times^{Z_G(t)} \mathrm{Uni}(t) \xrightarrow{\sim} \chi^{-1}(\chi(t))$$

from the bundle associated to the principal fibration $G \longrightarrow G/Z_G(t)$ and the adjoint action of $Z_G(t)$ onto the fiber $\chi^{-1}(\bar{t})$. In particular, the classification of the orbits and the description of the geometric structure (e.g. singularities) of a given fiber $\chi^{-1}(\bar{t})$ are completely reduced to the corresponding problems for the unipotent variety Uni(t) of a "smaller" semisimple group, i.e. the commutator subgroup of $Z_G(t)$.

**Remark:** The descriptions 2) and 3) for the adjoint quotient extend without change to the case of a general reductive group. In 1) one has to use a slightly modified notion of fundamental representations and one has to take into account the algebraic dependence of the corresponding characters in case the semisimple part of G is not any more simply connected.

## 2. Groups Attached to Kac-Moody Lie Algebras

Let $A \in M_\ell(\mathbb{Z})$ be a generalized Cartan matrix and $\mathfrak{g}$ the corresponding complex Kac-Moody Lie algebra generated by $3\ell$ generators $e_1, h_1, f_1, \ldots, e_\ell, h_\ell, f_\ell$. We have the usual decomposition

$$\mathfrak{g} = \mathfrak{u}^- \oplus \mathfrak{h} \oplus \mathfrak{u}^+,$$
$$\mathfrak{u}^\pm = \bigoplus_{\alpha \in \Sigma^+} \mathfrak{g}_{\pm\alpha},$$

where $\Sigma^+$ denotes the set of positive roots. The groups we want to consider will depend on slightly finer data than A or $\mathfrak{g}$:

A $\mathbb{Z}$-root basis with Cartan matrix A is a triple $(H, \nabla, \Delta)$ consisting of

- a free $\mathbb{Z}$-module H of finite rank r,
- a free indexed subset $\nabla = \{h_1, \ldots, h_\ell\}$ of H (the set of simple coroots),
- a free indexed subset $\Delta = \{\alpha_1, \ldots, \alpha_\ell\}$ in the dual $H^* = \mathrm{Hom}_{\mathbb{Z}}(H, \mathbb{Z})$ of H

such that

$$\alpha_i(h_j) = A_{ji} \text{ for } i,j \in \{1, \ldots, \ell\}.$$

If $\alpha = \alpha_i \in \Delta$ we shall also write $h_\alpha$ for $h_i$, and we put $\alpha^\vee = h_\alpha$, $h_\alpha^\vee = \alpha$.

Using one of the construction procedures as described in [2], [11], [12], [13], [19], [20] one can attach to $(H, \nabla, \Delta)$ a group G with subgroups B and N satisfying the axioms of a Tits-system (cf. [8]):

- G is generated by B and N
- the intersection $T = B \cap N$ is normal in N

- the quotient $W = N/T$ is generated by a set of involutions $S \subset W$ such that

$$sBw \subset BwB \cup BswB$$

and

$$sBs \neq B$$

for all $s \in S$, $w \in W$.

Moreover, G is linked to $(H,\nabla,\Delta)$ and the corresponding Kac-Moody Lie algebra $\mathfrak{g}$ in the following way:

- the group T is isomorphic to $H \otimes_{\mathbb{Z}} \mathbb{C}^*$

- the pair (W,S) is isomorphic to the pair (W',S') where S' is the set $\{s_\alpha \mid \alpha \in \Delta\}$ of reflections

$$s_\alpha: H \longrightarrow H, \; s_\alpha(h) = h - \alpha(h)h_\alpha,$$

  and $W' \subset \mathrm{Aut}(H)$ is the group generated by S',

- under the isomorphisms given above the action of $W = N/T$ on T is the one induced by the action of W' on H.

- the group B is a semidirect product

$$B = T \ltimes U$$

  where U is the prounipotent proalgebraic group whose Lie algebra is the completion of $\mathfrak{n}^+$ with respect to the natural root-height filtration. The action of T on U is induced by the natural action of T on $\mathfrak{u}^+$ (via the positive roots $\Sigma^+$).

For more details on this particular type of group one may consult [16] Ch. 5.

In the following we will always assume that $\nabla$ is cotorsionfree in H, i.e. that $H/\mathbb{Z}\nabla$ has no torsion. This corresponds to the situation that the commutator subgroup of G (in the topological sense) is simply connected. Properties of groups attached to other root bases are easily derived from this special case.

In the infinite-dimensional situation we face some obstacles when we try to extend the definition of the adjoint quotient from the finite-dimensional case:

i) First one has to clarify what kind of functions one wants to consider as regular functions on the group G.

ii) The characters of the infinite-dimensional fundamental representations of G are not defined on all elements of G. In particular, they do not belong to any class of regular functions on G introduced up to now.

iii) Concerning the "set-theoretical" definition of the adjoint quotient (cf. Section 1, 2) and 3)), it is no longer true that all elements of G are conjugate into the Borel subgroup B, nor does there exist a Jordan decomposition for all elements of G.

iv) Since the Weyl group W acts on T with infinite isotropy groups, the quotient T/W does not inherit a natural structure of an analytic space. (This is also related to point ii).)

To obtain a reasonable adjoint quotient of G we are forced on one side to restrict ourselves to subsets of G and T (because of ii) and iv)). On the other side we have to enlarge T by additional points (because of iii)) and -- as we believe -- we also have to enlarge G to get a satisfactory picture in connection with point i) (cf. [4]) and the deformation theory of certain surface singularities (cf. [16]).

Both, the restrictions and the enlargements will be described in terms of the so-called Tits-cone attached to the root basis $(H,\nabla,\Delta)$ and its Weyl group W.

## 3. The Tits-Cone and Its Boundary Components

The action of W on H induces natural W-actions on the real vector spaces $V^* = H \otimes_{\mathbb{Z}} \mathbb{R}$ and $V = H^* \otimes_{\mathbb{Z}} \mathbb{R}$. We let

$$C = \{\omega \in V \mid \omega(h) > 0 \text{ for all } h \in \nabla\}$$

respectively

$$\bar{C} = \{\omega \in V \mid \omega(h) \geq 0 \text{ for all } h \in \nabla\}$$

denote the open (resp. closed) fundamental chamber, and let

$$A = W.\bar{C}$$

be the union of W-translates of $\bar{C}$. For any subset $\Theta \subset \nabla$ we define the open (resp. closed) face of $\bar{C}$ of type $\Theta$ by

$$F_\Theta = \{\omega \in V \mid \omega(h) = 0 \text{ for all } h \in \Theta,\ \omega(h) > 0 \text{ for all } h \in \nabla \setminus \Theta\}$$

resp.

$$\bar{F}_\Theta = \{\omega \in V \mid \omega(h) = 0 \text{ for all } h \in \Theta,\ \omega(h) \geq 0 \text{ for all } h \in \nabla \setminus \Theta\}.$$

Then the following properties hold (cf. [8], [9], [21]):

i) A is a convex solid cone in V.

ii) $\bar{C}$ is a fundamental domain for the action of W on A, and if $w(F_\Theta) \cap \bar{C} \neq 0$ for some $w \in W$, $\Theta \subset \nabla$, then w fixes every point of $F_\Theta$.

This property ii) allows one to extend the "faceting" of $\bar{C}$ in a well-defined way to the whole of A such that A may be viewed as a geometrical realization of the Coxeter complex of W.

iii) The interior $\overset{\circ}{A}$ of A is the union of all faces with finite stabilizer in W. In particular, W acts properly discontinuously on $\overset{\circ}{A}$.

iv) We have $A = \overset{\circ}{A}$ if and only if $A = V$ if and only if W is finite if and only if the Cartan matrix is of finite type.

Because of property i) the set A is called the Tits cone of W in V.

A subset A' of A is called a boundary component of A if there exists an element $\varphi \in V^*$ such that $\varphi(\omega) \geq 0$ for all $\omega \in A$ and $A' = \{\omega \in A \mid \varphi(\omega) = 0\}$.

A subset $\Theta$ of $\nabla$ is called special if all connected components of $\Theta$ are of infinite type. For any subset $\Theta$ of $\nabla$ we define $\Theta^{\perp} = \{h \in \nabla \mid h^{\vee}(h')=0 \text{ for all } h' \in \Theta\}$. For any subset $\Theta \subset \nabla$ we let $W_\Theta$ be the subgroup of W generated by the reflections $s_\alpha$, $h_\alpha \in \Theta$.

The following proposition is essentially due to Looijenga (cf. [9], [16]):

**Proposition:**

1) Let $\Theta \subset \nabla$ be a special subset. Then $A(\Theta) = \{\omega \in A \mid \omega(h)=0 \text{ for all } h \in \Theta\}$ is a boundary component.

2) For any boundary component A' of A there is a special subset $\Theta \subset \nabla$ and an element $w \in W$ such that $A' = w.A(\Theta)$.

3) Every element in $A(\Theta)$ is $W_{\Theta^{\perp}}$-conjugate to an element of $\bar{F}_\Theta = \bar{C} \cap A(\Theta)$.

4) The stabilizer $N_W(A(\Theta)) = \{w \in W \mid wA(\Theta)=A(\Theta)\}$ of the

boundary component $A(\Theta)$ equals $W_\Theta \times W_{\Theta^\perp}$. The subgroup $W_\Theta$ is the centralizer $Z_W(A(\Theta)) = \{w \in N_W(A(\Theta)) \mid w|_{A(\Theta)} = id\}$ of $A(\Theta)$.

<u>**Examples:**</u>

1) Let the Cartan matrix be of type $A_2$, i.e. $\begin{bmatrix} 2 & -1 \\ -1 & 2 \end{bmatrix}$. Then we may choose V two-dimensional

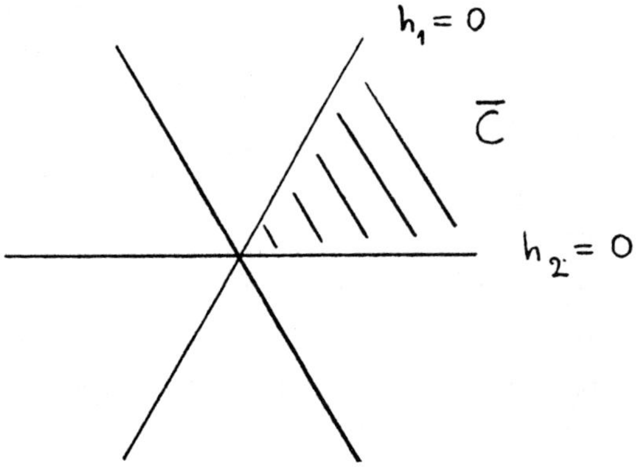


Fig. 1

Here $W = \sigma_3$ and $A = W.\bar{C} = V$. There is only one non-proper boundary component $A(\emptyset) = A = V$.

2) Next let us look at the Cartan matrix of type $A_1^{(1)}$:

$$\begin{bmatrix} 2 & -2 \\ -2 & 2 \end{bmatrix}.$$

Here we have to choose V of dimension 3.

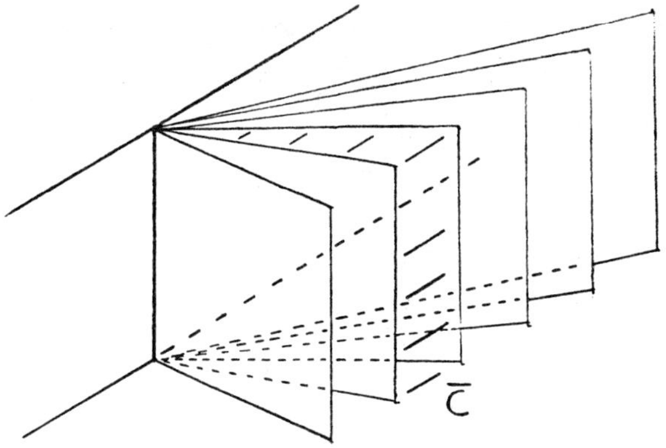


Fig. 2

The Weyl group W is the infinite dihedral group and the Tits cone A = W.$\bar{C}$ consists of an open half space (the open book) and a line (the spine of the book). There are two boundary components, A = A(∅) and A(∇) = the spine line.

3) The matrix

$$\begin{bmatrix} 2 & -3 \\ -3 & 2 \end{bmatrix}$$

is hyperbolic and may be realized by a two-dimensional root basis.

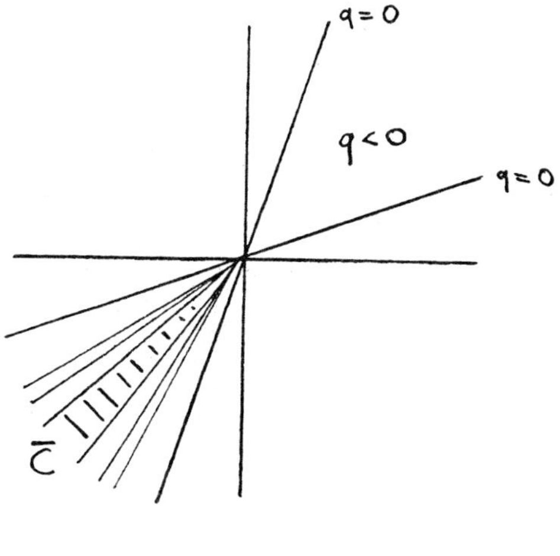


Fig. 3

There is a hyperbolic form q on V (given by the Cartan matrix with respect to a basis of simple roots) which is invariant under the action of the infinite dihedral Weyl group W. The Tits cone consists of one component of the open negative cone q < 0 and the point 0. There are two boundary components, $A = A(\emptyset)$ and $A(\nabla) = \{0\}$.

4) The Cartan matrix

$$\begin{bmatrix} 2 & -2 & 0 \\ -2 & 2 & -1 \\ 0 & -1 & 2 \end{bmatrix}$$

is hyperbolic and defines a hyperbolic form $q$ on the three-dimensional space $V$ spanned by the simple roots

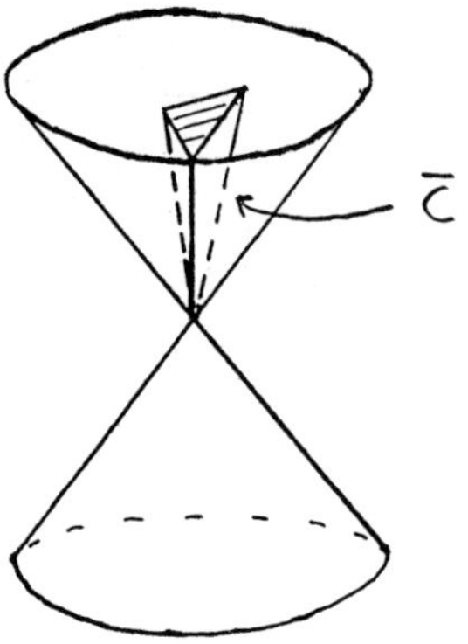

Fig. 4

The Weyl group $W$ may now easily be identified with $PGL_2(\mathbb{Z})$ (cf. [1]). The Tits cone $A$ consists of one component of the open negative cone $q < 0$, the isotropic rational half-lines in the closure of this component, and the point 0. The "Dynkin-diagram" looks as follows 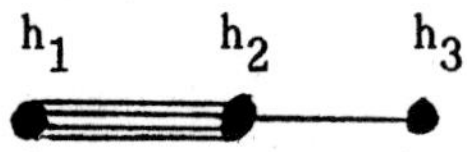

. There are three special subsets: $\emptyset$, $\Theta = \{h_1, h_2\}$, $\nabla = \{h_1, h_2, h_3\}$. Correspondingly we have three types of boundary components:

$A(\emptyset) = A$

$A(\nabla) = \{0\}$

all $W$-conjugates of $A(\Theta)$ = all rational isotropic half lines

**Remark**: Instead of working in V we can equally look at $V^*$ and define the dual fundamental chamber $\bar{C}^*$ and the dual Tits cone $A^*$. Then of course similar properties for the W-action on $A^*$ hold as for the W-action on A.

## 4. Looijenga's Partial Compactification

Let us fix the exponential map

$$\mathbb{C} \longrightarrow \mathbb{C}^*, \; z \longmapsto \exp(2\pi i z).$$

Tensoring with H gives then rise to an exponential map

$$\mathfrak{h} = H \otimes_{\mathbb{Z}} \mathbb{C} \longrightarrow H \otimes_{\mathbb{Z}} \mathbb{C}^* = T.$$

We denote the image of the tube domain $V + i\overset{\circ}{A}{}^* \subset \mathfrak{h}$ in T under this map by $\mathcal{T}$ and call it Looijenga's domain. Then $\mathcal{T}$ may be alternatively described as

$$\mathcal{T} = \{t \in T \mid |\gamma(t)| > 1 \text{ for all positive imaginary roots}\}.$$

By the property iii) of the W-action on $\overset{\circ}{A}{}^*$ we get that W acts properly discontinuously on $\mathcal{T}$. Hence the quotient $\mathcal{T}/W$ exists as an analytic space. In fact, since $\nabla$ is assumed to have no cotorsion in H the isotropy groups of W on $\mathcal{T}$ are generated by reflections and thus $\mathcal{T}/W$ is even a complex manifold (cf. [9]).

We now want to describe Looijenga's partial compactification of the orbit space $\mathcal{T}/W$. A convenient way to formulate this is the language of toroidal embeddings (not necessarily of locally finite type) (cf. [6]).

Let $\mathbb{C}[H^*]$ the complex group ring of the lattice $H^*$. Then T identifies with the maximal spectrum of $\mathbb{C}[H^*]$. Corresponding to the inclusion

$$A \cap H^* \hookrightarrow H^*$$

of semigroups we have an inclusion of semigroup rings

$$\mathbb{C}[A\cap H^*] \hookrightarrow \mathbb{C}[H^*]$$

which gives rise to an embedding

$$\begin{array}{ccc} \mathrm{Specm}\ \mathbb{C}[H^*] & \longrightarrow & \mathrm{Specm}\ \mathbb{C}[A\cap H^*] \\ \| & & \| \\ T & & \hat{T} \end{array}$$

Here $\hat{T}$ may be described as a union of homogeneous T-spaces (under "left translation")

$$\hat{T} = \bigcup_{\substack{A' \text{ boundary} \\ \text{component of } A}} T/\mathrm{Ann}(A') ,$$

where $\mathrm{Ann}(A') = \{t\in T \mid \omega(t)=1 \text{ for all } \omega\in H^*\cap A'\}$. If A' has the form $A' = A(\Theta)$ for some special subset $\Theta \subset \nabla$ we have

$$\mathrm{Ann}(A(\Theta)) = (\mathbb{Z}.\Theta) \otimes_{\mathbb{Z}} \mathbb{C}^* \subset H \otimes_{\mathbb{Z}} \mathbb{C}^* .$$

In this case we shall denote $T/\mathrm{Ann}(A(\Theta)) \cong (H/\mathbb{Z}.\Theta) \otimes_{\mathbb{Z}} \mathbb{C}^*$ by $T(\Theta)$.

Since W stabilizes A we get a W-action on $\hat{T}$. Using the proposition of section 3 we get the following set-theoretic picture for the quotient of $\hat{T}$ by W:

$$\hat{T}/W = \bigcup_{\substack{\Theta\subset\nabla \\ \text{special}}} T(\Theta)/W_{\Theta^\perp} .$$

Let $\mathcal{T}(A') \subset T/\mathrm{Ann}(A')$ denote the image of $\mathcal{T}$ in the quotient torus $T/\mathrm{Ann}(A')$. If $A' = A(\Theta)$ we simply put $\mathcal{T}(A') = \mathcal{T}(\Theta)$. Then one can equip

$$\hat{\mathcal{T}} = \bigcup_{\substack{A' \text{ boundary} \\ \text{component of } A}} \mathcal{T}(A')$$

with a W-invariant topology (of Satake type, cf. [9]) such that the quotient

$$\hat{\mathcal{T}}/W = \bigcup_{\substack{\Theta \subset \nabla \\ \text{special}}} \mathcal{T}(\Theta)/W_{\Theta^\perp}$$

is locally compact and Hausdorff. If one calls a continuous function on $\hat{\mathcal{T}}/W$ analytic, when it is analytic on every stratum $\mathcal{T}(\Theta)/W_{\Theta^\perp}$, then $\hat{\mathcal{T}}/W$ will be a Stein manifold with this analytic structure.

**Examples:** We will look at the examples of section 3 in the same numbering.

1) $A_2$: There is no proper boundary component, $\mathcal{T}$ equals T and $\mathcal{T}/W = \hat{\mathcal{T}}/W = T/W \cong \mathbb{C}^2$.

2) $A_1^{(1)}$: Here $\mathcal{T}$ may be identified with $(\mathbb{C}^*)^2 \times \mathbb{C}_{>1}$, where $\mathbb{C}_{>1}$ denoes the complex numbers of absolute value bigger than 1. The completion $\hat{\mathcal{T}}$ consists of $\mathcal{T}$ and $\mathcal{T}(\nabla) = \mathbb{C}_{>1}$. The action of the Weyl group W commutes with the projection of $\hat{\mathcal{T}}$ onto $\mathbb{C}_{>1}$, and the quotient $\hat{\mathcal{T}}/W$ is isomorphic to $\mathbb{C}^2 \times \mathbb{C}_{>1}$, $\mathcal{T}/W$ being identified with $(\mathbb{C}^2 \setminus 0) \times \mathbb{C}_{>1}$ and $\mathcal{T}(\nabla) = \mathcal{T}(\nabla)/W_{\nabla^\perp}$ with $\{0\} \times \mathbb{C}_{>1}$

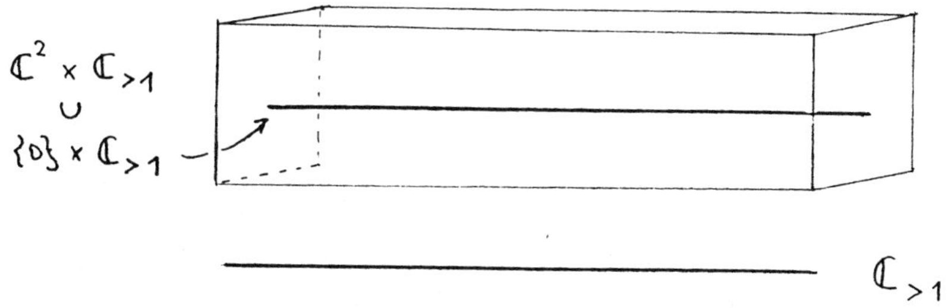


Fig. 5

3) In the rank-2-hyperbolic case $\mathcal{T}(\nabla)$ is a point and in a neighborhood of $\mathcal{T}(\nabla) \subset \hat{\mathcal{T}}/W$ we have a picture analogous to that of 0 in $\mathbb{C}^2$:

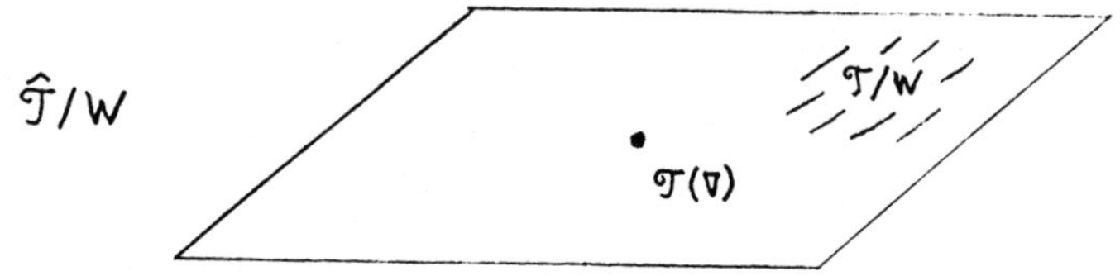


Fig. 6

4) In the last, rank-3-hyperbolic case $\hat{\mathcal{T}}/W$ is a three-dimensional manifold. In a neighborhood of the point $\mathcal{T}(\nabla) \in \hat{\mathcal{T}}/W$ the stratum $\mathcal{T}(\Theta)$ looks like a pointed disk $\mathbb{C}_{>1} \cong \mathbb{C}_{<1} \setminus \{0\}$:

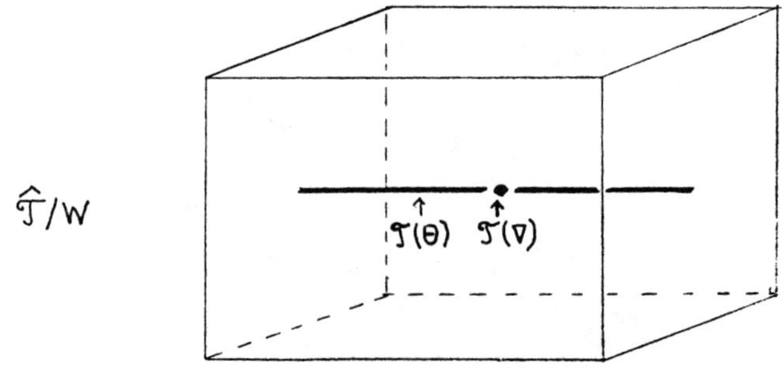


Fig. 7

In the general situation, the position of the boundary strata of $\hat{\mathcal{T}}/W$ in a neighborhood of the smallest stratum $\mathcal{T}(\nabla)$ (we assume $\nabla$ itself special, now) can be easily described in terms of local coordinates transversal to $\mathcal{T}(\nabla)$. These may be realized by the fundamental characters $\chi_\omega$, $\omega$ a fundamental dominant weight. It follows from [15] and [16], 6.9, Lemma 3, that the closure $\overline{\mathcal{T}(\Theta)/W_{\Theta^\perp}}$ of the stratum $\mathcal{T}(\Theta)/W_{\Theta^\perp}$ is locally given by

$$\{\hat{t}\in\hat{\mathcal{T}}/W \mid \chi_\omega(\hat{t})=0 \text{ for all } \omega \text{ with } \omega(\Theta)\neq\{0\}\}.$$

## 5. The Parabolic Partition of G

In the next section we are going to define a set-theoretic map

$$\chi: G \longrightarrow \hat{T}/W$$

which is invariant under conjugation. Parallel to the decomposition of $\hat{T}/W$ into the union of its boundary components $T(\Theta)/W_{\Theta^\perp}$, $\Theta \subset \nabla$ special, we now shall introduce a decomposition

$$G = \bigcup_{\substack{\Theta\subset\nabla \\ \text{special}}} G(\Theta)$$

of G into Ad(G)-stable subsets $G(\Theta)$ such that $G(\Theta)$ will be exactly the preimage $\chi^{-1}(T(\Theta)/W_{\Theta^\perp})$. This decomposition and the definition of $\chi$ are essentially due to Looijenga (cf. [16] 6.8, 7.7).

Let $\Theta \subset \nabla$ be an arbitrary subset of $\nabla$. Then the standard parabolic subgroup $P_\Theta$ of G is defined as the group generated by B and representatives in N of the reflections $s_\alpha$, $h_\alpha \in \Theta$. Any G-conjugate of $P_\Theta$ is called a parabolic subgroup of type $\Theta$.

For $\Theta \subset \nabla$ special let $G(\Theta)$ denote the set of elements $g \in G$ which are conjugate into $P_\Theta$ but not into a smaller parabolic subgroup $P_{\Theta'}$, $\Theta' \subset \Theta$, $\Theta' \neq \Theta$. An element $g \in G(\Theta)$ is called parabolic of type $\Theta$. The following theorem is crucial for the definition and analysis of $\chi$.

**Theorem**:

1) Any element $g \in G$ is contained in some $G(\Theta)$, $\Theta$ special.

2) If $\Theta \neq \Theta'$ are both special, then $G(\Theta) \cap G(\Theta') = \emptyset$.

3) Each element $g \in G(\Theta)$ lies in a unique parabolic subgroup of type $\Theta \cup \Theta^{\perp}$.

4) For all special $\Theta$, $G(\Theta)$ is not empty.

**Remarks**: Statements 1), 2), 3) are essentially due to Looijenga (cf. [16] 6.8). The proof of 2) and 3) uses the interpretation of these statements in terms of the building $\mathcal{P}$ attached to the Tits system (G, B, N, S). The geometry of the boundary components of the Tits-cone A (cf. section 3), i.e. of an apartment of $\mathcal{P}$, enters here in a decisive way. The proof of 4) uses the irreducible highest weight modules $L(\lambda)$, $\lambda$ a dominant weight in $H^*$, of G. In connection with this proof we were suggested the following (cf. [16] 6.9):

**Conjecture**: Let $\Theta \subset \nabla$ be special. Then $G(\Theta)$ is the set of all $g \in G$ which have an eigenvector in a module $L(\lambda)$ if and only if $\lambda(h) = 0$ for all $h \in \Theta$.

In loc. cit. we show that an element g with the above property is contained in the union of all $G(\Theta')$ with $\Theta' \supset \Theta$, and we reduce the conjecture to the following special case (we assume $\nabla$ itself special):

**Conjecture'**: The stabilizer in G of a vector $v \neq 0$ of a module $L(\lambda)$, $\lambda(\nabla) \neq \{0\}$, is contained in a proper parabolic subgroup.

One corollary of these conjectures would be the characterization of $G(\emptyset)$ as the set of all elements of G which act on the modules $L(\lambda)$ by locally finite transformations. There is some evidence for that (cf. [16] 6.9).

## 6. Definition of an Adjoint Quotient

We will define an Ad(G)-invariant map

$$\chi: G \longrightarrow \hat{T}/W$$

by piecing it together from its restrictions

$$\chi\big|_{G(\Theta)}: G(\Theta) \longrightarrow T(\Theta)/W_{\Theta^{\perp}}$$

for all special subsets $\Theta \subset \nabla$. For the definition of $\chi\big|_{G(\Theta)}$ we now fix $\Theta$. Let $P_\Theta$ the standard parabolic subgroup of type $\Theta$. Then there is a natural isomorphism

$$P_\Theta/DP_\Theta \cong T(\Theta),$$

where $DP_\Theta$ denotes the commutator subgroup of $P_\Theta$ in the topological sense (with respect to the topology on $P_\Theta$ induced by the proalgebraic structure of U).

Now let $g \in G(\Theta)$. Then g is conjugate to an element $p \in P_\Theta$. Let $\bar{t} \in T(\Theta)$ be its natural image in $T(\Theta)$. We put

$$\chi(g) = \bar{t} \bmod W_{\Theta^{\perp}} \in T(\Theta)/W_{\Theta^{\perp}} \subset \hat{T}/W .$$

Using the property 3) in the theorem of section 5 and the Bruhat decomposition one can see that $\chi(g)$ is unambiguously defined and, in particular, that $\chi$ is invariant under conjugation.

The proof for the non-emptyness of $G(\Theta)$ actually shows that $\chi$ maps $G(\Theta)$ surjectively onto $T(\Theta)/W_{\Theta^{\perp}}$.

We call $\chi: G \longrightarrow \hat{T}/W$ the adjoint quotient of G. Any fiber of $\chi$ will be called an adjoint fiber.

**Remark**: To call $\chi$ an adjoint quotient may seem premature at the moment. However, we believe that the following will provide sufficient justification for this terminology.

## 7. Adjoint Fibers of Classical Type

We first want to look at the fibers of the restriction

$$\chi\Big|_{G(\emptyset)} : G(\emptyset) \longrightarrow T(\emptyset)/W_{\Theta^{\perp}} = T/W ,$$

which we call of classical type. As in the finite-dimensional case the essential tool for their investigation is a Jordan decomposition.

Recall that by definition $G(\emptyset)$ consists of all elements of G which are conjugate into B, i.e. $G(\emptyset)$ is the union of all Borel subgroups of G.

We call an element $g \in B$ semisimple (resp. unipotent) if its image in any algebraic quotient group of B is semisimple (resp. unipotent). The Jordan decomposition in these quotients then lifts consistently to a Jordan decomposition in B, i.e. any element $b \in B$ admits a unique decomposition $b = s.u$ into the product of a semisimple element $s \in B$ and a commuting unipotent element $u \in B$. By conjugation with G we extend the notions of semisimplicity and of unipotence to all of $G(\emptyset)$. It follows essentially from the Bruhat decomposition in G that this definition is not ambiguous (cf. [16] 7.3). We can then prove (loc. cit.):

**Theorem:** Let $g \in G(\emptyset)$. Then there exist unique commuting elements $s,u \in G(\emptyset)$, such that s is semisimple, u is unipotent, and $g = s.u$.

To prove the uniqueness of the decomposition one can either invoke the representation theory of G or a fixed point theorem on the "flag manifold" G/B.

To determine the structure of the fibers $\chi^{-1}(\chi(t))$, $t \in T$, one can now proceed as in the classical situation (loc. cit. 7.5).

**Theorem:** The fiber $\chi^{-1}(\chi(t))$ is G-isomorphic to the associated bundle $G \times^{Z_G(t)} \mathrm{Uni}(t)$, where Uni(t) denotes the unipotent elements in $Z_G(t)$.

If t lies in the domain $\mathcal{T} \subset T$ then $Z_G(t)$ is a finite-dimensional reductive group and the structure of its unipotent variety is well-understood in this case (to some reasonable extent, at least). In general, the elements of Uni(t) can be shown to be $Z_G(t)$-conjugate to elements of $Z_G(t) \cap U$ (loc. cit. 7.4).

## 8. Arbitrary Adjoint Fibers

For simplicity we shall assume that $\nabla$ itself is special, i.e. that all connected components of $\nabla$ are of infinite type. Any fiber of the restriction

$$x\big|_{G(\nabla)}: G(\nabla) \longrightarrow T(\nabla)/W_{\nabla^{\perp}} = T(\nabla)$$

will be called a special fiber. We don't know too much about the structure of these fibers. If the Cartan matrix A is non-singular then all special fibers are G-isomorphic. Otherwise, the corank of A should be viewed as the number of "moduli" of the G-isomorphism classes of special fibers (cf. [16] 7.9).

Any progress in understanding the structure of the special fibers would immediately lead to a progress in the understanding of arbitrary fibers of the adjoint quotient as will be obvious from the following.

Let us now fix a special subset $\Theta \subset \nabla$ different from $\emptyset$ and $\nabla$. We want to look at the fibers of

$$x\big|_{G(\Theta)}: G(\Theta) \longrightarrow T(\Theta)/W_{\Theta^{\perp}} .$$

Let $P_{\Theta\cup\Theta^{\perp}}$ be the standard parabolic subgroup of type $\Theta \cup \Theta^{\perp}$ and $P_{\Theta\cup\Theta^{\perp}}(\Theta)$ the set of all elements in $P_{\Theta\cup\Theta^{\perp}}$ which are of parabolic type $\Theta$. Since every element $g \in G(\Theta)$ is lying in exactly one parabolic subgroup of type $\Theta \cup \Theta^{\perp}$ (cf. section 5) we obtain that $G(\Theta)$ may be rewritten as an associated bundle $G \times^{P_{\Theta\cup\Theta^{\perp}}} P_{\Theta\cup\Theta^{\perp}}(\Theta)$. The map from $G(\Theta)$ to $T(\Theta)/W_{\Theta^{\perp}}$ is now simply induced by an $\mathrm{Ad}(P_{\Theta\cup\Theta^{\perp}})$-invariant map

$$x': P_{\Theta\cup\Theta^{\perp}}(\Theta) \longrightarrow T(\Theta)/W_{\Theta^{\perp}}$$

which we will describe now.

We write the standard parabolic subgroups of type $\Theta$, $\Theta^{\perp}$,

$\Theta\cup\Theta^\perp$ as semidirect products $P = L \ltimes U$, i.e.

$$P_\Theta = L_\Theta \ltimes U_{(\Theta)}, \qquad P_{\Theta^\perp} = L_{\Theta^\perp} \ltimes U_{(\Theta^\perp)}$$

$$P_{\Theta\cup\Theta^\perp} = L_{\Theta\cup\Theta^\perp} \ltimes U_{(\Theta\cup\Theta^\perp)} .$$

Here L is a Kac-Moody group attached to a root basis $(H,\nabla',\Delta')$, $\nabla' = \Theta$, $\Theta^\perp$, or $\Theta\cup\Theta^\perp$, and U is the unipotent radical of P. We have natural embeddings

$$DL_\Theta \subset L_\Theta \subset L_{\Theta\cup\Theta^\perp} \supset L_{\Theta^\perp} \supset DL_{\Theta^\perp} ,$$

DL denoting the derived group of L. Let $M_{\Theta^\perp}$ (resp. $M_\Theta$) denote the quotient of $L_{\Theta\cup\Theta^\perp}$ by $DL_\Theta$ (resp. $DL_{\Theta^\perp}$). Then $M_{\Theta^\perp}$ (resp. $M_\Theta$) is a Kac-Moody group associated to the root base $(H/\mathbb{Z}.\Theta, \Theta^\perp, \Theta^{\perp\vee})$ (resp. $(H/\mathbb{Z}.\Theta^\perp, \Theta, \Theta^\vee)$).

Under the natural projection $P_{\Theta\cup\Theta^\perp} \xrightarrow{\pi} M_{\Theta^\perp}$ the set $P_{\Theta\cup\Theta^\perp}(\Theta)$ is mapped to $M_{\Theta^\perp}(\emptyset)$, and the map $\chi'$ is simply the composition of this projection with the adjoint quotient $\chi_{M_{\Theta^\perp}}$ of $M_{\Theta^\perp}$:

$$P_{\Theta\cup\Theta^\perp} \longrightarrow M_{\Theta^\perp}(\emptyset) \longrightarrow T(\Theta)/W_{\Theta^\perp} .$$

From this one finally can derive the following.

**Theorem**: Any fiber of $\chi|_{G(\Theta)}$ is G-isomorphic to a fiber bundle associated to the principal fibration $G \longrightarrow G/P_{\Theta\cup\Theta^\perp}$ and the natural action of $P_{\Theta\cup\Theta^\perp}$ on the product $U_{(\Theta\cup\Theta^\perp)} \times S_\Theta \times F_{\Theta^\perp}$, where $S_\Theta$ is a special adjoint fiber of $M_\Theta$ and $F_{\Theta^\perp}$ is an adjoint fiber of classical type in $M_{\Theta^\perp}$.

In [16] 7.9 one can find a more precise statement of this theorem.

## 9. Simultaneous Partial Resolutions

When G is semisimple and simply connected (as in section 1) the adjoint quotient $\chi: G \longrightarrow T/W$ admits a simultaneous resolution of its fibers, i.e. there is a commutative diagram

$$\begin{array}{ccc} G\times^{B}B & \xrightarrow{\varphi} & G \\ {\scriptstyle\tilde{\chi}}\downarrow & & \downarrow{\scriptstyle\chi} \\ T & \xrightarrow{\psi} & T/W \end{array}$$

with $\varphi(g*b) = gbg^{-1}$, $\tilde{\chi}(g*b)$ = T-part of B, and $\psi$ the natural quotient map, such that any fiber $\tilde{\chi}^{-1}(t)$, $t \in T$, is a resolution of the corresponding adjoint fiber $\chi^{-1}(\psi(t))$ (Grothendieck-Springer, cf. [14]).

From the point of view of the surface singularities related to certain Kac-Moody groups G it is excluded that the adjoint quotient $\chi: G \longrightarrow \hat{T}/W$ admits a simultaneous resolution. However, it is suggested that there exists a separate simultaneous resolution over each boundary component $T(\Theta)/W_{\Theta^{\perp}}$ of $\hat{T}/W$. In fact, the following diagram

$$\begin{array}{ccc} G\times^{P_{\Theta}}P_{\Theta}(\Theta) & \xrightarrow{\varphi} & G\times^{P_{\Theta\cup\Theta^{\perp}}}P_{\Theta\cup\Theta^{\perp}}(\Theta)=G(\Theta) \\ {\scriptstyle\tilde{\chi}}\downarrow & & \downarrow{\scriptstyle\chi} \\ T(\Theta) & \xrightarrow{\psi} & T(\Theta)/W_{\Theta^{\perp}} \end{array}$$

with $\tilde{\chi}(g*p)$ = $T(\Theta)$-projection of p, $\varphi$ and $\psi$ the natural maps, provides something like a simultaneous partial resolution of $\chi|_{G(\Theta)}$, i.e. when $\chi^{-1}(\psi(t))$, $t \in T(\Theta)$, has the form of an associated bundle $G\times^{P_{\Theta\cup\Theta^{\perp}}}(U_{(\Theta\cup\Theta^{\perp})} \times S_{\Theta} \times F_{\Theta^{\perp}})$ (see section 8), then the fiber

$\tilde{\chi}^{-1}(t)$ has the form $G \times^{P_{\Theta \cup \Theta^{\perp}}} (U_{(\Theta \cup \Theta^{\perp})} \times S_{\Theta} \times \tilde{F}_{\Theta^{\perp}})$ where $\tilde{F}_{\Theta^{\perp}}$ is a "resolution" of $F_{\Theta^{\perp}}$ and where the map $\varphi$ is induced by the natural map

$$U_{(\Theta \cup \Theta^{\perp})} \times S_{\Theta} \times \tilde{F}_{\Theta^{\perp}} \longrightarrow U_{(\Theta \cup \Theta^{\perp})} \times S_{\Theta} \times F_{\Theta^{\perp}} .$$

In the case that t is in the domain $\mathcal{T}(\Theta) \subset T(\Theta)$ then $F_{\Theta^{\perp}}$ is itself an associated bundle which has a finite-dimensional unipotent variety as its fiber. The term "resolution" can then be used in its honest sense (for more details see [16] 7.10).

## 10. Some Open Problems

In the case of simply connected semisimple groups G the adjoint quotient $\chi$: G $\longrightarrow$ T/W may be realized by means of the fundamental characters (cf. section 1). Naturally, one would like to have a similar realization for Kac-Moody groups G. Since the fundamental highest weight representations $L(\lambda)$ are now infinite-dimensional the corresponding characters will not be defined on all elements of G (e.g. the neutral element $e \in G$) and the definition itself requires analytical preparation.

When the Cartan matrix of G is symmetrizable, then all the modules $L(\lambda)$ are equipped with a structure of a Pre-Hilbert space which is invariant with respect to a compact form of G (cf. [5]). Relative these structures we can define the subset $\mathfrak{D} \subset G$ of all elements $g \in G$ which are of trace class in all representations $L(\lambda)$. We would like some of the following properties to hold, if not for the group G at least for some "reasonable" group intermediate between G and the minimal group $G^{min}$ studied by Kac and Peterson ([4], [13]):

1) There should be a subset $\mathcal{U} \subset \hat{T}/W$ which is an open neighborhood of the smallest stratum $\mathcal{T}(\nabla) \subset \hat{\mathcal{T}}/W \subset \hat{T}/W$ such that $\chi^{-1}(\mathcal{U})$ is contained in $\mathfrak{D}$.

2) Optimistically $\mathcal{U}$ might be the domain of convergence of the

characters (these may be naturally extended to a domain in $\hat{T}$ and $\hat{T}/W$, cf. [15], compare also [5] where the exact domain of convergence in T is described).

3) For all elements $g \in \mathfrak{D}$ and $x \in G$ one should have the same trace for g as for $xgx^{-1}$ on all modules $L(\lambda)$ (note that this is not trivial since the element x might act as an unbounded operator on some (or all) $L(\lambda)$.

4) For sufficiently small $\mathcal{U} \subset \hat{\mathcal{T}}/W$ (see 1) the fundamental characters $\chi_i$, i=1,...,r (where r = rank H $\geq \ell$ = card $(\nabla)$) induce an isomorphism of $\mathcal{U}$ with an open neighborhood of $\{0\} \times (\mathbb{C}^*)^{r-\ell}$ in $\mathbb{C}^\ell \times (\mathbb{C}^*)^{r-\ell}$ (cf. [9], [15]). With respect to this isomorphism the adjoint quotient

$$\begin{array}{cccc} \chi: & \chi^{-1}(\mathcal{U}) & \longrightarrow & \mathcal{U} \\ & \cap & & \cap \\ & \mathfrak{D} & & \hat{\mathcal{T}}/W \end{array}$$

should be identical to the character map

$$\bar{\chi}: \chi^{-1}(\mathcal{U}) \longrightarrow \mathbb{C}^\ell \times (\mathbb{C}^*)^{r-\ell}$$

$$g \longmapsto (\chi_1(g),\dots,\chi_r(g)) \; .$$

In [15] it is shown that $\chi$ and $\bar{\chi}$ coincide at least on the intersection $N \cap \mathfrak{D}$ and that $\chi(N\cap\mathfrak{D})$ is a neighborhood of the smallest stratum $\mathcal{T}(\nabla)$ in $\hat{\mathcal{T}}/W$.

Granted points 1) and 3) we can derive a proof of 4) from the conjectured characterization of the elements in $G(\Theta)$, $\Theta\subset\nabla$ special, via the nonexistence of eigenvectors in certain representations (cf. section 5).

Besides the problems above many more pose themselves naturally when one tries to interpret the adjoint quotient $\chi: G \longrightarrow \hat{T}/W$, or rather its "analytic" part, as a categorical quotient. We intend to come back to these questions at another occasion.

## References

[1] A. Feingold, I. Frenkel: A hyperbolic Kac-Moody algebra and the theory of Siegel modular forms of genus 2; Math. Ann. 263, 87-144 (1983).

[2] H. Garland: The arithmetic theory of loop groups; Publ. Math. IHES 52, 5-136 (1980).

[3] V. Kac, D. Peterson: Infinite-dimensional Lie algebras, Theta-functions, and modular forms; Advances in Math. 53, 125-264 (1984).

[4] ——————, ——————: Regular functions on certain infinite-dimensional groups; in "Arithmetic and Geometry", Vol. 2, 141-166, Ed. M. Artin, J. Tate, Birkhäuser, Boston 1983.

[5] ——————, ——————: Unitary structure in representations of infinite-dimensional groups and a convexity theorem; Inventiones Math. 76, 1-14 (1984).

[6] G. Kempf et alii: Toroidal Embeddings I, Lecture Notes in Math. 339, Springer, Berlin-Heidelberg-New York, 1974.

[7] B. Kostant: Lie group representations on polynomial rings; Amer. J. Math. 85, 327-404 (1963).

[8] N. Bourbaki: Groupes et algèbres de Lie, IV, V, VI, Hermann, Paris, 1968.

[9] E. Looijenga: Invariant theory for generalized root systems; Inventiones Math. 61, 1-32 (1980).

[10] ——————: Rational surfaces with an anti-canonical cycle; Annals of Math. 114, 267-322 (1981).

[11] R. Marcuson: Tits' systems in generalized nonadjoint Chevalley groups; J. Algebra 34, 84-96 (1975).

[12] R. V. Moody, K. L. Teo: Tits' systems with cristallographic Weyl groups; J. Algebra 21, 178-190 (1972).

[13] D. Peterson, V. Kac: Infinite flag varieties and conjugacy theorems; Proc. Natl. Acad. Sci. USA 80, 1778-1782 (1983).

[14] P. Slodowy: Simple singularities and simple algebraic groups; Lecture Notes in Math. 815, Springer, Berlin-Heidelberg-New York, 1980.

[15] ——————: A character approach to Looijenga's invariant theory for generalized root systems; Compositio Math. (to appear).

[16] ——————: Singularitäten, Kac-Moody Liealgebren, assoziierte Gruppen und Verallgemeinerungen; Habilitationsschrift Universität Bonn, Bonn, 1984.

[17] R. Steinberg: Regular elements in semisimple algebraic groups; Publ. Math. IHES 25, 49-80 (1965).

[18] ——————: Conjugacy classes in algebraic groups; Lecture Notes in Math. 366, Springer, Berlin-Heidelberg-New York, 1974.

[19] J. Tits: Définition par générateurs et relations de groupes avec BN-paires; C. R. Acad. Sc. Paris, 293, 317-322 (1981).

[20] ——————: Annuaire du Collège de France, 1980/81, 75-86, 1981/82, 91-106, Paris.

[21] E. B. Vinberg: Discrete linear groups generated by reflections; Math. USSR Izvestija 35, 1983-1190 (1971).

# ANALYTIC AND ALGEBRAIC ASPECTS OF THE KADOMTSEV-PETVIASHVILI HIERARCHY FROM THE VIEWPOINT OF THE UNIVERSAL GRASSMANN MANIFOLD

By

Kimio Ueno

Department of Mathematics
Yokohama City University
22-2 Seto, Kanazawa-ku
Yokohama 236, JAPAN

## §0. Introduction and Notations

In this note we shall study the Cauchy problems for the KP hierarchy and for the KP equation, commutative subrings of ordinary differential operators. Our central idea is to connect the characterization theorem for wave functions [1] to the language of the universal Grassmann manifold due to Professor Mikio Sato [5], [6], [7]. From this viewpoint, the problems stated above are resolved in an extremely simple manner.

The results in this note have been obtained during the stay of the author's at Mathematical Sciences Research Institute (MSRI), and have been announced at the workshop at MSRI, "Infinite Dimensional Lie Groups" (May 11-15, 1984). The author would like to thank Professor Victor Kac for giving him a chance to talk at the workshop.

The details will be published elsewhere.

List of notations used throughout this note.

$\mathcal{E}$; the ring of ordinary microdifferential operators of finite order with coefficients in scalor functions.

$\mathcal{D}$; the ring of ordinary differential operators with coefficients in scalor functions.

$\mathcal{E}^{(m)}$ (resp. $\mathcal{D}^{(m)}$); the subspace of ordinary microdifferential (differential) operators of order at most m.

$$\mathcal{E}^{(0),\text{monic}} = \{P \in \mathcal{E}^{(0)};\ \text{the leading coefficient is } 1\}$$

Under these notations, $\mathcal{E}$ has a direct sum decomposition $\mathcal{E} = \mathcal{D} \oplus \mathcal{E}^{(-1)}$, so that any operator $P = P(x,\partial_x) \in \mathcal{E}$ is uniquely expressed as $P = (P)_+ + (P)_-$, where $(A)_+ \in \mathcal{D}$, $(A)_- \in \mathcal{E}^{(-1)}$. As symbols of the derivation, we shall freely use $\partial_x (= \frac{d}{dx})$, $f_x (= \frac{\partial f}{\partial x})$ etc.

$\mathbb{N}^c = \{-1,-2,\ldots\}$.

$GL(\mathbb{N}^c)$; the group of invertible matrices of size $\mathbb{N}^c \times \mathbb{N}^c$.

## §1. Review of the Theory of KP Hierarchy

In the KP hierarchy, a monic, microdifferential operator of first order, L plays a role of an unknown function to be solved;

$$L = L(t,\partial_x) = \sum_{i=0}^{\infty} u_i(t)\partial_x^{1-i} \quad (u_0 = 1,\ u_1 = 0).$$

Here $t = (t_1,t_2,t_3,\ldots)$ denotes a set of an infinite number of time variables, and the coefficients $u_i(t)$'s are analytic functions or formal power series in t. Compared with the other time variables, $t_1$ plays a role equivalent to x, so that, from now on, we set $t_1 = x$.

The KP hierarchy is, by definition, a system of infinitely many Lax equations,

(1) $\partial_{t_n} L = [B_n, L]$, $n = 2,3,4,\ldots,$

describing the time evolution of L. Here $B_n = (L^n)_+ \in \mathcal{D}^{(n)}$. Thus, the KP hierarchy is thought of to be an infinite number of equations for the unknown functions $u_i(t)$ $(i=2,3,\ldots)$. It is well known that the totality of the Lax equations, (1) coincides with the totality of the following Zakharov-Shabat equations [5], [1],

(2) $\partial_{t_n} B_m - \partial_{t_m} B_n + [B_m, B_n] = 0$, for $n,m = 2,3,\ldots$ .

Especially, setting $n = 2$, $m = 3$ in (2), one gets a single non-linear equation,

$$\frac{3}{2} u_{yy} = (2u_t - 6uu_x - \frac{1}{2} u_{xxx})_x,$$

which is nothing but the celebrated Kadomtsev-Petviashvili (or the two-dimensional KdV) equations.

For a solution L to the KP hierarchy, one can find the so-called wave function, $w(t,k)$ of the form

(3) $$w(t,k) = \left[\sum_{j=0}^{\infty} w_j(t)\, k^{-j}\right] \exp \eta(t,k),$$

where $w_0 = 1$, and $\eta(t,k) = \sum_{n=1}^{\infty} t_n k^n$, and satisfying the following linear equations,

(4) $Lw = kw$, $\partial_{t_n} w = B_n w$, $n=2,3,\ldots$ .

The existence of a wave function is not unique. Actually it has ambiguity such as $w \longmapsto w(1 + \sum_{j=1}^{\infty} c_j k^{-j})$ where $c_j$ is an arbitrary constant. Taking the integrability condition for (4), one recovers the system of the Lax equations, (1) and the system of the Zakharov-Shabat equations (2). In this sense, we often call (4) the linearization problem for the hierarchy.

For a wave function $w(t,k)$, (3), let us introduce a wave operator $W = W(t,\partial_x) \in \mathcal{E}^{(0),\mathrm{monic}}$ through the following formula,

$$W(t,\partial_x) = \sum_{j=0}^{\infty} w_j(t)\partial_x^{-j} .$$

Then $w(t,k) = W(t,\partial_x)e^{\eta(t,k)}$ (note that $\partial_x^j e^{kx} = k^j e^{kx}$ for any $j \in \mathbb{Z}$). Now let us define the dual version of a wave function by

$$w^*(t,k) = (W^*)^{-1} e^{-\eta(t,k)},$$

where $W^*$ is the formal adjoint operator for W, which is, by definition,

$W^* = \sum_{j=0}^{\infty} (-\partial_x)^{-j} w_j(t)$. Consequently it takes the form

$$(5) \qquad w^*(t,k) = \left[\sum_{j=0}^{\infty} w_j^*(t)\, k^{-j}\right] \exp(-\eta(t,k)).$$

Furthermore one can easily check that it satisfies the dual linearization problem

$$L^* w^* = kw^*, \quad \partial_{t_n} w^* = -B_n^* w^*, \quad n=2,3,\ldots .$$

Here $L^*$ and $B_n^*$ denote the formal adjoint operators of $L$ and $B_n$, respectively.

A wave function and its dual version of the KP hierarchy are completely characterized by the "bilinear residue formula" which was first proved by Kashiwara [1], [8].

**Theorem**. Let $w(t,k)$ and $w^*(t,k)$ be functions of the form (3) and (5), respectively. Then they are a wave function and its dual version for the KP hierarchy if and only if they satisfy the following integral equation

$$(6) \qquad \oint w(t,k)\, w^*(t',k) \frac{dk}{2\pi(-1)^{1/2}} = 0, \text{ for any } t,t'.$$

Here the integration means to take the residue at $k = \infty$.

**Remark**. The theorem is valid even in the case that $w(t,k)$ is a wave function of finitely many Zakharov-Shabat equations.

From now on, we shall call (6) the <u>"bilinear residue formula"</u>. From this formula one can deduce the existence of a $\tau$ function such that

$$w(t,k) = \frac{\tau(t-\varepsilon[k^{-1}])}{\tau(t)} \exp \eta(t,k) \qquad \text{where } \varepsilon[k^{-1}] = (k^{-1}, \tfrac{1}{2}k^{-2}, \tfrac{1}{3}k^{-3}, \ldots)$$

and further one can derive Hirota's bilinear differential equations for $\tau$ functions whose generating functional expression is ([1], [8])

$$(7) \qquad \sum_{j=0}^{\infty} p_j(-2y)\, p_{j+1}(\tilde{D}) \exp\left[\sum_{n=1}^{\infty} y_n D_{t_n}\right] \tau \cdot \tau = 0.$$

Here $y_n$'s are indeterminate variables, and $D_{t_n}$'s are Hirota's bilinear differentiation and $\tilde{D} = (D_{t_1}, \frac{1}{2} D_{t_2}, \frac{1}{3} D_{t_3}, \ldots)$. $p_j(t)$'s are polynomials introduced through $\exp \eta(t,k) = \sum_{j=0}^{\infty} p_j(t)\, k^j$.

As to the bilinear residue formula, it will be profitable to note the following points: Let $w(x,k) = \left[\sum_{j=0}^{\infty} w_j(x)\, k^{-j}\right] e^{kx}$, $w^*(x,k) = \left[\sum_{j=0}^{\infty} w^*_j(x)\, k^{-j}\right] e^{-kx}$ with $w_0 = w^*_0 = 1$. Then they are expressed as $w(x,k) = W e^{kx}$, $w^*(x,k) = (W^*)^{-1} e^{-kx}$ by means of a certain microdifferential operator $W = W(x,\partial_x) \in \mathcal{E}^{(0),\text{monic}}$ if and only if they satisfy the bilinear residue formula

$$\oint w(x,k)\, w^*(x',k) \frac{dk}{2\pi(-1)^{1/2}} = 0 \qquad \text{for any } x,x'.$$

## §2. The Grassmann Equation and the Cauchy Problem of the KP Equations

After the theory of the universal Grassmann manifold (UGM) of Professor Sato's [5], [6], the totality of Hirota bilinear equations satisfied by $\tau$ functions of the KP hierarchy is equivalent to the totality of the Plücker's relations. More precisely one can prove the following statement: Let $\mathcal{Y}$ be the set of all the Young diagrams. Any function f(t) admits the expansion of the form,

$$f(t) = \sum_{Y\in\mathcal{Y}} c_Y \chi_Y(t),$$

where $\chi_Y(t)$ is the Schur polynomial corresponding to the diagram Y, and the coefficient $c_Y$ is given by $c_Y = \chi_Y(\tilde{\partial}_t) f(t)\big|_{t=0}$. Then, f(t) is a $\tau$ function of the KP hierarchy if and only if the coefficients $c_Y$'s satisfy the Plücker relations. In other words, the Hirota bilinear differential equations (6) are obtained upon eliminating such coefficients.

The first purpose in this section is to, without using Sato's results, take out the Grassmann variables which are hidden behind the bilinear residue formula. From now on, we assume that the unknown functions $u_i(t)$ (the coefficients $w_i(t)$ of a wave function as well) are analytic functions in t near t = 0, or formal power series in t.

Before presenting our result, let us give the definition of an $\mathbb{N}^c$-frame.

**Definition.** Let $\underline{\xi} = (\ldots\xi_{-3},\xi_{-2},\xi_{-1})$ be a rectangular matrix of size $\mathbb{Z} \times \mathbb{N}^c$ (each $\xi_\nu \in \mathbb{C}^{\mathbb{Z}}$). $\underline{\xi}$ is called an $\mathbb{N}^c$-frame when vectors $\xi_\nu$ are linearly independent to each other.

**Theorem 1.** Let w(t,k),(3) be a wave function of the KP hierarchy, and $\vec{w}(t)$ an infinite dimensional vector formed by the coefficients of (3); $\vec{w}(t) = (\ldots w_2(t), w_1(t), 1)$. Then there exists an $\mathbb{N}^c$-frame $\underline{\xi}$ satisfying the following conditions:

$$(G_1)\qquad \underline{\xi} \text{ is the form, } \underline{\xi} = \left(\begin{array}{cccccc} & & & & & \\ & & & 0 & & \\ & \cdots & 1 & & & \\ & \cdots & * & 1 & & \\ * & \cdots & * & * & 1 & \\ & \cdots & * & * & * & \\ & \cdots & * & * & * & \\ & & \vdots & \vdots & \vdots & \\ & & & & & \end{array}\right| \begin{array}{c} -\infty \\ \vdots \\ -2 \\ -1 \\ 0 \\ 1 \\ \vdots \\ +\infty \end{array} .$$

$(G_2)$ $\vec{w}(t)$ solves the following linear algebraic equation,

$$\vec{w}(t)\, A_1^* \exp(\eta(t,\Lambda))\underline{\xi} = 0. \tag{8}$$

Here $\Lambda^n = (\delta_{\mu+n,\nu})_{\mu,\nu\in\mathbb{Z}}$ is the n-th shift matrix,

$$\Lambda^n = \begin{bmatrix} \ddots & & & & & \\ & 1 & \phantom{\cdot} & & & \\ & & \ddots & & 0 & \\ & & & 1 & & \\ & & & & 1 & \\ \hline & & & & & 1 \\ & 0 & & & & \ddots \\ & & & & & \end{bmatrix} \begin{matrix} \\ \\ -n \\ \\ \\ \\ \nu \\ \\ \end{matrix},$$

$$\phantom{xxxx}\mu\phantom{xx}n$$

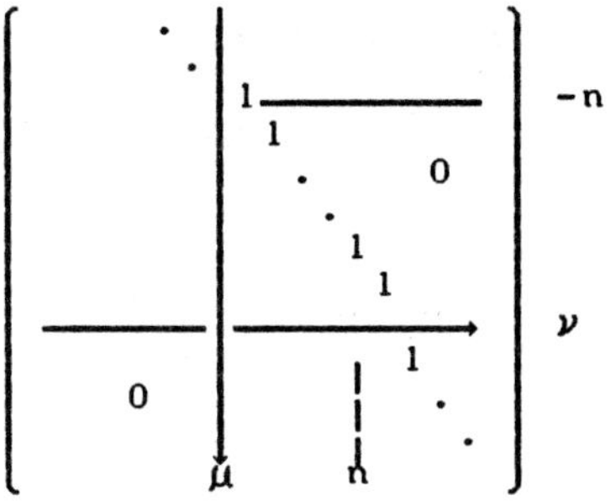

and $A_1^* = (\delta_{\mu\nu})_{\mu\leq 0, \nu\in\mathbb{Z}}$ is a rectangular matrix of the form,

$$A_1^* = \begin{bmatrix} \ddots & 0 & & & \\ & \ddots & & 0 & \\ & & 1 & & \\ & 0 & 1 & & \\ & & 1 & & \end{bmatrix} \begin{matrix} -\infty \\ \\ \\ \\ \\ \end{matrix}$$

$$-\infty \phantom{xxxxxxxx} +\infty$$

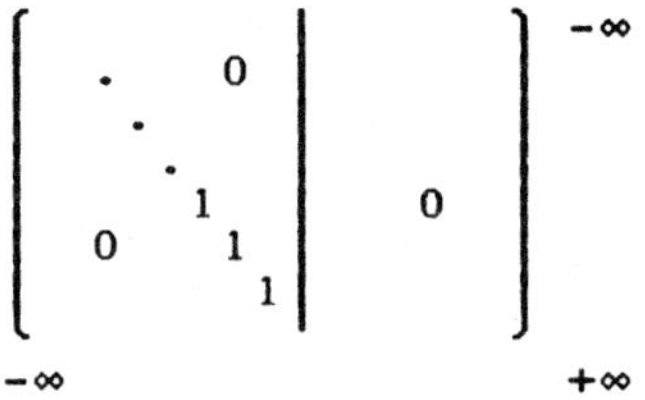

Sketch of the proof. Let ${}^t\vec{w}^*(t)$ be an infinite dimensional column vector formed by the coefficients of a dual wave function, ${}^t\vec{w}^*(t) = {}^t(\ldots w_2^*(t), w_1^*(t), 1)$. Let $A_1 = {}^t(A_1^*)$, and $J$ an anti-diagonal matrix, $J = (\delta_{\mu,-\nu})_{\mu,\nu\in\mathbb{Z}}$. The bilinear residue formula (6) reads as

(9) $\quad \vec{w}(t) \, A_1^* \, \exp(\eta(t,\Lambda)) \, \Lambda J \, \exp(-\eta(t',\Lambda^{-1})) \, A_1 \, {}^t\vec{w}^*(t') = 0.$

We set

$$\xi'_{S_\nu} = (-\delta_{x'})^{-\nu-1} \exp(-\eta(t',\Lambda^{-1})) \, A_1 \, {}^t\vec{w}^*(t') \Big|_{t'=0}, \quad (\nu\in\mathbb{N}^c),$$

and further set $\xi_{S_\nu} = \Lambda J \xi'_{S_\nu}$. Then one can easily verify that $\xi_{S_\nu}$ takes the form,

$$\xi_{S_\nu} = \left[\begin{array}{c} \cdot \\ \cdot \\ 0 \\ 1 \\ * \\ \cdot \\ \cdot \\ * \\ \hline * \\ \cdot \\ \cdot \end{array}\right] \begin{array}{l} \\ \\ \\ \nu \\ \\ \\ \\ -1 \\ 0 \\ \\ \\ \end{array}$$

so that $\xi_{S_\nu}$'s turn out to be linearly independent to each other. Setting $\underline{\xi}_S = (\ldots \xi_{S_{-3}}, \xi_{S_{-2}}, \xi_{S_{-1}})$, one sees that $\underline{\xi}_S$ is an $\mathbb{N}^c$-frame (we shall call this "<u>the standard $\mathbb{N}^c$-frame</u>") and satisfies $\vec{w}(t)\, A_1^*\, \exp(\eta(t,\Lambda))\, \underline{\xi}_S = 0$ from (9). Q.E.D.

**<u>Remark</u>.** Let $\alpha = (\alpha_1, \alpha_2, \ldots)$ be an arbitrary multi-index. Then the vector $\xi'_{S_\alpha} = (-\partial_{t'})^\alpha \exp(-\eta(t',\Lambda^{-1}))\, A_1\, {}^t\vec{w}^*(t')\big|_{t'=0}$ is expressed as a linear combination of $\xi'_{S_\nu}$'s.

If an $\mathbb{N}^c$-frame $\underline{\xi}$ satisfies the equation (8) for a wave vector $\vec{w}(t)$, so does $\underline{\xi} \cdot h$ for $h \in GL(\mathbb{N}^c)$. Thus one should consider $\underline{\xi}$, an $\mathbb{N}^c$-frame in the theorem, as a point of the infinite dimensional Grassmann manifold. In more precise statement, letting $\underline{\xi}$ be an $\mathbb{N}^c$-frame satisfying the condition $(G_1)$ in the theorem, one has

$$\underline{\xi} \bmod GL(\mathbb{N}^c) \in UGM^{\emptyset}.$$

$UGM^{\emptyset}$ is the largest cell in <u>the universal Grassmann manifold UGM</u> introduced by Sato [5], [6] where $\emptyset$ denotes the empty Young diagram (cf. §5 Appendix). For this reason, we call (8) <u>the Grassmann equation</u>.

Next we shall consider the converse assertion to the theorem. For an $\mathbb{N}^c$-frame $\underline{\xi}$ satisfying $(G_1)$, the Grassmann equation (8) can be uniquely solved by means of an elementary method, Cramér's formula [7].

Let $A_1^* \exp(\eta(t,\Lambda))\ \underline{\xi} = (\xi_{\mu\nu}(t))_{\mu\leq 0,\nu<0}$, and let us consider the following equation which is a slight modification of (8);

$$(\ldots\tilde{w}_2(t),\tilde{w}_1(t),\tilde{w}_0(t))\cdot(\xi_{\mu\nu}(t))_{\mu\leq 0,\nu<0} = 0.$$

making use of Cramér's formula, one has

$$\tilde{w}_i(t) = (-)^i \det \left[\frac{\xi_{\mu\nu}(t)_{\substack{\mu<-i\\ \nu<0}}}{\xi_{\mu+1,\nu}(t)_{\substack{-i\leq\mu<0\\ \nu<0}}}\right],$$

where the determinant in the right-hand side is the minor determinant obtained by extracting the (-i)-th row, the 0-th column from $(\xi_{\mu\nu}(t))_{\mu\leq 0,\nu<0}$. We should note that $\tilde{w}_i(t)$ makes sense as a formal power series in t, and that, especially, $\tilde{w}_0(t)$ is an invertible formal power series. Thus, setting $w_i(t) = \tilde{w}_i(t)/\tilde{w}_0(t)$, one sees that this is again a formal power series and solves the Grassmann equation. Furthermore one can find a differential operator of order n, $B_n = \partial_x^n + \sum_{j=0}^{n-2} b_{n,j}(t)\ \partial_x^j$ such that

$$(\partial_{t_n} - B_n)\ \{\vec{w}(t)\ A_1^* \exp(\eta(t,\Lambda))\ \underline{\xi}\}$$

$$= (\ldots v_{-2}(t), v_{-1}(t), 0)\ A_1^* \exp(\eta(t,\Lambda))\ \underline{\xi}.$$

The right side is, of course, equal to the null vector. The uniqueness of a solution to the Grassmann equation implies that $v_j(t) = 0$ for any j. Introducing a wave function w(t,k) or a wave operator $W(t,\partial x)$ through $\vec{w}(t)$, one sees that

$$\left[\sum_{j=0}^{\infty} v_j(t)\ k^{-j}\right] e^{\eta(t,k)} = (\partial_{t_n} - B_n)\ w(t,k) = 0.$$

Let $L(t,\partial_x) = W(t,\partial_x)\ \partial_x W(t,\partial_x)^{-1}$. Summing up our arguments, the wave function w(t,k) solves the linearization problem for the KP hierarchy;

$$Lw = kw, \ \partial_{t_n} w = B_n w.$$

Hence $L(t,\partial_x)$ is a solution to the hierarchy.

Now let us give two remarks.

(i) The first minor determinant, $\tilde{w}_0(t)$ is nothing but a $\tau$ function.

(ii) In the above discussion, the resulting solution $L(t,\partial_x)$ $= \sum_{i=0}^{\infty} u_i(t)\partial_x^{1-i}$ is formal power series. If one wants to get an analytic solution, one must start from $\underline{\xi}$ in the analytic universal Grassmann manifold, $UGM^{ana(r)}$ (as for the precise definition, the readers should refer to [7]).

Next we consider the Cauchy problems for the KP hierarchy and the KP equation, as an application of the Grassmann equation. The KP hierarchy is, as in §1, a system of infinitely many equations for infinitely many unknown functions, while the KP equation is a single equation for only one unknown function. Consequently, there exists a very big gap between the analytic aspects of the both systems.

First let us resolve the Cauchy problem for the hierarchy. It is formulated as follows:

For a given microdifferential operator $L_0(x,\partial_x) \in \partial_x + \mathcal{E}^{(-1)}$, find a solution $L(t,\partial_x)$ to the KP hierarchy such that

$$L(t,\partial_x)\Big|_{t_2=t_3=\ldots=0} = L_0(x,\partial_x).$$

This problem can be solved through the following five steps:

(i) Find an initial wave operator $W_0(x,\partial_x) \in \mathcal{E}^{(0),\mathrm{monic}}$ such that $L_0(x,\partial_x) = W_0(x,\partial_x)\partial_x W_0(x,\partial_x)^{-1}$, and set $w_0^*(x,k) = (W_0(x,\partial_x)^*)^{-1}e^{-kx}$.

(ii) Compute the standard $\mathbb{N}^c$-frame $\underline{\xi}_S$ corresponding to $w_0^*(x,k)$ (see the proof of Theorem 1).

(iii) Solve the Grassmann equation, $\vec{w}(t)\ A_1^*\ \exp(\eta(t,\Lambda))\ \underline{\xi}_S = 0$.

(iv) Construct a wave operator $W(t,\partial_x) \in \mathcal{E}^{(0),\text{monic}}$ from the solution $\vec{w}(t)$.

(v) Then $L(t,\partial_x) = W(t,\partial_x)\partial_x W(t,\partial_x)^{-1}$ is a solution to the problem. The unique solvability can be easily verified.

The Cauchy problem for the KP equation can be resolved in the following, more generalized version.

**Theorem 2 (Extension Theorem).** Let $\tilde{L} = \tilde{L}(x,t_2,\ldots,t_N,\partial_x) \in \partial_x + \mathcal{E}^{(-1)}$, and $\tilde{B}_n = \tilde{B}_n(x,t_2,\ldots,t_N,\partial_x) = (\tilde{L}^n)_+ \in \mathcal{D}^{(n)}$. Suppose that they satisfy the Zakharov-Shabat equations,

$$\partial_{t_n}\tilde{B}_m - \partial_{t_m}\tilde{B}_n + [\tilde{B}_m,\tilde{B}_n] = 0,\ n,m = 2,\ldots,N.$$

Then there exist differential operators $B_n = B_n(x,t_2,\ldots,t_{N+1},\partial_x) \in \mathcal{D}^{(n)}$, $(n = 2,\ldots,N+1)$, which solve the Zakharov-Shabat equations

$$\partial_{t_n}B_m - \partial_{t_m}B_n + [B_m,B_n] = 0,\ n,m = 2,\ldots,N+1,$$

and have initial values of $B_n\big|_{t_{n+1}=0} = \tilde{B}_n$ $(n = 2,\ldots,N)$.

In order to prove this theorem, we need the following crucial proposition.

**Proposition 3.** Under the same hypothesis as in Theorem 2, one can find a wave function of the form

$$\tilde{w}(x,t_2,\ldots,t_N,k) = \left[\sum_{j=0}^{\infty} \tilde{w}_j(x,t_2,\ldots,t_N)k^{-j}\right] \exp\left[xk + \sum_{n=2}^{N} t_n k^n\right]$$

with $\tilde{w}_0 = 1$ such that

$$\partial_{t_n} \tilde{w} = \tilde{B}_n \tilde{w}, \quad n = 2,\ldots,N.$$

Making use of this proposition, one can prove Theorem 2 in the same way as in the hierarchy. As a corollary, we obtain the solvability theorem of the Cauchy problem for the KP equation. We remark that the uniqueness of the solvability does not hold now because of the big ambiguity of wave functions for the finitely many Zakharov-Shabat equations.

## §3. The Reduced Hierarchies and Commutative Subrings of Ordinary Differential Operators

The purpose in this section is to introduce certain commutative subrings which characterize subholonomic hierarchies such as the KdV hierarchies etc., or special solutions such quasi-periodic, soliton and rational solutions in the KP hierarchy. Introducing and studying such commutative rings was originally due to Professor Sato's deep insights, however his results have not been published unfortunately. The results in this section are closely connected with the theory of commutative subrings of ordinary differential operators studied by Burchnal-Chaundy, Krichver and Mumford [2], [4]. And they are also deeply relevant to the results in Mulase [3].

First of all, we shall consider how the $\ell$-reduced hierarchies are described in terms of the Grassmann variables ($\ell$ is a positive integer).

<u>Definition</u>. A solution $L = L(t,\partial_x) \in \partial_x + \mathcal{E}^{(-1)}$ to the KP hierarchy is called <u>$\ell$-reduced</u> if the condition

(10) $L^n \in \mathcal{D}$ for any $n \equiv 0 \pmod{\ell}$

holds.

Let $w(t,k)$ be a wave function, and $\vec{w}(t)$ a wave vector for the solution $L(t,\partial_x)$. Then the condition (10) equivalently reads as

$$
(11) \qquad
\begin{aligned}
&\partial_{t_n} w(t,k) = k^n w(t,k) \text{ for any } n \equiv 0 \pmod{\ell}, \\
&\text{or, } \partial_{t_n} \vec{w}(t) = 0 \text{ for any } n \equiv 0 \pmod{\ell}.
\end{aligned}
$$

<u>**Definition**</u>. An $\mathbb{N}^c$-frame $\underline{\xi}$ is called $\ell$-<u>periodic</u> if

$$\Lambda^{\ell}\underline{\xi} = \underline{\xi}\tilde{\Lambda}^{\ell}$$

holds, where

$$
\tilde{\Lambda}^{\ell} = (\delta_{\mu+\ell,\nu})_{\mu,\nu\in\mathbb{N}^c} =
\begin{array}{c}
\left[\begin{array}{cccc}
\ddots & & & \\
 & 1 & & \\
 & & 1 & \\
 & & & 1 \\
 & & & \\
\end{array}\right]
\begin{array}{l}
-\infty \\ \\ \\ -\ell-1 \\ \\ -1
\end{array} \\
\begin{array}{ccc} -\infty & \qquad\qquad & -1 \end{array}
\end{array} .
$$

From now on, we shall denote by $\underline{\xi}_S$ a standard $\mathbb{N}^c$-frame formed by a wave function $w(t,k)$ or $w(x,k)$.

The following proposition completely characterizes the $\ell$-reduced hierarchy.

<u>**Proposition 4**</u>. Let $\underline{\xi}$ be an $\mathbb{N}^c$-frame satisfying the condition $(G_1)$ in Theorem 1. Let $\vec{w}(t)$ be a solution to the Grassmann equation,

$$\vec{w}(t)\ A_1^{*}\ \exp(\eta(t,\Lambda))\ \underline{\xi} = 0,$$

and let $W(t,\partial_x) \in \mathcal{E}^{(0),\text{monic}}$ be the corresponding wave operator. We set $L = L(t,\partial_x) = W(t,\partial_x)\partial_x W(t,\partial_x)^{-1}$. Then we have;

(i) If there exists an $\ell$-periodic $\mathbb{N}^c$-frame $\underline{\xi}^{(\ell)}$ such that $\underline{\xi} \equiv \underline{\xi}^{(\ell)} \bmod GL(\mathbb{N}^c)$, L belongs to the $\ell$-reduced hierarchy.

(ii) If L belongs to the $\ell$-reduced hierarchy, the standard $\mathbb{N}^c$-frame $\underline{\xi}_S$ is deformed to an $\ell$-periodic $\mathbb{N}^c$-frame $\underline{\xi}^{(\ell)}$ by a succession of elementary operations with respect to columns of matrices. Consequently $\underline{\xi}_S \equiv \underline{\xi}^{(\ell)} \bmod GL(\mathbb{N}^c)$. Furthermore it turns out that $\underline{\xi}^{(\ell)}$ takes the form

$$\underline{\xi}^{(\ell)} = (\ldots,\Lambda^{\ell}\xi_{S_{-\ell-1}},\ldots,\Lambda^{\ell}\xi_{S_{-1}},\xi_{S_{-\ell-1}},\ldots,\xi_{S_{-1}}),$$

where we set $\underline{\xi}_S = (\xi_{S_\nu})_{\nu\in\mathbb{N}^c}$.

Sketch of the proof. We shall prove only (i). Suppose that $\underline{\xi} = \underline{\xi}^{(\ell)}\cdot G$ for some $G \in GL(\mathbb{N}^c)$. Then one sees that

$$\exp(\eta(t,\Lambda))\underline{\xi} = \exp\left[\sum_{n\not\equiv 0 (\mathrm{mod}\ \ell)} t_n\tilde{\Lambda}^n\right] G^{-1}.$$

Hence the Grassmann equation reduces to

$$\vec{w}(t)\ A_1^*\ \exp\left[\sum_{n\not\equiv 0 (\mathrm{mod}\ \ell)} t_n\Lambda^n\right] \underline{\xi} = 0,$$

so that the resulting solution $\vec{w}(t)$ satisfies the condition (11). Q.E.D.

**Remark**. A similar statement is valid even for the case that the time evolution is not considered.

Next let us study the relation between commutative subrings of ordinary differential operators and the Grassmann variables.

Let P and Q be differential operators of the form

$$P = \partial_x^{\ell} + \sum_{j=0}^{\ell-1} p_j(x)\partial_x^j,\quad Q = \partial_x^m + \sum_{j=0}^{m-1} q_j(x)\partial_x^j.$$

We assume that they commute with each other, i.e. $[P,Q] = 0$. Then there exists a simultaneous eigenfunction $w = w(x,k) = \left[\sum_{j=0}^{\infty} w_j(x)k^{-j}\right]e^{kx}$ for the operators P,Q. That is, it satisfies the following equations,

(12) $\quad Pw = k^{\ell}w,\ Qw = \alpha(k)w,$

where $\alpha(k) = k^m + \sum_{j=-\infty}^{m-1} \alpha_j k^j \in \mathbb{C}((k^{-1}))$.

Abusing the terminology, we call such an eigenfunction "a wave function," as well.

**Definition.** Let $\underline{\xi}$ be an $\mathbb{N}^c$-frame, and $\alpha(k) = \alpha^m + \sum_{j=-\infty}^{m-1} \alpha_j k^{-j} \in \mathbb{C}((k^{-1}))$. $\underline{\xi}$ is called an $\alpha$-quasi periodic if it satisfies the condition,

$$\alpha(\Lambda)\underline{\xi} = \underline{\xi}\tilde{\Lambda}^m.$$

The eigenfunction w(x,k), (12) is characterized as follows.

**Proposition 5.**

(i) Let $\underline{\xi}_S$ be the standard $\mathbb{N}^c$-frame for the wave function w(x,k), (12). Then there exist an $\ell$-periodic $\mathbb{N}^c$-frame $\underline{\xi}^{(\ell)}$ and an $\alpha$-quasi periodic $\mathbb{N}^c$-frame $\underline{\xi}^{(\alpha)}$ such that

(13) $\underline{\xi}_S \equiv \underline{\xi}^{(\ell)}$, and $\underline{\xi}_S \equiv \underline{\xi}^{(\alpha)} \mod GL(\mathbb{N}^c)$.

Here $\underline{\xi}_S$ is deformed to $\underline{\xi}^{(\ell)}$ or $\underline{\xi}^{(\alpha)}$ by a succession of elementary operations with respect to columns of matrices. Furthermore $\underline{\xi}^{(\ell)}$ and $\underline{\xi}^{(\alpha)}$ are explicitly given from the standard frame.

(ii) Suppose that, in the Grassmann equation

$$\vec{w}(x)\ A_1^* \exp(\eta(x,\Lambda))\underline{\xi} = 0,$$

(the frame $\underline{\xi}$ is assumed to admit the condition $(G_1)$) $\underline{\xi}$ satisfies the same condition as (13). Then the resulting wave function w(x,k) is a simultaneous eigenfunction, (12).

Sketch of the proof. The frame $\underline{\xi}^{(\alpha)}$ is chosen as follows. Let $\xi'_{S_\nu} = (-\partial_{x'})^{-\nu-1} \{\exp(-x'\Lambda^{-1})\ A_1{}^t\vec{w}^*(x')\}\big|_{x'=0}$, $(\nu \in \mathbb{N}^c)$,

and let $\underline{\xi}^{(\alpha)} = (\underline{\xi}_\nu^{(\alpha)})_{\nu\in\mathbb{N}^c}$. Then $\underline{\xi}_\nu^{(\alpha)}$'s are defined by the following recursive relation;

$$\xi_{\nu}^{(\alpha)} = \Lambda J \xi'_{\nu}, \ (\nu = -1,\dots,-m),$$
$$\xi_{-sm+\nu}^{(\alpha)} = (\alpha(\Lambda))^{s} \ \underline{\xi}_{\nu}^{(\alpha)}, \ (s = 0,1,\dots). \quad \text{Q.E.D.}$$

Let A be a subring in $\mathbb{C}((\Lambda^{-1}))$ generated by $\Lambda^{\ell}$ and $\alpha(\Lambda)$, i.e. $A = \mathbb{C}[\Lambda^{\ell},\alpha(\Lambda)]$. The condition (13) says that $\underline{\xi}_S$ mod $GL(\mathbb{N}^c)$ is stable point in UGM. It is rather easy to check that A satisfies the condition,

(14) $\quad A \cap \mathbb{C}[[\Lambda^{-1}]]\cdot\Lambda^{-1} = \{0\},$

(cf. Mulase [3]). Now let us consider such commutative rings.

E.g. We give some simple examples.

$A = \mathbb{C},$
$\;\; = \mathbb{C}[\Lambda^{\ell}]$, ($\ell$ is a positive integer)
$\;\; = \mathbb{C}[\Lambda^{\ell},\Lambda^{m}]$, ($\ell$,m are positive integers)

The final statement in this section is the following.

<u>**Theorem 6.**</u> For a wave function w(t,k) for the KP hierarchy, there exists a unique maximal subring $A_w \subset \mathbb{C}((\Lambda^{-1}))$ such that $A_w \cap \mathbb{C}[[\Lambda^{-1}]]\cdot\Lambda^{-1} = \{0\}$ and the standard $\mathbb{N}^c$-frame $\underline{\xi}_S$ is $A_w$-stable.

## §4. Discussions

Concluding this note, I would like to present some problems to be left unsolved. Besides those problems discussed here, there are many interesting, stimulating open problems on the KP hierarchy, the universal Grassmann manifold, the Fock representation of the Clifford algebra.

(i) (Classification of the orbits.) In the final theorem, I showed that, for any $\mathbb{N}^c$-frame $\underline{\xi}_S$, there exists a maximal subring $A_W$ in $\mathbb{C}((\Lambda^{-1}))$ which characterizes the orbit from time evolution with $\underline{\xi}_S$ mod $G(\mathbb{N}^c)$ as an initial point. What I want to know is the converse. That is, "for a given A satisfying (14), does there exist an $\mathbb{N}^c$-frame $\underline{\xi}$ such that $\underline{\xi}$ is A-stable?". If one solves this problem, it turns out that the set of the orbits of time evolution corresponds, one to one, to the set of such rings.

(ii) (Schottoky problem and Novikov conjecture.) The Schottoky problem is to characterize the Jacobian variety among the obelian varieties. This problem has been solved independently by several mathematicians, though I would remark that, if one combines my results plus something, the Schottoky problem can be automatically solved. As for the Novikov conjecture, Mulase will give detailed explanation in this proceeding, so I will not mention here anymore. Let me remark only that the extension theorem (Theorem 2) plays an important role in his discussion.

## §5. Appendix

Let us denote by $\mathcal{E}_{\mathbb{R}}$ the ring of microdifferential operators with the coefficients in $\mathbb{R} = \mathbb{C}[[x]]$, and introduce a left $\mathcal{E}_{\mathcal{R}}$-module $\mathcal{V}$ by

$$\mathcal{V} = \mathcal{E}_{\mathbb{R}}/\mathcal{E}_{\mathbb{R}}\cdot x.$$

Setting $e_\nu = \partial_x^{-1-\nu}$ mod $\mathcal{E}_{\mathbb{R}}\cdot x$ $(\nu \in \mathbb{Z})$, one sees that as a vector space over $\mathbb{C}$,

$$\mathcal{V} = \{\sum_{\nu\in\mathbb{Z}} \xi_\nu e_\nu;\ \xi_\nu \in \mathbb{C},\ \xi_\nu = 0 \text{ for } \nu \ll 0\},$$

and that $\mathcal{V}$ has the direct sum decomposition

$$\mathcal{V} = V^{\phi} \oplus \mathcal{V}^{(0)}$$

where $V^{\phi} = \bigoplus_{\nu\in\mathbb{N}^c} \mathbb{C}e_\nu$, $\mathcal{V}^{(0)} = \prod_{\nu\in\mathbb{N}} \mathbb{C}e_\nu$.

The universal Grassmann manifold, UGM, is by definition,

$$\mathrm{UGM} = \{U \subset V;\ \dim(U \cap V^{(0)}) = \dim V/(U + V^{(0)}) < +\infty\}.$$

Now we shall present some main theorems on UGM.

(i) Let $\mathcal{Y}$ be the set of all the Young diagrams. UGM has cellular decomposition associated with $\mathcal{Y}$,

$$\begin{aligned}\mathrm{UGM} &= \coprod_{Y \in \mathcal{Y}} \mathrm{UGM}^{Y} \\ &= \mathrm{UGM}^{\emptyset} \cup \mathrm{UGM}^{\square} \cup \mathrm{UGM}^{\square\square} \cup \dots .\end{aligned}$$

Especially $\mathrm{UGM}^{\emptyset}$ is the largest cell, defined by

$$\mathrm{UGM}^{\emptyset} = \{U \subset V:\ V = U \oplus V^{(0)}\}.$$

In terms of $\mathbb{N}^c$-frames, one has

$$\mathrm{UGM}^{\emptyset} = \{\underline{\xi} \bmod GL(\mathbb{N}^c);\ \underline{\xi} = \left[\begin{array}{ccc} \ddots & & 0 \\ & 1 & \\ * & & 1 \\ & & 1 \\ \hline & * & \end{array}\right]\begin{array}{l} \\ \\ \mathbb{N}^c \\ \\ \mathbb{N} \end{array}\}.$$

(ii) UGM is embedded into the projective space $\mathbb{P}(\mathbb{C}^{\mathcal{Y}})$ as the zero-locus of all the Plücker relations.

(iii) Let $K$ be the quotient field of $\mathbb{R}$, and let

$$W = \{W \in \mathcal{E}_K^{(0),\mathrm{monic}};\ x^m W,\ W^{-1}x^n \in \mathcal{E}_{\mathcal{R}} \text{ for some } m,n \in \mathbb{N}\}$$

where $\mathcal{E}_K^{(0),\mathrm{monic}}$ is the set of monic microdifferential operators of 0-th order with the coefficients in $K$. We define a map $\gamma$: $W \longrightarrow$ UGM by

$$\Upsilon(W) = (W^{-1}x^n)V^{\emptyset} \subset \mathcal{V} \text{ for } W \in \mathcal{W}.$$

Then $\Upsilon$ is a bijection.

For the details, the readers should refer to [6], [7].

## References


[1] Date, E., Jimbo, M., Kashiwara, M. and Miwa, T., Transformation groups for soliton equations, Nonlinear integrable systems - Classical Theory and Quantum Theory, Ed by M. Jimbo and T. Miwa (World Scientific Publishing Company, Singapore, 1983).

[2] Krichever, I. M., Functional Anal. Appl. 11 (1977), 12-26.

[3] Mulase, M., Cohomological structure of solutions of soliton equations, iso-spectral deformations of ordinary differential operators and a characterization of Jacobian varieties, MSRI Preprint 003-84 (to appear in J. Diff. Geometry).

[4] Mumford, D., An algebro-geometric construction of commuting operators and of solutions to the Toda lattice equation, Korteweq-de-Vries equation and related non-linear equations, Intl. Symp. on Algebraic Geometry, Kyoto (1977), 115-153.

[5] Sato, M., RIMS Kôkyûroku 439, Kyoto Univ. (1981), 30.

[6] , KP hôteishiki-kei to fuhen Grassmann tayô tai, Lecture Note Series delivered at Sophia Univ., Japan (1984) (Notes by M. Noumi, in Japanese).

[7] Sato, M., and Sato, Y., Soliton Equations as Dynamical Systems on Infinite Dimensional Grassmann Manifold, Preprint (1983).

[8] Ueno, K., and Takasaki, K., RIMS Preprint 425, Kyoto Univ. (1983), to appear in Adv. Studies in Pure Math.

# COMMENTS ON DIFFERENTIAL INVARIANTS

By

B. Weisfeiler
MSRI and Pennsylvania State University

The subject of differential invariants is possibly as old as the algebraic invariant theory itself. The first differential invariant discovered was the Schwarzian derivative $(y'''/y') - (3/2)(y''/y')^2$ of a function y(t) of one variable. It has several invariance properties; two of them are under a fractional linear change of an independent variable and, separately, under a fractional linear change of the dependent variable.

The theory of differential invariants has never achieved the degree of maturity of the algebraic invariant theory. There seem to be several reasons for that. One of them is that by the turn of the century the theory of differential invariants was able to tackle (almost) only functions of one variable. The development of differential geometry required handling functions of more than one variable and such were developed in particular cases. The theory of torsion and curvature of a connection is an example.

Another possible reason is that by the turn of the XX-th century mathematical values changed. In particular, Hilbert's finiteness theorem became a worthier model than explicit results in particular cases. The topic of differential invariants was not ready for a similar kind of conceptualization.

We give below several examples representing, we hope, some important features of the subject. Then we prove a differential analog of Hilbert's theorem on finite generation of invariants. We conclude by pointing out several problems. In some of our comments we are rather informal.

---

Professor Boris Weisfeiler disappeared on January 5, 1985, in a mountain zone near Chillan in Chile in circumstances which are still not completely explained.

Conversations with J. Bernstein, R. Herman, V. Kac, D. Kazhdan, and V. S. Varadarajan helped me enormously in understanding the subject of this paper. I am grateful to them.

## 1. Examples.

To start with let us give several examples. More examples can be found in literature, see annotated bibliography in the end of this paper. We always assume that we are working over $\mathbb{C}$, the complex numbers.

**Example 1.** Consider a system of ordinary linear differential equations of the first order:

$$X' = AX \text{ where } X = \begin{bmatrix} x_1(t) \\ \vdots \\ x_n(t) \end{bmatrix}, \quad A = (a_{ij}(t))_{i,j=1,\dots,n},$$

where the $x_i(t)$ and the $a_{ij}(t)$ are functions of t from a given class $\mathcal{F}$ and $X' = dX/dt$.

A substitution $X = BY$, where $B = B(t)$ is an invertible $n \times n$ matrix and Y is the column of new variables, gives rise to

$$Y' = (B^{-1}AB - B^{-1}B')Y$$

Thus to describe the equivalence classes of systems with respect to linear changes of dependent variables is the same as describing orbits of matrices $A \in \mathrm{Mat}_n(\mathcal{F})$ under the action of $GL_n(\mathcal{F})$ given by

$$B(A) = B^{-1}AB - B^{-1}B' \text{ for } A \in \mathrm{Mat}_n(\mathcal{F}),\ B \in GL_n(\mathcal{F}).$$

This is a classical problem, see [BV1] for a historical sketch. The group $GL_n(\mathcal{F})$ is sometimes called the group of gauge transformations. We assume, of course, that $\mathcal{F}$ is closed under subtraction, multiplication, taking the inverse, and differentiation; in other words we assume that $\mathcal{F}$ is a differential field.

As in the geometric invariant theory to describe the orbits we need, in particular, to construct functions P of the $a_{ij}$ and their derivatives such that the values of P for A and B(A) are the same. We restrict our attention now to the case when the P are polynomials. In other words we are looking at the ring $R = \mathcal{F}[a_{ij}, a'_{ij}, a''_{ij}, \ldots]_{i,j=1,\ldots,n}$ of the so called differential polynomials in the $a_{ij}$, see [K] and [R], and the action of $GL_n(\mathcal{F})$ on it, and we want to describe $R^{GL_n(\mathcal{F})}$, the ring of invariants of $GL_n(\mathcal{F})$ in R.

This problem admits a geometric interpretation. Recall that a connection (of class $\mathcal{F}$) on the principle $GL_n(\mathbb{C})$-bundle $\mathbb{C}^1 \times GL_n(\mathbb{C})$ over $\mathbb{C}^1$ is a map $A: \mathbb{C}^1 \longrightarrow \text{Lie } GL_n(\mathbb{C}) = Mat_n(\mathbb{C})$, i.e. $A \in Mat_n(\mathcal{F})$. Given a section $B: \mathbb{C}^1 \longrightarrow \mathbb{C}^1 \times GL_n(\mathbb{C})$ one can compare it pointwise with the section $c \longmapsto c \times Id$ to obtain a function $B: \mathbb{C}^1 \longrightarrow GL_n(\mathbb{C})$, i.e. $B \in GL_n(\mathcal{F})$. This new section B determines a new map $B(A) \in Mat_n(\mathcal{F})$ given by $B(A) = B^{-1}AB - B^{-1}B'$ (see [BV]). Thus the question we are looking at can also be considered as a question of classifying connections in the principal $GL_n(\mathbb{C})$-bundles over a line.

This is the problem we are concerned with here. The answer to this problem is trivial in our case (see [NW, Theorem 5.2] which is applicable because in the one-dimensional case curvature is always zero): *If* $\mathcal{F}$ *is differentially closed then* $GL_n(\mathcal{F})$ *acts transitively on* $Mat_n(\mathcal{F})$ (action as above) *and, therefore,* $R^{GL_n(\mathcal{F})} = \mathbb{C}$.

However, if $\mathcal{F}$ is not differentially closed there may exist more than one orbit of $GL_n(\mathcal{F})$ on $Mat_n(\mathcal{F})$. In the classically most interesting cases where $\mathcal{F}$ is the field of formal Laurent series in t or the field of convergent (outside 0) series the problem was solved recently by D. G. Babitt and V. S. Varadarajan (see [BV1], [BV2] and the forthcoming papers). They first classify the orbits for the action of $GL_n(\mathcal{F}_f)$, where $\mathcal{F}_f = \mathbb{C}((t))$ is the field of the formal Laurent power series. Then they study the action of $GL_n(\mathcal{F}_{conv})$, where $\mathcal{F}_{conv}$ is the subring of $\mathbb{C}((t))$ consisting of the series $s = s(t)$

convergent for $0 < |t| < \mathcal{E} = \mathcal{E}(s)$ some $\mathcal{E}(s) > 0$, on the convergent parts of orbits of $GL_n(\mathcal{F}_f)$. In other words, they study the orbits of $GL_n(\mathcal{F}_{conv})$ on the sets $\{B(A),\ B \in GL_n(\mathcal{F}_f)\} \cap Mat_n(\mathcal{F}_{conv})$, that is, the equivalence classes of germs of linear differential meromorphic (with at most an isolated pole in 0) equations at 0.

**Example 2.** Consider now a system of ordinary linear differential equations of the second order: $X'' = 2A_1X' + A_2X$. Setting $X = BY$ we obtain that $Y'' = 2(B^{-1}A_1B - B^{-1}B')Y' + (B^{-1}A_2B + 2B^{-1}A_1B' - B^{-1}B'')Y$. Thus the action of $GL_n(\mathcal{F})$ on $M = Mat_n(\mathcal{F}) \oplus Mat_n(\mathcal{F})$ is given now by

$$B\begin{bmatrix} A_1 \\ A_2 \end{bmatrix} = \begin{bmatrix} B^{-1}A_1B - B^{-1}B' \\ B^{-1}A_2B + 2B^{-1}A_1B' - B^{-1}B'' \end{bmatrix}$$

If $\mathcal{F}$ is a differentially closed field with the field of constants $\mathbb{C}$ then, again by [NW, Theorem 5.2], we see that $GL_n(\mathcal{F})$ acts transitively on $Mat_n(\mathcal{F})$ by $B(A_1) = B^{-1}A_1B - B^{-1}B'$. Thus to study the action of $GL_n(\mathcal{F})$ on M is the same as to study the action of the stabilizer of the subset $N := 0 \oplus Mat_n(\mathcal{F})$ on N. The stabilizer is

$$\{B \in GL_n(\mathcal{F}) \mid B^{-1}B' = 0\} = GL_n(\mathbb{C}).$$

The action of $GL_n(\mathbb{C})$ on $Mat_n(\mathcal{F})$ is given by $B(A_2) = B^{-1}A_2B$ and we are led to considering the invariants $R^{GL_n(\mathbb{C})}$ where R is the algebra of differential polynomials in the entries of $A_2$, as in the previous example.

A more detailed study of this case and some of its generalizations is conducted in [W,Ch. IV] and Section 2 of the present paper.

In the nextexample we give a description of a result from [W] which is, undoubtedly, the most interesting study of differential invariants from the "algebraic" point of view. A more modern formulation of this result is given in [M1], [M2].

**Example 3.** Consider the linear differential operators

$$L(y) = y^{(n)} + a_1(t)y^{(n-1)} \dots a_n(t)y$$

The changes of variables (both dependent and independent) of the form

$$\begin{cases} y_1 = \lambda(t)y \\ t_1 = \mu(t) \end{cases}$$

form a group if $\lambda \neq 0$ and $\mu' \neq 0$. This group preserves the equation $L(y) = 0$. We are interested in the polynomials in $a_1, a_2, \dots, a_n$ and their derivatives which are invariant under this group. It turns out that such invariants are very few. We are led then to considering relative differential invariants. These are, in essence, differential forms, i.e. such differential polynomials P in the $a_i$ that the substitution $y_1 = \lambda(t)y$, $t_1 = \mu(t)$ gives P times some power of the Jacobian of our substitution. Using our substitution (and solving a couple of differential equations, the procedure corresponding to an extension of the differential field) one can bring the equation $L(y) = 0$ into the form where $a_1 = 0 = a_2$. Since the most interesting action occurs in this case we assume now that $a_1 = a_2 = 0$. The only transformations from our group which preserve linearity of $L(y)$, the equation $L(y) = 0$, and the condition $a_1 = a_2 = 0$ are of the form $y_1 = ky/(ct+d)^{n-1}$, $t_1 = (at+b)/(ct+d)$ with $k \neq 0$, $ad-bc \neq 0$, $a,b,c,d,k \in \mathbb{C}$. The group $G(\mathbb{C})$ of such transformations is isomorphic to a quotient of $SL_2(\mathbb{C}) \times \mathbb{C}^*$ by a finite central subgroup. If we write $(gL)(y) = 0$ for $g \in G(\mathbb{C})$ as $y_1^{(n)} + (g(a_3))y_1^{(n-3)} + \dots + (g(a_n))y_1 = 0$ then the problem of finding (relative) differential invariants is the same as finding for all $m \in \mathbb{Z}$ the differential polynomials $P(a_3,\dots,a_n)$ such that $P(a_3,\dots,a_n)dt^m = P(g(a_3),\dots,g(a_n))dt_1^m$. Thus we consider the algebra $R: = \bigoplus_{m\geqslant 0} \mathcal{F}\{a_3,\dots,a_n\}dt^m$ (where $\mathcal{F}\{a_3,\dots,a_n\}$ is the algebra of differential polynomials) and we want a description of $R^{G(\mathbb{C})}$. Set for $i\geqslant 3$, $i\leqslant n$

$$\Theta_i: = \frac{1}{2} \sum_{i\geqslant s\geqslant 0} \frac{(-1)^s(i-2)!\,i!\,(2i-s-2)!}{(i-s-1)!\,(i-s)!\,(2i-3)!\,s!} a_{i-s}^{(s)}$$

$$\Theta_{j\cdot 1}: = 2j\Theta_j\Theta_j'' - (2j+1)(\Theta_j')^2$$

and, for $P \in \mathcal{F}\{a_3,\dots,a_n\}dt^m$,

$$[\Theta_3,P]: = 3\Theta_3P' — m\Theta_3'P.$$

**Theorem** ([W,p. 53] and also [M1, Lemma 2]). *The* $\Theta_i$, $\Theta_{i\cdot 1}$, $i=3,\dots,n$ *and their repeated brackets with* $\Theta_3$ *generate the algebra* $R^{G(\mathbb{C})}$.

Morikawa gives in [M1, Lemma 2] also relations between the $2(n-2)$ relative invariants $\Theta_i$ and $\Theta_{ik1}$. One also has

**Theorem** (see [W, p. 39]). *Let* $R\cdot R^{-1}$ *be the quotient field of* R. *Then* $(R\cdot R^{-1})^{G(\mathbb{C})}$ *is the differential field of rational functions in the* $\Theta_i$, $i=3,\dots,n$, *and* $\Theta_{3\cdot 1}$ *with respect to derivation* $\frac{\Theta_3}{\Theta_4}\frac{d}{dt}$.

There are many results in [W] which describe in much detail decompositions of the level sets of the differential invariants into orbits under $G(\mathbb{C})$.

Finally, let us note that the invariants $\Theta_i,\Theta_{j\cdot 1}$ were used (first, I believe, by G. H. Halphen) to study geometry of the curves in $\mathbb{P}^{n-1}(\mathbb{C})$. Halphen used them to describe the moduli space of such curves in $\mathbb{P}^3(\mathbb{C})$; his proof, though, had gaps.

Let L(y) be as above and let $y_1(t),\dots,y_n(t)$ be a basis of solutions of $L(y) = 0$. Then the curve $C: \mathbb{C}^1 \longrightarrow \mathbb{C}^n$ is given by the parametric representation $(y_1(t),\dots,y_n(t))$. If $\bar{y}_s = \Sigma\, a_s^i y_i$ is another basis then $\det |a_s^i| \neq 0$ and, therefore, $L(y) = 0$ is the same for all curves A(C) where $A \in GL_n(\mathbb{C})$. Conversely, if a curve C is given parametrically by $(y_1(t),\dots,y_n(t))$ and is not contained in any $\mathbb{C}^{n-1}$ then

$$L(y): = \begin{bmatrix}\text{Wronskian of} \\ y_1,\dots,y_n\end{bmatrix}^{-1} \cdot \begin{bmatrix}\text{Wronskian of} \\ y,y_1,\dots,y_n\end{bmatrix}$$

gives rise to C.

Thus the classes of linear equivalence of curves $C \subseteq \mathbb{C}^n$ which are not contained in any $\mathbb{C}^{n-1}$ correspond bijectively to equations $L(y) = 0$ of n-th order.

The image of our curve in $\mathbb{P}^{n-1}(\mathbb{C})$ is given by the homogeneous coordinates $(y_1(t): y_2(t): \ldots: y_n(t))$. Then it is clear that the curve in $\mathbb{P}^{n-1}(\mathbb{C})$ does not change if we replace y by $\lambda(t)y$ and t by $\mu(t)$. Thus we have

**Theorem.** *The equivalence classes of the curves in* $\mathbb{P}^{n-1}(\mathbb{C})$ (not contained in any $\mathbb{P}^{n-2}(\mathbb{C})$) *correspond bijectively to the equivalence classes of* $L(y) = 0$ *under the group of variables' changes* $y \longmapsto \lambda(t)y$, $t \longmapsto \mu(t)$.

The geometric properties of such curves are, therefore, expressible through the properties of the invariants $\Theta_i$, $\Theta_{i\cdot 1}$.

**Example 4.** Consider the differential invariants of the action of $G \simeq \mathbb{Z}/2$, $G = \{1,\sigma\}$, on $\mathbb{C}$ given by $\sigma(y) = -y$. Thus we consider $\mathcal{F}\{y\}^G$ where $\mathcal{F}$ is an apropriate (differential) field of functions on $\mathbb{C}$. It is clear that $\mathcal{F}\{y\}^G = \mathcal{F}[y^2,(y')^2,\ldots,(y^{(n)})^2,\ldots]$.

**Claim.** $\mathcal{F}\{y\}^G$ *is not a finitely generated differential algebra.*

Indeed, if $z_1,\ldots,z_m$ are (differential) generators of $\mathcal{F}\{y\}^G$ then we can assume that $z_1,\ldots,z_m$ have no constant term. Let $\bar{R}$ be the differential subalgebra without 1 of $\mathcal{F}\{y\}^G$ generated by the $z_i$. Then the ideal $\bar{R}\cdot\mathcal{F}\{y\}$ of $\mathcal{F}\{y\}$ will be a finitely generated (by the $z_i$) ideal of $\mathcal{F}\{y\}$. This, however, is not the case by [R, §I.15].

## 2. Formalism and finite generation theorems.

Consider a commutative algebra K over $\mathbb{C}$. The algebra $\mathrm{Der}_{\mathbb{C}} K$ is a vector space over K and a Lie algebra over $\mathbb{C}$ with the bracket defined by

$$[p_1,p_2]\ (k) = p_1(p_2(k)) - p_2(p_1(k))$$

for $k \in K$, $p_1, p_2 \in \mathrm{Der}_{\mathbb{C}} K$.

We fix a pair (K,P) where K is a commutative algebra over $\mathbb{C}$ and P is a K-subspace of $\mathrm{Der}_{\mathbb{C}} K$ such that $[P,P] \subseteq P$. The algebra $K[P]$ of differential operators is, by definition, the associative algebra over $\mathbb{C}$ generated by its subalgebra K and the K-subspace P with relations

$$pk = p(k) + kp$$
$$pp_1 - p_1p = [p,p_1]$$

for $k \in K$, $p, p_1 \in P$.

Let $K\{x_1,\ldots,x_n\}$ denote the algebra of differential polynomials over K, i.e. the symmetric K-algebra over the free $K[P]$-module $\bigoplus_{1 \leq i \leq n} K[P]x_i$. $K[P]$ acts on $K\{x_1,\ldots,x_n\}$ in such a way that for $a,b \in K\{x_1,\ldots,x_n\}$ and $p \in P$ we have

$$p(ab) = p(a)b + ap(b)$$

Generally, any K-algebra A with an action of $K[P]$ on it is called a differential algebra if it satisfies the above condition.

Even a finitely generated differential algebra over a differential field need not be Noetherian. An example exists already in $K\{x\}$ when $\dim_K P = 1$; it is described in [R, §I.15]. (Actually a stronger counter-example is established in [R, §I.15].) However $K\{x\}$ has a weaker Noetherian property. Recall that for an ideal I of a ring A its radical, Rad(I), is the ideal of A given by

$$\mathrm{Rad}(I) := \{a \in A \mid a^m \in I \text{ for some } m = m(a) \in \mathbb{N}\}.$$

**Definition.** A differential K-algebra A will be called rdN (: = radically differentially Noetherian) if every radical differential ideal is the radical of a finitely generated ideal.

In the terminology of [K] radical ideals are called "perfect" and rdN above is expressed in [K] by saying that perfect differential ideals form a "conservative Noetherian system". As with the

Noetherian property, rdN implies that a strictly increasing sequence of radical differential ideals is finite.

As Example 4 shows, passing to the rings of invariants may result in non-finitely generated rings even if one starts with the finitely generated rings. We still, however, have that the weaker finiteness property, rdN, is preserved if the action of the group is sufficiently simple.

**Theorem.** *Let* A *be an* rdN *algebra over a differential field* K (with derivations P) *and let* G *be a reductive (possibly disconnected) algebraic group over* $\mathbb{C}$. *Suppose that* G($\mathbb{C}$) *acts on* A *by* K-*algebra automorphisms and that*

(i) $pP(g(a)) = g(p(a))$ *for* $g \in G(\mathbb{C})$, $p \in P$, $a \in A$.

*Assume also that* A *has a filtration* $A_0 = K \subseteq A_1 \subseteq \ldots \subseteq A_s \subseteq \ldots$ *such that*

(ii) $A_i \subseteq A_{i+1}$, $A = \cup A_i$,

(iii) $G(\mathbb{C})(A_i) \subseteq A_i$

(iv) $\dim_K A_i < \infty$, $i=0,1,\ldots$

*Then* $A^{G(\mathbb{C})}$ *is a differential subalgebra of* A *and it is* rdN.

Note that this result applies to Example 4 and to Example 2 (see the concluding remarks in this example). We shall give a wide class of examples satisfying the above theorem later on.

That $A^{G(\mathbb{C})}$ is a differential subalgebra of A follows directly from (i). The proof that it is rdN is essentially the same as the proof of Hilbert's theorem on finite generation of invariants. We shall outline this proof.

Since G($\mathbb{C}$) is Zariski-dense in the algebraic group G we see that G itself acts on A by K-automorphisms. Since G is reductive (and char $K = 0$) it is completely reducible on each $A_i$ and there is a canonical projection $E_i\colon A_i \longrightarrow A_i^G = A_i^{G(\mathbb{C})}$ of $A_i$ on its subspace of fixed vectors. We have that $E_i | A_j = E_j$ if $j \leq i$. Thus we obtain a canonical $E\colon A \longrightarrow A^{G(\mathbb{C})}$. We have

$$E(ab) = E(a)b \text{ for } a \in A,\ b \in A^G.$$

Indeed if $a,b \in A_j$ then we write $a = \bar{a} + a_0$, $a_0 \in A_j^{G(\mathbb{C})}$ and $\bar{a}$ belongs to the (canonical again) complement $C_j$ of $A_j^{G(\mathbb{C})}$ in $A_j$. Then $ab = \bar{a}b + a_0b$. Clearly $a_0b \in A^{G(\mathbb{C})}$ and $\bar{a}b \in C_j \cdot A_j^{G(\mathbb{C})}$. The latter is (as a $G(\mathbb{C})$-module) a quotient of $C_j \otimes A_j^{G(\mathbb{C})}$ and since the latter is, as a $G(\mathbb{C})$-module, a multiple of $C_j$ it follows that $C_j \otimes A_j^{G(\mathbb{C})}$ has no trivial $G(\mathbb{C})$-submodules whence $(C_j \cdot A_j^{G(\mathbb{C})})^{G(\mathbb{C})} = \{0\}$ whence the required property of E.

A similar argument based on commutativity of actions of $G(\mathbb{C})$ and P shows that

$$pE(a) = E(pa) \text{ for } p \in P,\ a \in A.$$

Now let J be a radical differential ideal of $A^G$. Then Rad(JA) is a radical differential ideal of A. rdN implies (in the same way as in the Noetherian case) that a finite set of radical generators of Rad(AJ) can be chosen from J. Let it be $a_1,\dots,a_d \in J$. Let $a \in J$. Then $a \cdot 1 \in \mathrm{Rad}(AJ)$ and, therefore, $a^t \in \Sigma_{1\leq i\leq d}\ A\cdot K[P]a_i$. Write $a^t = \Sigma\, b_j f_j(p)(a_{i_j})$. Then by the above $a^t = E(a^t)$

$= \Sigma_j E(b_j)\cdot f_j(p)(e(a_{i_j}))$ $\subseteq \Sigma A^{G(\mathbb{C})} K[P] a_i$ whence

$a^t \in \mathrm{Rad}(\Sigma A^{G(\mathbb{C})} K[P] a_i)$, i.e. $J = \mathrm{Rad}(\Sigma A^{G(\mathbb{C})} K[P] a_i)$. Thus J is the radical of a finitely generated ideal of $A^{G(\mathbb{C})}$ as claimed.

A version of the above result for the field of rational differential invariants is much simpler.

**Theorem.** *Let* $\Gamma$ *be a group acting on a finitely generated differential field extension* L *of* K (with derivations P) *by automorphisms. If* $p(\gamma(\ell)) = \gamma(p(\ell))$ *and*

$\gamma(k) = k$ *for all* $\gamma \in \Gamma$, $p \in P$, $\ell \in L$, *and* $k \in K$, *then* $L^{\Gamma}$ *is a finitely generated differential field extension of* K.

**Proof.** Since the action of P commutes with that of $\Gamma$, $L^{\Gamma}$ is a differential subfield of L. But by [K,Proposition II.14] any subfield of a finitely generated differential field is a finitely generated differential field, whence our claim.

It is proper now to recall again that a very interesting study of the field of differential rational and algebraic functions on the differential manifolds of $\infty$-jets of maps of $U \subseteq \mathbb{C}^1$ into $M(\mathbb{C})$, where M is a homogeneous under G algebraic variety, was undertaken by M. Green [G]. His results give, in particular, a more detailed information on the structure of the differential quotient field of $A^{G(\mathbb{C})}$ for some A from the next to last Theorem. In addition, M. Green computes explicitly differential invariants in a large number of cases.

We shall now describe a class of examples to which the above theorems apply.

First, let us consider the algebra $\mathcal{H}$ of germs at 0 of the holomorphic functions $f\colon U \longrightarrow \mathbb{C}$ defined on a neighborhood U of 0 in $\mathbb{C}^n$. Set $\mathbb{C}^n := \oplus\, \mathbb{C}t_i$ and $\partial_i := \partial/\partial t_i$. Then $f(t_1,\dots,t_n) = \Sigma_{\bar{i}\geq 0}$

$a_{\bar{i}}\, t^{\bar{i}}/\bar{i}!$ where, as usual, $\bar{i} := (i_1,\dots,i_n)$, $t^{\bar{i}} := t_1^{\,i_1}\dots t_n^{\,i_n}$, $\bar{i}! = i_1!\dots i_n!$ and

$\bar{i} \geq 0$ means that the $i_j \geq 0$. We consider the $a_{\bar{i}}$ as the coordinates of f, i.e., as functions on $\mathcal{H}$. Then the coordinates of $\partial f/\partial t_m$ are $a_{i_1\dots i_{m-1},i_m+1,i_{m+1},\dots,i_n}$. Thus the coordinates of f are acted upon by $\mathbb{C}[\partial_1,\dots,\partial_n]$ as the elements of the free $\mathbb{C}[\partial_1,\dots,\partial_n]$-module generated by $a_{\bar{0}}$. To make our considerations independent of the choice of the uniformizers $t_1,\dots,t_n$ we are led to considering $P_{\mathcal{H}} := \oplus\, \mathcal{H}\partial_i \subseteq \mathrm{Der}_{\mathbb{C}}\, \mathcal{H}$ and the algebra $\mathcal{H}\{a_0\}$ of differential polynomials in one variable.

Note now that in actuality the $a_{\bar{i}}$ are also coordinates on the ring $\mathbb{C}[[t_1,\dots,t_n]]$ of formal power series. Therefore an algebraic study of objects arising in the algebra $\mathcal{H}\{a_{\bar{0}}\}$ will, by necessity, be only a study of <u>formal</u> objects. We thus replace $\mathcal{H}$ by $\mathcal{F}: = \mathbb{C}[[t_1,\ldots,t_n]]$ and $P_{\mathcal{H}}$ by $P_{\mathcal{F}}: = \oplus\, \mathcal{F}\partial_i$ and then $\mathcal{F}$ by its field of quotients K and $P_{\mathcal{F}}$ by $P: = \oplus\, K\partial_i$. (Note that there is a difficulty with our definition of K as the notion of formal Laurent series in many variables is not straightforward.)

Now let G be an algebraic $\mathbb{C}$-group acting, say, linearly on $M: = \mathbb{A}^m$. We consider $\mathcal{M}: = \mathcal{M}_n: = \{$the set of $\infty$-jets at 0 of holomorphic maps $f: U \longrightarrow M(\mathbb{C})$, U open in $\mathbb{C}^n$, $0 \in U\}$. The action of $G(\mathbb{C})$ on M gives an action of $G(\mathbb{C})$ on $\mathcal{M}$ by

$$(g(m))(x) = g(m(x)) \text{ for } g\in G(\mathbb{C}),\ m\in\mathcal{M},\ x\in U\subseteq\mathbb{C}^n.$$

This action is differential which in our case means that $p(g(m)) = g(p(m))$ for $p \in P_{\mathcal{H}}$, $g \in G(\mathbb{C})$, $m \in \mathcal{M}$. Our action gives rise to a differential action of $G(\mathbb{C})$ on the algebra $\mathcal{H}\{x_1,\ldots,x_m\}$. As before, we pass over from $\mathcal{H}$ to $\mathcal{F}: = \mathbb{C}[[t_1,\ldots,t_n]]$, and then to $K: = \mathcal{F}\cdot\mathcal{F}^{-1}$. Our Theorems apply to such actions (with G reductive for the first Theorem). They also apply to the case when M is taken to be just an algebraic variety over $\mathbb{C}$.

As a more concrete example, consider the case $G = SL_2$ acting via the natural representation on $M: = \mathbb{A}^2$. We take $n < \infty$ and consider the action of $G(\mathbb{C})$ on $A: = K\{x,y\}$. As $G(\mathbb{C})$ commutes with $P: = \bigoplus_{1\leqslant i\leqslant n} K\partial_i$ we are looking at invariants of $G(\mathbb{C})$ acting on direct sums $(= \oplus_{\bar{i}<m} K\partial^{\bar{i}}(Kx \oplus Ky))$ of natural $SL_2(\mathbb{C})$-modules. By the classical invariant theory we know then that $A^{G(\mathbb{C})}$ is generated (as a K-algebra) by the determinants

$$D_{\bar{i};\bar{j}} = \det \begin{bmatrix} \partial_{\bar{i}}x & \partial_{\bar{j}}x \\ \partial_{\bar{i}}y & \partial_{\bar{j}}y \end{bmatrix}, \quad \bar{i} \neq \bar{j},\ \bar{i},\bar{j} \geqslant 0.$$

## 3. <u>Problems</u>

The above results most probably extend to the case (including Example 1 and its analogs) when an action of the group is such that P still acts on the algebra or the field of the invariants. However some actions of interesting groups on interesting algebras do not have this latter property, see, e.g., Example 3. I do not know how to define such actions, whence

**Problem 1.** Give a definition of a differential action covering Example 3 and other examples appearing in the literature (say in [V,pp. 143–152]).

The first two Theorems quoted in our discussion of Example 3 give some kind of finite generation of the type one should be looking for.

Such a definition should be given, it seems, in terms of factors (i.e., cocycles) associated to an action of a group in question on the base differential field K.

Also it may be more convenient to consider actions of the differential Lie algebras instead of groups. A formalism is introduced in [NW] for classification of certain differential Lie algebras; it should be applicable here.

Another argument for considering Lie algebras instead of groups is that the differential Lie algebras of Cartan type (see [NW]) do not seem to have any group analog, except in the class of convergent or holomorphic maps, which seem to be of different flavour, see comments after Problem 3.

Once we have a definition of the general differential action we may look again at differential invariants. These, however, will not be generally preserved by some differentiations from P.

**Problem 2. (J. Bernstein)** Show that for an appropriate class of differential actions of differential groups on rdN-algebras the rings of invariants are again rdN with respect to derivations from P which normalize the action of our group or its Lie algebra.

Example 3 seems to confirm this conjecture: the bracket with

$\Theta_3$ plays the role of the desired derivation.

The algebraic differential invariants of the type we consider here are generally formal invariants, i.e. invariants of germs of functions, equations, groups, etc. at some point.

<u>**Problem 3**</u>. Find a way of putting the invariants together to obtain global results.

Some results of this type are given by M. Green [G, §3].

We return to the interesting question of describing even at some point the decomposition of formal orbits into orbits under the action of the group of analytic-, $C^\infty$ etc. transformations. This has been addressed on a number of occasions, but the discussion of this question is beyond the scope of the present note (see however concluding remarks in Example 3).

[BV] contains several references to problems of such kind.

<u>**References**</u> (with the library call numbers for books)


[BV1] D. G. Babbitt, V. S. Varadarajan, Formal reduction theory of meromorphic differential equations: a group-theoretic view, Pacific J. Math 109(1983), 1-80. [Describes the orbits of systems of linear ordinary differential equations X' = AX with coefficients from the field of formal Laurent series $\mathbb{C}((t))$ under the action of $GL_n(\mathbb{C}((t)))$, see remarks toward the end of Example 1.]

[BV2] D. G. Babbitt, V. S. Varadarajan, Local moduli for meromorphic differential equations, to appear. [Considers the systems of ordinary linear differential equations X' = AX with convergent outside 0 Laurent series coefficients and studies the decomposition of the set of such systems in the same formal orbit into the orbits under the group of convergent transformations, see concluding remarks to Example 1.]

[G] M. L. Green, The moving frame, differential invariants, and rigidity theorems for curves in homogeneous spaces, Duke Math. J. 45(1978), 735-779. [For a real Lie group H and a homogeneous space

M for H the paper studies the action of H on curves on M (under suitable genericity conditions on such curves). Many interesting explicit differential invariants of such action are computed.]

[K] E. R. Kolchin, Differential algebra and algebraic groups, Academic Press, New York, 1973(call no. QA3/P8/vol 54). [A fundamental introduction into differential algebra, contains many basic theorems and notions which can not be found in any other book.]

[M1] H. Morikawa, On differential invariants of holomorphic projective curves, Nagoya Math. J. 77(1980), 75-87. [Restates some results of [W] below and gives relations between the invariants.]

[M2] H. Morikawa, Some analytic and geometric applications of the invariant-theoretic method, Nagoya Math. J. 80(1980), 1-47. [Gives relations between the differential invariant theory of [W] and [M1] to ordinary invariants of $SL_2(\mathbb{C})$ and to automorphic forms.]

[NW] W. Nichols, B. Weisfeiler, Differential formal groups of J. F. Ritt, Amer. J. Math. 104(1982), 943-1003. [Describes a formalism needed for a classification of formal differential groups and gives a classification of their Lie algebras.]

[R] J. F. Ritt, Differential algebra, Amer. Math. Soc. Coll. Publ., Vol 33, Amer. Math. Soc., Providence, R.I., 1950(call no. QA1/A54/vol 33). [A terse and fast introduction into differential algebra.]

[V] E. Vessiot, Méthodes d'intégration elementaire. Études des équations differentielles ordinaires au point de vue formel. Section II.16 in Encyclopédie des sciences mathematiques et appliqués, Tom II, Vol 3, Fasc 1, Gauthier-Villars, Paris, 1904(call no. QA37/E62). [On pp. 143-152, V. gives a survey of results up to 1907.]

[W] E. L. Wilczynski, Projective differential geometry of

curves and ruled surfaces, Teubner, Leipzig, 1906(call no. QA660/W5). [Studies differential invariants of one ordinary lin. diff eq$^n$ of order n with application to the study of curves in a projective space. Also studies a system of 2 eq$^{ns}$ in 2 unknowns with application to ruled surfaces in $\mathbb{P}^3$.]

# THE VIRASORO ALGEBRA AND THE KP HIERARCHY

By

Hirofumi Yamada
Hiroshima University, Hiroshima 730, Japan

## §0. Introduction

A certain central extension of the Lie algebra of vector fields on the circle is called by physicists "the Virasoro algebra". Mathematicians started to develop a representation theory of this algebra quite recently. The study of the highest weight representations of the Virasoro algebra was started by V. Kac ([4, 5]).

In this talk we construct the Fock representations of the Virasoro algebra and decompose them to the irreducible highest weight representations. In the procedure, the Schur polynomials play an essential role. We also give a relationship with a hierarchy of non-linear differential equations of a soliton type, the KP hierarchy.

## §1. The Virasoro algebra and its highest weight modules

The Virasoro algebra is an infinite dimensional complex Lie algebra defined as follows:

$$\mathfrak{g} := \bigoplus_{k\in\mathbb{Z}} \mathbb{C}\ell_k \oplus \mathbb{C}c$$

with the bracket relations for the basis

$$\begin{cases} [\ell_k, \ell_m] = (k-m)\ell_{k+m} + \frac{1}{12}(k^3-k)\,\delta_{k+m,0}\,c \\ [\mathfrak{g}, c] = \{0\}. \end{cases}$$

We denote the universal enveloping algebra of $\mathfrak{g}$ by $U(\mathfrak{g})$. We put $\mathfrak{h} := \mathbb{C}\ell_0 \oplus \mathbb{C}c$ and $\mathfrak{n}^{\pm} := \bigoplus_{k>0} \mathbb{C}\ell_{\pm k}$. The subalgebra $\mathfrak{h}$ is called the Cartan subalgebra of $\mathfrak{g}$.

A $\mathfrak{g}$-module M is called "a highest weight module" if and only if there exists a non-zero vector $v_0 \in M$ such that 1) $U(\mathfrak{g})v_0 = M$, 2) there exists $\lambda \in \mathfrak{h}^*$ (the dual space of $\mathfrak{h}$) such that $Hv_0 = \lambda(H)v_0$ for any $H \in \mathfrak{h}$, 3) $\mathfrak{n}^+ v_0 = 0$. For such a module M, $\lambda \in \mathfrak{h}^*$ is called "the highest weight" and $v_0 \in M$, "the highest weight vector". For an arbitrarily given $\lambda \in \mathfrak{h}^*$, there is a universal highest weight module $M(\lambda)$, the Verma module, which is defined by $M(\lambda) := U(\mathfrak{g}) \otimes_{U(\mathfrak{h}\oplus\mathfrak{n}^+)} \mathbb{C}$, where the action of $\mathfrak{h} \oplus \mathfrak{n}^+$ on $\mathbb{C}$ is given by $(H + X) \cdot 1 = \lambda(H) \cdot 1$ for any $H \in \mathfrak{h}$ and $X \in \mathfrak{n}^+$. Any highest weight module with highest weight $\lambda \in \mathfrak{h}^*$ is a quotient module of the Verma module $M(\lambda)$. There is a unique irreducible $\mathfrak{g}$-module with highest weight $\lambda$, which is denoted by $L(\lambda)$. If $\lambda(\ell_0) = \xi \in \mathbb{C}$ and $\lambda(c) = \eta \in \mathbb{C}$, we write $\lambda = (\xi, \eta)$ and $L(\lambda) = L(\xi, \eta)$.

## §2. Fock representations of the Virasoro algebra

We prepare the infinite dimensional vector space $V = \mathbb{C}[x_1,x_2,x_3,\ldots]$ of the polynomials of infinitely many variables. Let $a_j$'s be the operators on V defined by, for positive integer j, $a_j = (2)^{1/2}\partial_j$ ($\partial_j = \partial/\partial x_j$), $a_{-j} = \frac{1}{(2)^{1/2}} jx_j$, and $a_0 = \frac{1}{(2)^{1/2}}\mu$ ($\mu \in \mathbb{R}$). Now, using $a_j$'s, define the operators

$$L_k^{(\mu)} := \frac{1}{2} \sum_{j\in\mathbb{Z}} : a_{-j}a_{j+k}:$$

for $k \in \mathbb{Z}$. Here : : is the normal ordering:

$$:a_ia_j: = \begin{cases} a_ia_j & \text{if } i \leq j \\ a_ja_i & \text{if } i > j. \end{cases}$$

We write down some $L_k^{(\mu)}$'s. For example,

$$\begin{cases} L_0^{(\mu)} = \frac{1}{4}\mu^2 + x_1\partial_1 + 2x_2\partial_2 + 3x_3\partial_3 + \cdots \\ L_1^{(\mu)} = \mu\partial_1 + x_1\partial_2 + 2x_2\partial_3 + 3x_3\partial_4 + \cdots \\ L_2^{(\mu)} = \partial_1^2 + \mu\partial_2 + x_1\partial_3 + 2x_2\partial_4 + \cdots \end{cases}$$

We can calculate the commutation relations as the differential operators on V, namely,

$$[L_k^{(\mu)}, L_m^{(\mu)}] = (k-m) L_{k+m}^{(\mu)} + \frac{1}{12} (k^3-k) \delta_{k+m,0} \text{ id}.$$

Hence the application $\pi_\mu$: $\ell_k \longmapsto L_k^{(\mu)}$, $c \longmapsto$ id defines a representation of $\mathfrak{g}$ on V. Considered as the representation space of $\pi_\mu$, the vector space V is denoted by $V_\mu$. We call it "the Fock representation".

The Fock representation $(\pi_\mu, V_\mu)$ itself is not necessarily a highest weight representation. However, it includes a highest weight module with highest weight vector $1 \in V_\mu$. Obviously the highest weight is $(\frac{1}{4}\mu^2, 1)$. Therefore if $(\pi_\mu, V_\mu)$ is irreducible, then $V_\mu \simeq L(\frac{1}{4}\mu^2, 1)$. About the irreducibility of the Fock representations, we have a theorem of Kac. One can find a proof of this theorem in the article of Feigin and Fuks.

<u>**Theorem**</u> ([2, 4, 5]). The Fock representation $(\pi_\mu, V_\mu)$ is irreducible if and only if $\mu \notin \mathbb{Z}$. If $\mu = n \in \mathbb{Z}$, then $(\pi_n, V_n)$ is completely reducible.

Let $N(\lambda)$ be a highest weight module with the highest weight vector $v_0 \in N(\lambda)$. For a non-negative integer m, we put

$$N(\lambda)_m := \text{linear span of } \{\ell_{-\nu_1} \cdots \ell_{-\nu_k} v_0;\ k \geq 0,\ \nu_i > 0,\ \nu_1 + \ldots + \nu_k = m\}.$$

We consider the dimensions of $N(\lambda)_m$'s and their generating function, which is called "the formal character",

$$\text{ch } N(\lambda) := \sum_{m=0}^{\infty} \dim N(\lambda)_m \, q^m$$

where q is an indeterminate.

For the Verma module $M(\lambda)$, it is easy to see that dim

$M(\lambda)_m = p(m)$, the partition number of m. Therefore ch $M(\lambda) = \frac{1}{\phi(q)}$, where $\phi(q) = \prod_{n=1}^{\infty} (1-q^n)$, the Euler's function. The determination of the formal character of the irreducible module $L(\lambda)$ is, generally, a difficult problem (cf. [6, 8, 9]). For our purpose, however, we can use the following theorem of Kac.

**Theorem** ([4, 5]).

$$\mathrm{ch}\, L\left(\frac{1}{4} n^2, 1\right) = \frac{1}{\phi(q)} (1-q^{n+1}), \qquad \text{for } n = 0,1,2,\ldots .$$

## §3. Irreducible decompositions

First let us recall "the Schur polynomial" of $x_1,x_2,x_2,\ldots$ for a given Young diagram. Let Y be a Young diagram of size N. For a set of non-negative integers $(\nu_1,\nu_2,\ldots,\nu_N)$ with $\nu_1 + 2\nu_2 + \ldots + N\nu_N = N$, let $\pi_Y(1^{\nu_1},2^{\nu_2},\ldots)$ be the value of the irreducible character of $S_N$, the symmetric permutation group of N letters, labeled by the Young diagram Y, and evaluated at the conjugacy class consisting of $\nu_1$ cycles of size 1, $\nu_2$ cycles of size 2, and so on. It is an integer. The Schur polynomial for Y is by definition

$$\chi_Y(x) := \sum_{\substack{\nu_1+2\nu_2+\ldots=N \\ \nu_i \geq 0}} \pi_Y(1^{\nu_1},2^{\nu_2},\ldots) \frac{x_1^{\nu_1} x_2^{\nu_2}\ldots}{\nu_1!\nu_2!\ldots}$$

It is a weighted homogeneous polynomial ($\deg x_j = j$) with rational coefficients. Many formulas are known for Schur polynomials, for example, the orthogonal relations. We give some examples of the Schur polynomials.

$$\chi_\phi = 1,\ \chi_{\square} = x_1,\ \chi_{\square\square} = \frac{1}{2} x_1^2 + x_2$$

$$\chi_{\boxminus} = \frac{1}{2} x_1^2 - x_2,\ \chi_{\boxplus} = \frac{1}{12} x_1^4 - x_1x_3 + x_2^2.$$

For Young diagrams $\underbrace{\boxed{\cdot\ \cdot\ \cdot}}_{n}$, the Schur polynomials have the simple generating function:

$$\sum_{n=0}^{\infty} \chi_{\underbrace{\boxed{\cdot\ \cdot\ \cdot}}_{n}}(x)\, p^n = e^{\xi(x,p)}$$

where $\xi(x,p) = \sum_{j=1}^{\infty} x_j p^j$. This is a consequence of the fact that the representation of $S_n$ labeled by the Young diagram $\underbrace{\boxed{\cdot\ \cdot\ \cdot}}_{n}$ is trivial.

Our result is to describe the highest weight vectors of the irreducible components of the Fock representation in terms of the Schur polynomials.

<u>Theorem</u> **([11], see also [12]).** Fix a non-negative integer n. For a non-negative integer r, let $Y_{r,n+r} = \underbrace{\boxed{\phantom{xxxx}}}_{n+r}\, r$ (a rectangular Young diagram). Then $L_k^{(-n)} \chi_{Y_{r,n+r}} \equiv 0$ for all positive k.

Applying $\sum_{j=1}^{\infty} jx_j\partial_j$ to a weighted homogeneous polynomial, we get the weighted homogeneous degree as the eigenvalue. Hence

$$\begin{aligned} L_0^{(-n)} \chi_{Y_{r,n+r}} &= \left[\frac{1}{4} n^2 + r(r+n)\right] \chi_{Y_{r,n+r}} \\ &= \frac{1}{4} (n+2r)^2 \chi_{Y_{r,n+r}} \end{aligned}$$

Summing up, the Schur polynomial $\chi_{Y_{r,n+r}}$ is the highest weight vector of highest weight $\frac{1}{4}(n+2r)^2$. Therefore if $n \in \mathbb{N}$, $V_{-n} \supset \bigoplus_{r=0}^{\infty} L\left[\frac{1}{4}(n+2r)^2, 1\right]$. We compute the formal character of the right hand side.

$$\sum_{r=0}^{\infty} \text{ch } L\left[\frac{1}{4}(n+2r)^2, 1\right] q^{r(r+n)}$$

$$= \sum_{r=0}^{\infty} \frac{1}{\phi(q)} (1-q^{n+2r+1}) q^{r(r+n)}$$

$$= \frac{1}{\phi(q)}$$

Hence the irreducible decomposition of the Fock representation $(\pi_{-n}, V_{-n})$ is complete.

<u>**Corollary**</u>. For a non-negative integer n, we have $V_{-n} \simeq \bigoplus_{r=0}^{\infty} L\left[\frac{1}{4}(n+2r)^2, 1\right]$.

For the proof of the theorem we use some combinatorial relations of the Schur polynomials.

## §4. <u>KP hierarchy</u>.

Finally we mention the relation between the Fock representations of the Virasoro algebra and the bilinear differential equations of Hirota form.

We introduce the Hermitian form on $V = \mathbb{C}[x_1,x_2,x_3,\ldots]$ by $\langle f,g\rangle := f(\tilde{\partial})\bar{g}(x)\big|_{x=0}$, where $\tilde{\partial} = (\partial_1,\frac{1}{2}\partial_2,\frac{1}{3}\partial_3,\ldots)$. The Fock representation $(\pi_{-n}, V_{-n})$ $(n \in \mathbb{N})$ is decomposed as $V_{-n} = \Omega_{-n} \oplus \Omega_{-n}^{\perp}$, where $\Omega_{-n}$ is the subrepresentation isomorphic to $L\left[\frac{1}{4}n^2, 1\right]$, and $\Omega_{-n}^{\perp}$ is the orthogonal complement of $\Omega_{-n}$ with respect to $\langle\ ,\ \rangle$. For $f \in V_{-n}$ put $f(\tilde{D}) := f(D_1,\frac{1}{2}D_2,\frac{1}{3}D_3\ldots)$.

<u>**Theorem**</u>. For any $f \in \Omega_{-n}^{\perp}$, $f(\tilde{D})\tau\cdot\tau' = 0$ is one of the Hirota's bilinear equations of the n-th modified KP hierarchy. Conversely, any Hirota's bilinear equation of the n-th modified KP hierarchy corresponds to some $f \in \Omega_{-n}^{\perp}$.

For example, let $n = 0$ and take $f = x_{\boxplus} = \frac{1}{12} x_1^4 - x_1x_3 + x_2^2$. Then $f(\tilde{D}) = \frac{1}{12}(D_1^4 - 4D_1D_3 + 3D_2^2)$. This is the Hirota form of the original KP equation.

We give only a sketch proof of this theorem. This proof was obtained jointly with V. Kac and K. Ueno. For the sake of simplicity we restrict our attention to the case n = 0. The fundamental idea is making use of the vertex operator of the algebra $\mathfrak{gl}(\infty)$ (cf. [1], [7: §14.9]).

Let us denote $\tilde{V}(\Lambda) = \mathbb{C}[x_1,x_2,x_3,\ldots]$ the space of Fock representation of $\mathfrak{gl}(\infty)$ equipped with the pairing $\langle\ ,\ \rangle$. We take the tensor product:

$$\tilde{V}(\Lambda) \otimes \tilde{V}(\Lambda) = \mathbb{C}[x_1^{(1)},x_2^{(1)},\ldots] \otimes \mathbb{C}[x_1^{(2)},x_2^{(2)},\ldots]$$

$$= \tilde{V}(2\Lambda) \oplus \tilde{V}(2\Lambda)^{\perp}$$

where $\tilde{V}(2\Lambda)$ is the highest irreducible component. We consider the subspace

$$\tilde{\Omega} := \left\{ v \in \tilde{V}(2\Lambda);\ \left(\frac{\partial}{\partial x_j^{(1)}} + \frac{\partial}{\partial x_j^{(2)}}\right) v \equiv 0 \text{ for any } j \geq 1 \right\}$$

$$= \left\{ v \in \tilde{V}(2\Lambda);\ v \text{ is a polynomial of } y \right\}$$

where we have set $x_j = x_j^{(1)} + x_j^{(2)}$, $y_j = x_j^{(1)} - x_j^{(2)}$. We define the space of "Hirota polynomials" by

$$\text{Hir} := \mathbb{C}[y_1,y_2,\ldots] \cap \tilde{V}(2\Lambda)^{\perp}$$

$$= \tilde{\Omega}^{\perp} \text{ (= the orthogonal complement in the space } \mathbb{C}[y_1,y_2,\ldots]).$$

Denote by $H_m$ the space of Hirota polynomials of weighted homogeneous degree m, so that $\text{Hir} = \bigoplus_{m=1}^{\infty} H_m$. Results of [1] show that $\dim H_m = p(m-1)$. Hence $\sum_{m=0}^{\infty} (\dim \tilde{\Omega}_m)\, q^m = \frac{1}{\phi(q)}(1-q)$, exactly the formal character of L(0,1) for the Virasoro algebra.

The spaces $\tilde{\Omega}$ and Hir are invariant under the homogeneous components of the following "vertex operator":

$$Z(u,v) := \frac{1}{(u-v)^2}\left[e^{\frac{1}{2}(\xi(y,u)-\xi(y,v))}e^{-(\xi(\tilde{\partial}_y,u^{-1})-\xi(\tilde{\partial}_y,v^{-1}))}\right.$$

$$\left.+\,e^{-\frac{1}{2}(\xi(y,u)-\xi(y,v))}e^{\xi(\tilde{\partial}_y,u^{-1})-\xi(\tilde{\partial}_y,v^{-1})}-2\right].$$

Now if we take the diagonal part $Z(u,u) = \sum_{k\in\mathbb{Z}} L_k u^{-k-2}$, then the operators $L_k$'s are the Virasoro operators. Thus the theorem is proved for the case $n = 0$.

According to the theory of M. Sato, the KP hierarchy is much related to the infinite dimensional Grassmann variety, especially the Plücker's identities. There must be a deep connection between the Virasoro algebra and the Grassmann variety.


**Acknowledgments**

I am very grateful to Professors Victor G. Kac, Etsuro Date and Kimio Ueno for many stimulating discussions at MSRI.


**References**


[1] E. Date, M. Kashiwara, M. Jimbo and T. Miwa: Transformation groups for soliton equations, Proc. RIMS Symp. "Non-Linear Integrable Systems -- Classical Theory and Quantum Theory", M. Jimbo and T. Miwa ed., World Scientific Publishing Co., 1983, 39-119.

[2] B. L. Feigin and D. B. Fuks: Invariant skew-symmetric differential operators on the line and Verma modules over the Virasoro algebra, Funct. Anal. Appl. 16 (1982), 114-126.

[3] ——————: Verma modules over the Virasoro algebra, ibid.

17 (1983), 241-242.

[4] V. G. Kac: Highest weight representations of infinite dimensional Lie algebras, in Proc. of ICM, 299-304, Helsinki, 1978.

[5] ——————: Contravariant form for infinite dimensional Lie algebras and superalgebras, Lecture Notes in Physics 94, 1978, 441-445.

[6] ——————: Some problems on infinite dimensional Lie algebras and their representations, Lecture Notes in Mathematics 933, 1982, 117-126.

[7] ——————: Infinite Dimensional Lie Algebras, Birkhäuser, 1983.

[8] A. Rocha-Caridi and N. R. Wallach: Characters of irreducible representations of the Lie algebra of vector fields on the circle, Inventiones Math. 72 (1983), 57-75.

[9] ——————: Characters of irreducible representations of the Virasoro algebra, Math. Z. 185 (1984), 1-21.

[10] G. Segal: Unitary representations of some infinite dimensional groups, Commun. Math. Phys. 80 (1981), 301-342.

[11] M. Wakimoto and H. Yamada: Irreducible decompositions of Fock representations of the Virasoro algebra, Letters in Math. Phys. 7 (1983), 513-516.

[12] N. R. Wallach: Classical invariant theory and the Virasoro algebra, preprint.

[13] H. Yamada: The basic representation of the extended affine

Lie algebra of type $A_1^{(1)}$ and the BKP hierarchy, to appear in Letters in Math. Phys.